Spatial Analysis along Networks

Statistics in Practice

Series Advisors

Statistics in Practice is an important international series of texts which provide detailed coverage of statistical concepts, methods and worked case studies in specific fields of investigation and study.

With sound motivation and many worked practical examples, the books show in down-to-earth terms how to select and use an appropriate range of statistical techniques in a particular practical field within each title's special topic area.

The books provide statistical support for professionals and research workers across a range of employment fields and research environments. Subject areas covered include medicine and pharmaceutics; industry, finance and commerce; public services; the earth and environmental sciences, and so on.

The books also provide support to students studying statistical courses applied to the above areas. The demand for graduates to be equipped for the work environment has led to such courses becoming increasingly prevalent at universities and colleges.

It is our aim to present judiciously chosen and well-written workbooks to meet everyday practical needs. Feedback of views from readers will be most valuable to monitor the success of this aim.

A complete list of titles in this series can be found at www.wiley.com/go/statisticsinpractice

Spatial Analysis along Networks

Statistical and Computational Methods

Atsuyuki Okabe

School of Cultural and Creative Studies
Aoyama Gakuin University
Emeritus Professor, University of Tokyo, Japan

Kokichi Sugihara

Graduate School of Advanced Mathematical Sciences
Meiji University
Emeritus Professor, University of Tokyo, Japan

A John Wiley & Sons, Ltd., Publication

This edition first published 2012

Registered office
John Wiley & Sons Ltd, The Atrium, Southern Gate, Chichester, West Sussex,
PO19 8SQ, United Kingdom

For details of our global editorial offices, for customer services and for information about how to apply for permission to reuse the copyright material in this book please see our website at www.wiley.com.

Library of Congress Cataloging-in-Publication Data
Okabe, Atsuyuki, 1945-
Spatial analysis along networks : statistical and computational methods / Atsuyuki Okabe and Kokichi Sugihara.
p. cm.
Includes bibliographical references and index.
ISBN 978-0-470-77081-8 (cloth)
1. Spatial analysis (Statistics) 2. Spatial analysis (Statistics)–Data processing. 3. Geography–Network analysis. I. Sugihara, Kokichi, 1948- II. Title.
QA278.2.O359 2011
519.5'36–dc23

2011040047

A catalogue record for this book is available from the British Library.

ISBN: 978-0-470-77081-8

Set in 10.25/12pt Times by Thomson Digital, Noida, India
Printed and bound in Singapore by Markono Print Media Pte Ltd

Contents

Preface

As its title indicates, this book is devoted to spatial analysis along networks, referred to as *network spatial analysis*, or more explicitly, statistical and computational methods for analyzing events occurring on and alongside networks. Network spatial analysis is of practical use for analyzing, among other things, the occurrence of traffic accidents on highways, the incidence of crime on streets, the location of stores alongside roads, and the contamination of rivers (Chapter 1 introduces many applications). This usefulness is the main reason we focus on network spatial analysis in this volume. However, there is also a more general and somewhat more ambitious justification for this work. That is, when viewed from a broader perspective, we expect that network spatial analysis will prove to be a first step toward next-generation spatial analysis.

Having reviewed the extant literature on spatial analysis, we note that most empirical studies incorporate spatially aggregated data across subareas, such as administrative districts, census tracts, and postal zones. We refer to this type of spatial analysis as *subarea-based spatial analysis* or *meso-scale spatial analysis*. One of the earliest and most notable examples of this type of spatial analysis is included in a compilation titled *The City* (Park, Burgess, and McKenzie, 1925), written by sociologists at the Chicago School (sometimes described as the Ecological School). More specifically, Burgess (1925) surveyed land use of subareas in Chicago and formulated the concentric-zone model, subsequently followed by Hoyt's (1939) sector model and the Harris–Ullman multiple nuclei model (Harris and Ullman, 1945).

Since then, subarea-based spatial analysis has become one of the most important approaches to empirical spatial analysis. Even today, we frequently employ subarea-based spatial analysis for empirical studies because subarea data, including population and other census-related data, are widely available and because it is generally straightforward to apply ordinary statistical techniques, including regression analysis, to the attribute values of subareas. Unlike the empirical literature, we find that the development of most theoretical work on spatial analysis has assumed an 'ideal space', that is, real space is represented by unbounded homogeneous space with Euclidean distance. This ideal space is convenient for developing pure theories of spatial analysis or spatial stochastic processes; indeed, the derivations of many useful theorems employ this assumption (see, e.g., Illian *et al.*, (2008)). However, ideal space is far from the real world.

In the late twentieth century, the availability of detailed spatial data increased dramatically thanks to rapid progress in data acquisition technologies, such as the global positioning system (GPS) and many kinds of geosensors. Better data availability potentially enables us to analyze spatial events in detail by representing individual entities in the real world in terms of geometric objects in two- or three-dimensional Cartesian space instead of aggregating them into subareas (see Chapter 2 for this representation). We describe this possible form of spatial analysis as *object-based spatial analysis* or *micro-scale spatial analysis*, in contrast to the well-established subarea-based spatial analysis or meso-scale spatial analysis. At present, however, the methods for micro-scale spatial analysis are at an early stage. We believe that one clue to micro-scale spatial analysis would be to represent real space by networks embedded in two- or three-dimensional Cartesian space. This is because many kinds of events or activities in the real world are constrained by networks, such as streets, railways, water and gas pipe lines, rivers, electric wires, and communication networks. A first step toward micro-spatial analysis would thus appear to be network-constrained spatial analysis, which is the main concern of this volume.

In network spatial analysis, we measure the shortest-path distance. Unfortunately, its computation is much more difficult than that of Euclidean distance because it requires the management of network topology. Therefore, network spatial analysis becomes practical only when efficient computational methods are available. Dijkstra (1959) developed a key algorithm for this purpose in the middle of last century. Since then, there has been extensive study of location problems on networks by a variety of researchers, mainly in operations research (Handler and Mirchandani, 1979; Daskin, 1995; for a review, see Labbe, Peeters, and Thisse (1995)). We should note that the focus in these studies has been locational optimization or the computing of network characteristics (e.g., Kansky, 1963; Haggett and Chorley, 1969), with rather less attention paid to the statistical analysis of events on networks.

To fill this gap in the literature, we develop statistical and computational methods for network spatial analysis by introducing computational methods originally developed for operations research and computational geometry (Preparata and Shamos, 1985; Chapter 3 in this volume presents some basic computational methods). In this sense, the network spatial analysis presented in this volume is a first step toward micro-scale spatial analysis. However, we cannot present real world space by either network or Euclidean space alone as it is a complex hybrid system with elements of both. The next step, then, would be object-based spatial analysis in a hybrid space consisting of a discrete network space with shortest-path distance and a continuous space with Euclidean, or more generally, geodesic distance. An initial attempt is Cressie *et al.* (2006).

We are now in the midst of an ongoing revolution brought about by information and communication technologies. In the future, microcomputer tips, tags, and geosensors will be embedded in almost every entity (including moving objects) in our environment, and the integration of these devices with communication systems (e.g., the Internet) will establish an intellectual system joining the virtual world of computers and the global real world. This system will then realize a society we refer to as the *ubiquitous computing society*, in which at any time and in any place, people can receive the most appropriate personalized information for action given their particular circumstances in time and space (Sakamura and Koshizuka, 2005). To construct this system, micro-scale spatial analysis is expected to extend to *real-time spatial analysis*, that is, spatial analysis in which the circumstances of an acting body (including a person, a group of persons, a company, or possibly a robot) are analyzed and appropriate personalized information for action is derived almost instantaneously (Okabe, 2009a, 2009b). We intend that this volume, in presenting state-of-the-art methodology for network spatial analysis, will contribute a first step toward micro-scale spatial analysis and encourage our readers to further develop micro-scale spatial analysis and, from there, tackle the challenge of real-time spatial analysis.

Atsuyuki Okabe
Kokichi Sugihara
March 2012

Acknowledgements

When we first thought of the concept underlying this book in June 2007, we consulted Noel Cressie on possible publication. In turn, he was kind enough to introduce our proposal to the statistics and mathematics section at John Wiley. A positive response meant that our long project could begin in September 2007. Since then, so very many people have helped us in different ways in developing and presenting this book that it would be impossible to acknowledge all of them individually.

To start with, we are very grateful to those who have read our drafts and offered useful comments, particularly Ikuho Yamada on the general concepts underpinning network spatial analysis (Chapters 1 and 2) and spatial autocorrelation (Chapter 8), Toshiaki Satoh on kernel density estimation (Chapter 9) and GIS-based tools (Chapter 12), and Kei-ichi Okunuki on the Huff model (Chapter 11). Our special thanks also go to those with whom we discussed related subjects and who in turn provided us with inspiration. These especially include Mike Tiefelsdorf and Barry Boots on spatial autocorrelation, Yuzo Maruyama and Yonghe Li on kriging, Shino Shiode on inverse-distance weighting and cell counting, and Hisamoto Hiyoshi on spatial interpolation. They also include Atsuo Suzuki, Takehiro Furuta, and Shinji Imahori on equal cell splitting, Kei-ichi Okunuki and Masatoshi Morita on the *K* function method, and Yasushi Asami and Yukio Sadahiro on urban analysis.

We would also like to express our thanks to those who helped us to run the necessary programs, particularly Toshiaki Satoh, Kayo Okabe, Akiko Takahashi, and the staff at the Center for Spatial Information Science (CSIS) at the University of Tokyo and the Information Science Research Center at Aoyama Gakuin University. We are also indebted to Ayako Teranishi for collecting the more than 500 related papers, entering them in our database, and editing the references and compiling the index, and to Tsukasa Takenaka for constructing the online database with which we could develop our book while we were away. We also thank Masako Yoshida for the retrieval program used for the references, Tetsuo Kobayashi for collecting research articles, and Aya Okabe for designing the website, along with the web crew members involved in its management at CSIS, through which we received many practical comments on the GIS-based toolbox known as SANET from users across 51 countries.

We are thankful to the staff at John Wiley, particularly Richard Davies, Ilaria Meliconi, Heather Kay, Susan Barclay, Kathryn Sharples, and Prachi Sinha Sahay for their helpful assistance. We also acknowledge a grant-in-aid by the Japan

Society for the Promotion of Science for a project entitled 'Development of methods, algorithms, and GIS-based tools for statistical spatial analysis on networks' (#20300098), and data provision by the Chiba Prefectural Police, NTT Data, and CSIS. Finally, we thank our respective partners, Kayo Okabe and Keiko Sugihara, for their lifelong encouragement and invaluable support before and during the writing of this book.

1

Introduction

This book presents statistical and computational methods for analyzing events that occur on or alongside networks. To this end, the first three chapters are concerned with preparations. This chapter shows the scope of this book, Chapter 2 fixes a general framework for spatial analysis, and Chapter 3 describes computational methods commonly used throughout the subsequent chapters. In this introductory chapter, we first describe the events under consideration, i.e., events that occur on and alongside networks, termed *network events*. Second, we show that if traditional spatial analysis assuming a plane with Euclidean distances, referred to as *planar spatial analysis*, is applied to network events, then it is likely to lead to false conclusions. Third, to overcome this shortcoming, we propose a new type of spatial analysis, namely *network spatial analysis*, which assumes a network with shortest-path distances. Fourth, we review studies on network events in the related literature and show how to apply network spatial analysis to those studies. Last, we describe the structure of the twelve chapters of the book and suggest how to read them according to the reader's interests. Note that network spatial analysis viewed from a board perspective is described in the preface of this volume.

1.1 What is network spatial analysis?

To introduce this new type of spatial analysis, we first define a key concept, *network events*, and next consider typical questions about network events to be solved by network spatial analysis. We then describe the salient features of network spatial analysis in contrast to the traditional planar spatial analysis.

Spatial Analysis along Networks: Statistical and Computational Methods, First Edition.
Atsuyuki Okabe and Kokichi Sugihara.

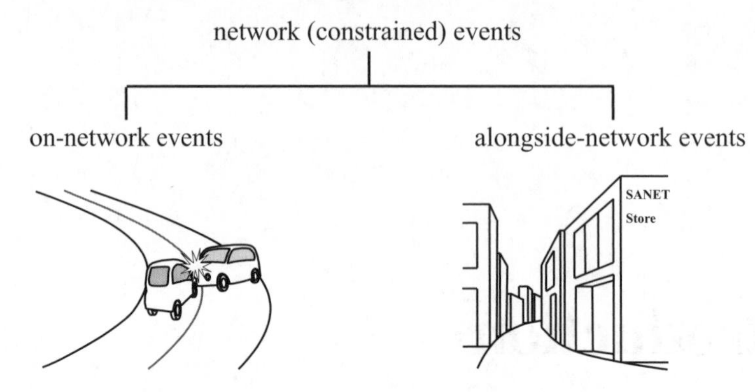

Figure 1.1 Network (constrained) events consisting of on-network events and alongside-network events.

1.1.1 Network events: events on and alongside networks

In the real world, there are numerous and various events that are strongly constrained by networks, such as car crashes on roads and fast-food shops located alongside streets. We call them *network-constrained events* (Yamada and Thill, 2007) or *network events* for short. Network events can be classified into two classes: events that occur directly on a network (e.g., car crashes on a road), and events that occur alongside a network rather than directly on it (e.g., fast-food shops located alongside a street). We refer to the former as *on-network events* and the latter as *alongside-network events*. Consequently, network events consist of on-network events and alongside-network events (Figure 1.1). Note that we sometimes use 'along' for both 'on' and 'alongside.'

Figure 1.2 illustrates an actual example of on-network events, where each dot represents a traffic accident around Chiba station, Japan. As with this example, many types of network event have been reported in the related literature, including pedestrian and motor vehicle street accidents, roadkills of animals on forest roads, street crime sites, tree spacing along the roadside, seabirds located along a coastline, beaver lodges in watercourses, levee crevasse distribution on river banks, leakages in gas and oil pipelines, breaks in a wiring network, disconnections on the Internet, and blood clots in a vascular network (studies on network events including these examples will be reviewed in Section 1.2).

Figure 1.3 depicts an actual example of alongside-network events, where the black dots indicate advertisement agency sites alongside streets in Shibuya ward, one of the subcentral districts in Tokyo. There are many facilities that are located alongside street networks within densely inhabited areas. In fact, the entrances to almost all facilities in a city are adjacent to streets and users access amenities through these (Figure 1.1). Consequently, the locations of almost all facilities within an urbanized area can be regarded as alongside-network events.

Figure 1.2 Sites of traffic accidents around Chiba station, Japan (private roads are not shown).

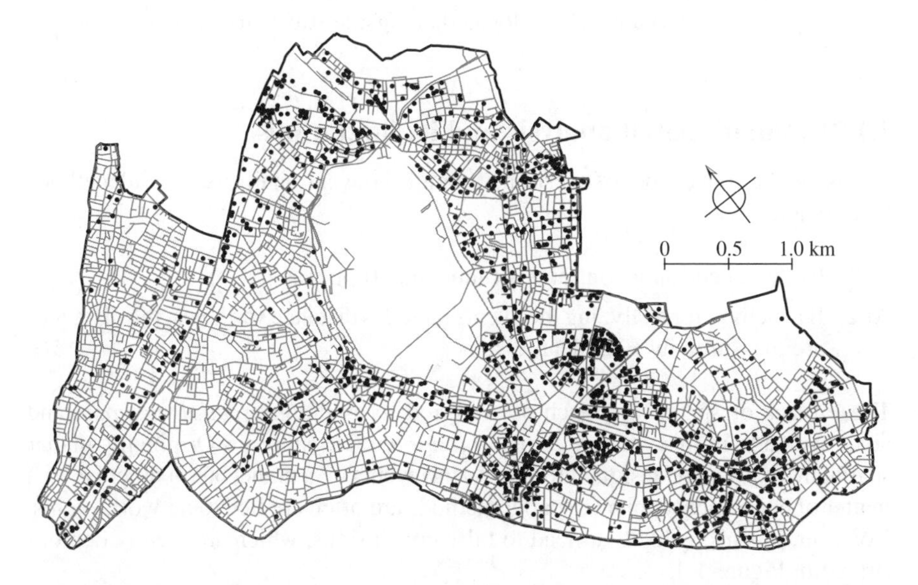

Figure 1.3 The distribution of advertisement agency sites (the black points) alongside streets (the gray line segments) in Shibuya ward, one of the subcentral districts in Tokyo.

On- and alongside-network events such as those in the above examples are the major concern of this book. More specifically, this book primarily focuses on spatial distributions and relationships of such events on and alongside networks. Typical questions to be discussed in this volume are as follows:

Q1: How can we obtain the catchment areas of parking lots in a downtown area including one-way streets, assuming that drivers access their nearest parking lots?

Q2: Do boutiques tend to stand side-by-side alongside streets in a downtown area?

Q3: Do street burglaries tend to take place near railway stations?

Q4: Is the roadside land price of a street segment similar to those of the adjacent street segments?

Q5: How can we locate clusters of fashionable boutiques alongside downtown streets?

Q6: How can we estimate the density of traffic accidents and street crimes incidence, and how can we identify locations where the densities of those occurrence are high, referred to as *black spots* and *hot spots*?

Q7: How can we spatially interpolate an unknown NO_x (nitrogen oxides) density at an arbitrary point on a road using known NO_x densities at observation points in a high-rise building district, such as Midtown Manhattan?

Q8: How can we estimate the probability of a consumer choosing a specific fast-food shop among alternative shops located alongside streets in a downtown area?

1.1.2 Planar spatial analysis and its limitations

To answer the above types of question, we might conventionally use spatial methods that assume:

AP1: Events occur on a continuous (unbounded) plane.

AP2: If a method for analyzing the events includes distance variables, the distances are measured by Euclidean distance.

These types of spatial approach are referred to as *planar spatial methods*, and analyses made in this way are termed *planar spatial analyses*. Originally, planar spatial methods were designed for analyzing events on a plane, but in practice, as a matter of convenience, planar spatial methods are often applied to network events. However, this use is likely to lead to false conclusions, which are clearly demonstrated in Figure 1.4.

Having assessed the distribution of points in Figure 1.4a, nobody would consider that the points are randomly distributed. This view is true if the points are considered as being distributed on a plane; however, this becomes false when the points are seen to be located on a network indicated by the line segments in

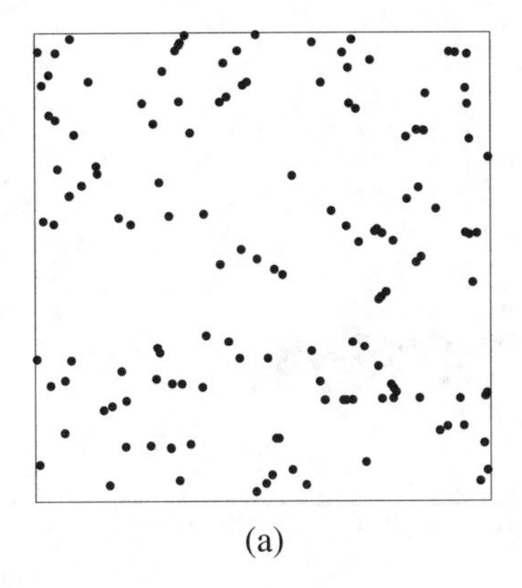

(a)

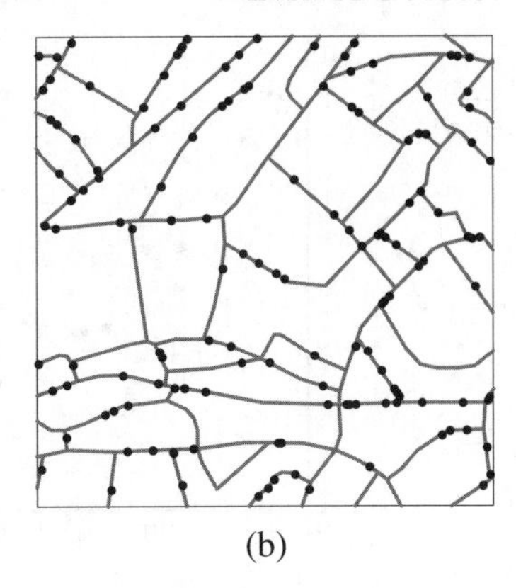

(b)

Figure 1.4 Point distributions: (a) nonrandomly distributed points on a bounded plane, (b) randomly distributed points on a network (note that the point distributions in (a) and (b) are the same).

Figure 1.4b. In fact, the points in this figure are randomly generated according to the uniform distribution across the network (for details, see Section 2.4.2 in Chapter 2). Figure 1.4 provides the following warning: analysis of network events using a planar spatial method is likely to lead to false conclusions. We shall show examples in subsequent chapters.

The second assumption AP2, i.e., the Euclidean distance assumption, is also arguable. The reasons for making this assumption are:

- it is much easier to compute Euclidean distance on a plane than the shortest-path distance on a network; and
- it is believed that the shortest-path distance is approximated by Euclidean distance.

The first reason remains true, although the difficulty is nowadays reduced because the use of geographical information systems (GIS) makes it easy to manage network data and to calculate shortest-path distances (a concise introduction to GIS is provided by Okabe (2004, 2005, Chapter 1)). The second reason may be true over a large region, but the validity of this concept is questionable across a small area or within a city. For example, Maki and Okabe (2005) report that in Kokuryo, a suburb of Tokyo, the difference between shortest-path distances and their corresponding Euclidean distances is significant if the Euclidean measurement is less than 400 m (see Figure 1.5). In addition, as shown in Table 1.1, the average radii of the service areas of many types of downtown store, exemplified by Shibuya ward in Tokyo, are less than 400 m. Planar spatial methods may be inappropriate therefore for analyzing alongside-network location events affected by trip behavior (for a further discussion, see Section 6.3 in Chapter 6).

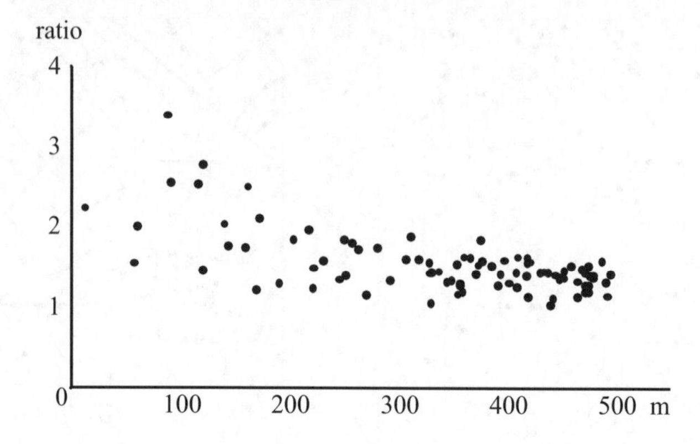

Figure 1.5 Ratio of the shortest-path distance to its corresponding Euclidean distance for the street network in Kokuryo, a suburb in Tokyo (data source: Maki and Okabe (2005)).

1.1.3 Network spatial analysis and its salient features

To overcome the above limitations of planar spatial methods, we now introduce a new type of spatial analysis that assumes:

AN1: Events occur on and alongside a network.

AN2: If a method for analyzing the events includes distance variables, the distances are shortest-path distances.

Corresponding to the planar spatial methods mentioned above (AN1 and AN2 correspond to AP1 and AP2, respectively), we call these methods *network spatial*

Table 1.1 Average radii of service areas in Shibuya ward, Tokyo.

Store type	Average radius (m)
Aromatherapy shop	282
Bag shop	271
Interior design shop	249
Daily necessities store	217
Preparatory school	216
Apartment estate agent	175
Printing store	167
Cafe	130
Japanese-style restaurant	106
Clothing store	85
Beauty shop	73

methods, and analyses that use network spatial methods, we call *network spatial analyses*. It should be noted that network spatial analysis does not imply the analysis of a network itself, such as geographical network analysis (Haggett and Chorley, 1969), communication network analysis (Kesidis, 2007), and circuit network analysis (Stanley, 2003). To avoid this confusion, we could use the terms *on-* or *alongside-network spatial analysis*, *network-constrained spatial analysis* (Yamada and Thill, 2004), *network-based spatial analysis* (Downs and Horner, 2007a, 2007b; Shiode, 2008) or more strictly, *spatial analysis on and alongside networks*. In this text, we use *network spatial analysis* for short.

We make a few remarks on the above two assumptions, AN1 and AN2. The first assumption AN1 describes places where events occur. The *on-network relation* is obvious. Events occur exactly on a network, such as traffic accidents. The *alongside-network relation* includes fairly broad spatial relations. It implies that the physical unit of an event (e.g., a store located at a site) has an access point on a network (the entrance of the store indicated by the black circle in Figure 1.6a) or the physical unit (the lot of the store) shares a common boundary line segment with a network (the bold line segment in Figure 1.6b). In addition, the alongside-network relation includes relations in which the physical unit of an event may intersect a network, for instance, a river intersects a road (Figure 1.6c) or a network goes through a forest area (Figure 1.6d). Computational treatments of these alongside-network relations are developed in Chapter 3 in detail.

The second assumption, AN2, specifies distance variables included in spatial methods. Consider, for instance, the analysis of boutique clusters in a downtown area using cluster analysis (for details, see Chapter 8). Because boutiques in clusters in a downtown area are located side-by-side alongside streets and customers access boutiques from entrances facing streets, it is natural to measure the closeness in terms of the shortest-path distance along streets. If a river separates two boutiques, it is not natural to assume that the boutiques belong to the same cluster even if the Euclidean distance between them is short. Underlying activities that result in boutique clusters are trips through streets, for example, window-shopping on sidewalks. In addition, many kinds of activities in a city are achieved through a street network, and so the configuration of activities may be influenced by trip

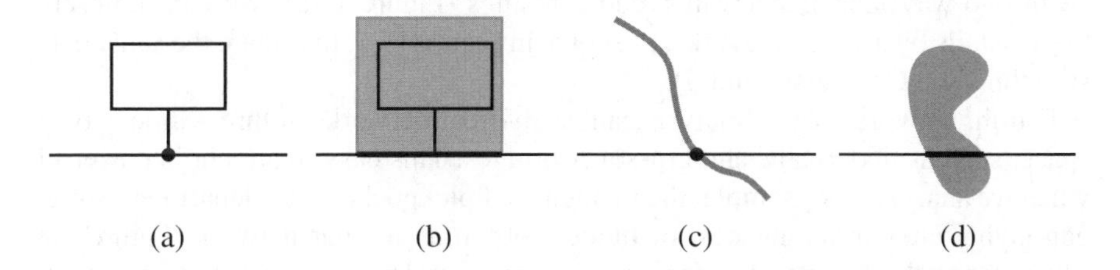

Figure 1.6 Alongside-network relations: (a) an access point (the black circle) of a polygon to a network (the horizontal line segment), (b) a boundary line segment of a polygon shared with a network (the bold line segment) (c) an intersection point of two networks (the black circles), (d) a network intersecting an area (the bold line segment).

behavior constrained by a street network. Consequently, network events may be best analyzed in terms of the shortest-path distance.

It should be noted, however, that there may be cases in which the shortest-path distance is not appropriate even if events occur on a network. For instance, consider the service area of a cell phone antenna. Although cell phone antennas stand on the edge of a street, their service areas are determined by Euclidean distance, because electric waves go straight through the air. The reader who wants to use a network spatial method should confirm whether or not the network spatial method is appropriate even when events are network events.

We notice from the above definition of network spatial analysis that it has salient features distinct from those of planar spatial analysis. First, by definition, network spatial analysis can properly analyze events occurring on and alongside a network. As a result, we can avoid the misleading conclusion illustrated by Figure 1.4. It is apparent from that figure that the selected points inevitably form clusters on a plane, because the points can exist only on a network. In fact, Yamada and Thill (2004) claimed that a planar spatial method (the *K* function method) overestimates clusters of traffic accidents in Buffalo (for details, see Chapter 6). Lu and Chen (2007) gave similar warning when analyzing urban crime distributed along streets. Such an overestimation is likely to happen not only for on-network events but also for alongside-network events. Therefore, clusters of stores in a city examined by planar spatial methods should be reexamined by network spatial methods.

Second, network spatial analysis can easily take account of directions, such as directions of current in a river and traffic flow regulation on a street network. In cities, particularly in downtown areas, many streets are one-way. In fact, about one third of streets in the downtown area of Kyoto are one-way (Okabe *et al.*, 2008). This implies that we cannot precisely estimate the delivery service areas of retail stores (e.g., pizza delivery stores) with Euclidean distances. Alternatively, we estimate the service areas in terms of the shortest-path distance on a directed network, and this estimation is investigated in detail in Chapter 4 (deterministic service areas) and Chapter 11 (probabilistic service areas).

Third, network spatial analysis can treat detailed networks using a common data structure. In a simple case, we represent a street by a line segment, but the street may consist of several components. For example, a street consists of vehicular roads (with two-way lanes), sidewalks, and crossings (Figure 1.7a). We can represent these details by a set of networks, as shown in Figure 1.7b, that share the same data structure (see Chapters 2 and 3).

Fourth, network spatial analysis can easily treat networks in three-dimensional space, such as underpaths and crossover bridges. This easy treatment is powerful when we analyze, for example, the incidence of pickpockets in a department store, egg-laying sites in an ant nest or blood clots in a vascular network. Figure 1.8 illustrates walkways, stairs, up/down escalators, and elevators in a department store, which are represented by a directed network.

Fifth, as will be shown in Section 2.3, network spatial analysis can treat nonuniform activities on a network more easily than planar spatial analysis can. While traditional spatial analysis methods are mostly designed to test the null

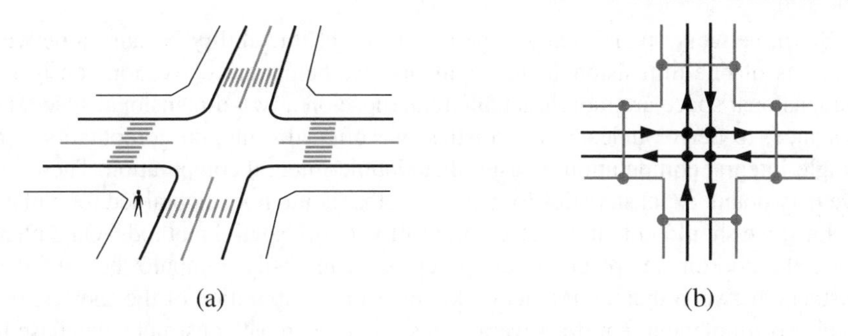

Figure 1.7 Entities represented by networks: (a) sidewalks, vehicular roads and crossing (entities), (b) the networks representing those entities.

hypothesis that events are uniformly distributed over a plane or network, this assumption is often violated in real-world phenomena. Consider, for example, traffic accidents on a road network. Obviously, traffic accidents do not occur uniformly across the network. Traffic accidents result from many factors, one of which is traffic volume (see Section 1.2.2). It is likely that the density of traffic accidents is proportional to traffic volume which naturally varies over a road network. Therefore, we cannot directly apply the traditional methods that assume uniform traffic volume to the distribution of traffic accidents resulting from nonuniform traffic volume. It is difficult to incorporate such nonuniformity in planar spatial analysis. Fortunately, however, we have good 'magic' that transforms a nonuniform density of an activity to a uniform density of the activity (to be shown in Section 2.4 in Chapter 2), to which we can apply traditional spatial methods assuming a uniform density. Through this transformation, we can easily analyze nonuniform activities on networks.

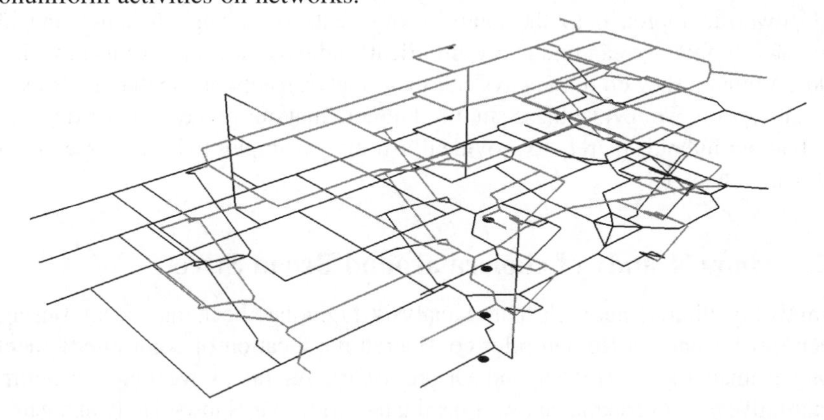

Figure 1.8 Walkways, stairs, up/down escalators, and elevators in a department store in Tokyo, where the circles indicate toilets (provided by T. Satoh). The subnetworks in different gray colors indicate the Voronoi cells of the three-dimensional network Voronoi diagram generated by the toilets (for definition, see Chapter 4).

Sixth, network spatial analysis gains analytical tractability because a network consists of one-dimensional line segments. Mathematical derivations on a one-dimensional space are more tractable than those on a two-dimensional space. For instance, to derive indexes or statistics, we often do integral computation, and single-integral computation is easier than double-integral computation. Therefore, we may obtain exact statistics for a network that could not be obtained for a plane.

Last, we should note the shortcomings of network spatial methods. On a plane, once the coordinates of points are given, we can easily compute the Euclidean distance between them. On a network, however, computation of the shortest path is not so simple and requires several steps. First, we must construct a database for managing a network. In practice, point data and network data are obtained from different sources and points that are supposed to be on a network are likely to be off the network rather than exactly on the network. Therefore, second, we must assign the points to the network. Third, we must use an algorithm for computing the shortest path on a network. In addition, we must perform many kinds of geometrical computation inherent in network spatial analysis. As a result, it is not straightforward in practice to extend statistical methods for planar spatial analysis to those for network spatial analysis. Network spatial analysis becomes practical only when its computation is possible. That is why the subtitle of this book is *Statistical and Computational Methods*. The computational methods in each chapter show how to solve difficult geometric computations encountered in network spatial methods in practice.

1.2 Review of studies of network events

As the above salient features indicate, network spatial analysis provides a suitable and powerful approach to the analysis of events occurring on and alongside networks. In fact, we can find many empirical studies of network events in various fields, although they do not always call their analyses network spatial analysis. In this subsection, we review these studies, but note that our review is not exhaustive and that our intent is merely to provide illustrative examples to be discussed in the following chapters.

1.2.1 Snow's study of cholera around Broad Street

Primitive qualitative network spatial analysis might date back many centuries ago when, for instance, a Roman ruler considered the location of colony settlements along Roman roads (Hodder and Orton, 1976). As far as we know, scientific quantitative network spatial analysis originated from John Snow's study in the mid-nineteenth century (Snow, 1855, 1936). John Snow's cholera map (Figure 1.9), which he called a diagram of the *topography of the outbreak* (Snow, 1855), illustrated one of the worst outbreaks of cholera that occurred around Broad Street and Golden Square in London in the mid-nineteenth century. The black bars along streets in Figure 1.9 indicate the number of victims. To find the source of the

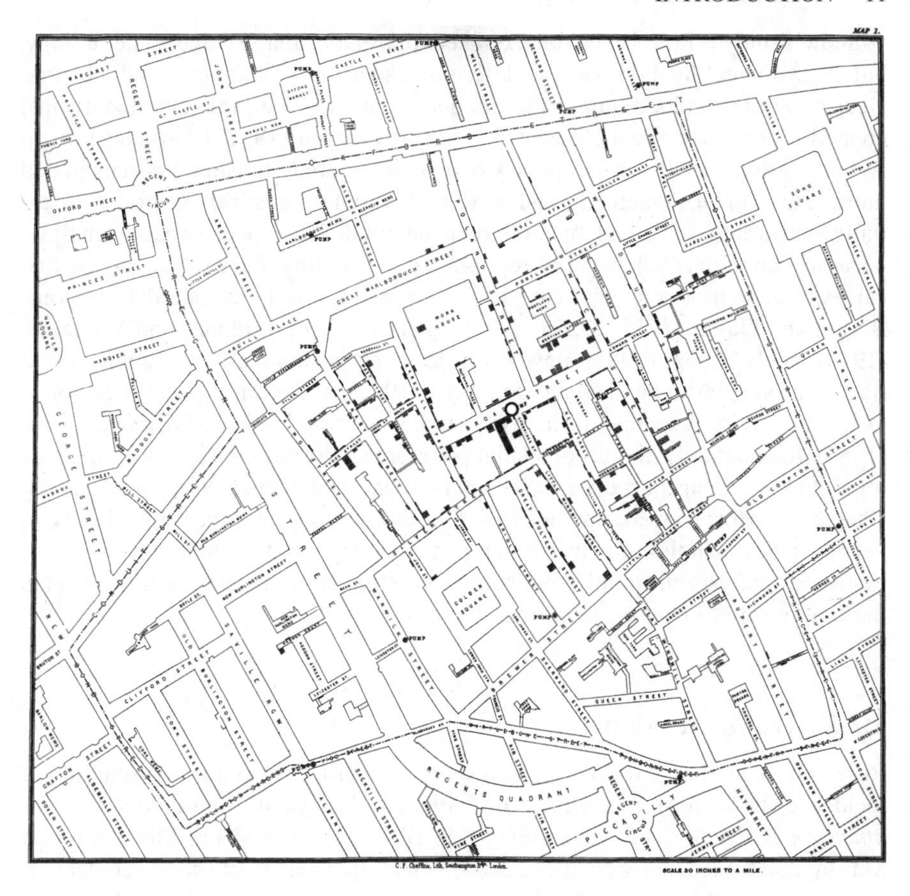

Figure 1.9 Snow's cholera map (Snow, 1855).

infections, he demarcated the area for which the Broad Street pump (the white circle in Figure 1.9) was the nearest pump among the 13 water pumps around this region in terms of the shortest-path distances along streets. Noticing that almost all victims were found in this demarcated area, he came to the conclusion that the cause of the victims' illness was contamination of the water from the Broad Street pump (his detection process was described in a book, *Ghost Map*, by Johnson (2006); see also Gilbert (1958), Smith (1982), Tufte (1997), McLeod (2000), and Koch (2004)). Snow called the demarcated area the *cholera area* (Snow, 1855). In modern terms, the area is a Voronoi subnetwork of the *network Voronoi diagram* (Okabe *et al.*, 2000, Section 3.8; Chapter 4 in this volume). Cliff and Haggett (1988, Figure 1.16D) explicitly applied the network Voronoi diagram to Snow's map, and Shiode (2012) deepened their study. Nakaya (2001) and Koch (2005) also reexamined Snow's map, although they used the planar Voronoi diagram.

Snow's study is indeed a landmark of network spatial analysis and epidemiology, but his statistical method was the descriptive statistics of the nineteenth century. Deepening his results required the 'modern' statistics (i.e., inferential statistics) created in the early twentieth century by Karl Pearson (1851–1936) and Ronald Aylmer Fisher (1890–1962), among others. Inferential statistics first developed univariate statistical methods and next multivariate statistical methods. Potentially, univariate statistical methods can be used for network spatial analysis because a univariate value can be represented by a point on a real line and the line can be regarded as a special case of a network. Using univariate distributions, Barton and David (1955, 1956a, 1956b), Pearson (1963), Pinder and Witherick (1973), and Young (1982) developed tests of randomness for points on a line. Watson (1961), Pearson and Stephens (1962), and Kuiper (1962) developed tests on a circle. A review of the early studies is provided by Selkirk and Neave (1984). In general, textbooks of spatial point processes refer to statistical methods on a line in their introductory chapters (e.g., Daley and Vere-Jones, 2003, Chapter 3). However, the methods used in those studies can be applied only to very specific examples of networks, i.e., one link between two nodes or one circle. Statistical analysis on complex networks consisting of more than one link began in the late twentieth century in various fields.

1.2.2 Traffic accidents

Researchers studying traffic accidents are concerned with the distribution of incidence spots on road networks. In particular, they want to find *black spots* or *black zones* (Maher and Mountain, 1988; Black, 1992; Flahaut, Mouchart, and Martin, 2003; Steenberghen *et al.*, 2004), where the density of traffic accidents is significantly high. In the early phase of traffic accident studies, researchers used spatially aggregated data to find black spots; specifically, the number of accidents was spatially aggregated with respect to road segments and the density of accidents on each road segment obtained from the aggregated number was used to detect black spots. The problem with that approach is how to divide a road network into road segments. Some researchers used unequal-length road segments (e.g., Ceder and Livneh, 1978; Turner and Thomas, 1986; Ng and Hauer, 1989; Stern and Zehavi, 1990; Miaou, 1994); and some researchers used equal-length road segments (e.g., Golob, Recker, and Levine, 1990; Thomas, 1996; Black, 1991; Erdogan *et al.*, 2008; Yamada and Thill, 2010). The result may vary according to the lengths of the road segments. This variation is a notorious known problem in spatial analysis, termed the *modifiable area unit problem* (abbreviated to MAUP) (Openshaw and Taylor, 1979, 1981; Openshaw, 1984).

For statistical tests, equal-length road segments are preferable, because statistical adjustments required by differences in size are not necessary, and several statistical methods assume equal-size spatial units, for instance, the *cell-count method*, i.e., the analysis using the numbers of road segments (cells in general) having certain numbers of accidents (see Section 9.1.3 in Chapter 9 for the precise definition). Nicholson (1989) and Black (1992) used a similar method using accident counts on equal-size road segments. In practice, however, it is not easy to divide a road network into equal-length road segments, because the length of a link between nodes varies from link to link and it is impossible to find a common divisor for all the links. We discuss this problem in depth in Chapter 9, and show a few practical methods in Section 9.3.1.

Once the numbers of accidents are given with respect to road segments (whether they are equal length or not), we can regard the numbers as the attribute values of statistical data units. In this general context, various methods developed in ordinary statistics can be used for finding high-density road segments, for example, the goodness-of-fit test in Erdogan *et al.* (2008), and the Gini coefficient in Nicholson (1989) for pedestrian accidents. However, if attribute values of spatially close (or distant) units are correlated, which is referred to as *spatial autocorrelation* (see Chapter 7 for the precise definition), then statistical indexes that take spatial autocorrelation into account should be used. Black (1991, 1992), Black and Thomas (1998), and Yamada and Thill (2010) analyzed traffic accidents using such indexes. In Chapter 7, we describe a few types of spatial autocorrelation defined on networks.

The use of aggregated data in ordinary statistical analyses produces the MAUP. Alternatively, researchers can use disaggregated data, i.e., the data of individual incidence spots on a network. There are several statistical methods based on disaggregated data in the traffic accident literature. One such method is the *kernel density estimation method* (for the precise definition, see Chapter 9). Intuitively speaking, this method estimates the density of accidents by generating 'mole hills' (bell-shaped density functions with small domains; see Figure 9.3a in Chapter 9) at incidence spots and estimates the density by heaping those 'hills' (Figure 9.3b). Sabel *et al.* (2006) and Erdogan *et al.* (2008) employed this method to study traffic accidents in Christchurch in New Zealand and Afyokarahisar in Turkey, respectively.

A second method is the *nearest-neighbor distance method*, i.e., a statistic defined in terms of the distance from each point to its nearest neighbor (see Chapter 5 for the precise definition). Black (1992) applied this method to traffic accidents in a time–space context. Modifying the nearest-neighbor distance method, Steenberghen *et al.* (2004) used a statistic defined in terms of the number of accidents within a certain radius centered at each incidence spot. A third method is an extension of their method, called the *K function method*, in which the radius varies from zero to infinity (see Chapter 6 for the precise definition). Applications of the *K* function method to traffic accidents are found in Jones, Langford, and Bentham (1996), Yamada and Thill (2004), and Erdogan *et al.* (2008). A fourth method is *cluster analysis*, which finds sets of points that are spatially clustered on a network. Levine, Kim, and Nitz (1995) applied this method to traffic accidents in Honolulu. We develop cluster analysis for points on networks in Chapter 8.

Once black zones are detected by the above methods, the next problem is to identify the factors that produce those zones. Examples of influential factors reported in the literature are: traffic density (Pfundt, 1969; Ceder and Livneh, 1978, 1982; McGuigan, 1981; Ceder, 1982; Brodsky and Hakkert, 1983;

Lassarre, 1986; Ng and Hauer, 1989; Golias, 1992; Steenberghen *et al.*, 2004); geometry of junctions (Tanner, 1953; McGuigan, 1981; Nicholson, 1989), signal control (McGuigan, 1981; Braddock *et al.*, 1994), points of access and egress (Black, 1991), trees (Mok, Landphair, and Naderi, 2006), access to the nearest emergency services (Brodsky and Hakkert, 1983), population density along roads (Levine, Kim, and Nitz, 1995), lane and shoulder width (Ng and Hauer, 1989; Klop and Khattak, 1999), median strips (Bellis and Graves, 1971), speed limits (Lassarre, 1986; Abdel-Aty, Cunningham, and Gayah, 2008), curvature (Fink and Krammes, 1995), horizontal and vertical alignments (Shankar, Mannering, and Barfield, 1995; Kim and Boski, 2001), light conditions (Aerts *et al.*, 2006), line-of-sight distances (Van Kirk, 2000; Kim and Boski, 2001), bicycles traveling with or against traffic (Wachtel and Lewiston, 1994), and natural environmental factors, for example, fog (Black, 1991).

Analysis of the spatial relationship between black spots (more generally, network events) and the above influential factors differs according to whether data are aggregated or disaggregated. When the number of traffic accidents and the values of explanatory variables (e.g., traffic volume, speed limit) are spatially aggregated with respect to road segments, we can simply use a regression model, for example, the numbers of traffic accidents of road segments are regressed on the traffic volumes and speed limits of these road segments. In this case, as mentioned above, we may encounter the MAUP. To avoid it and to deepen the results of aggregated data analysis, we can alternatively use the disaggregated data of accident incidence spots on a road network, and treat the above influential factors in the following disaggregated manner (full developments are shown in Chapter 2).

The factors (in the form of entities) are classified in four types according to their geometrical forms. First, junctions, points of access/egress, traffic signs, traffic lights, street lights and other elements are represented by points on a network with their attribute values (the black circles in Figure 1.10a). Second, road segments with specific discrete or categorical characteristics, such as 50 m width roads and national roads, are represented by line segments with their attribute values (e.g., 50 m and national; the bold line segment in Figure 1.10b). Third, areas with specific characteristics, say foggy areas are represented by polygons, and the foggy areas are treated as foggy road segments that are included in the polygons or the minimum distance between a point in the polygon and a point on the road (the bold line segments or the broken line segment in Figure 1.10c). Last, traffic density, population density, illumination intensity, curvature, line-of-sight distance, access to the nearest emergency services and other attribute values that vary continuously along roads are represented by a function with a network as its domain (called a *field function* defined in Section 2.1.2 in Chapter 2, as in Figure 1.10d). Geometrical formulations and computational methods are developed in Chapters 2 and 3.

1.2.3 Roadkills

Animals can be traffic accident victims as well as humans. The former accidents are referred to as *roadkills of animals* or *roadkills* for short. It is reported in the

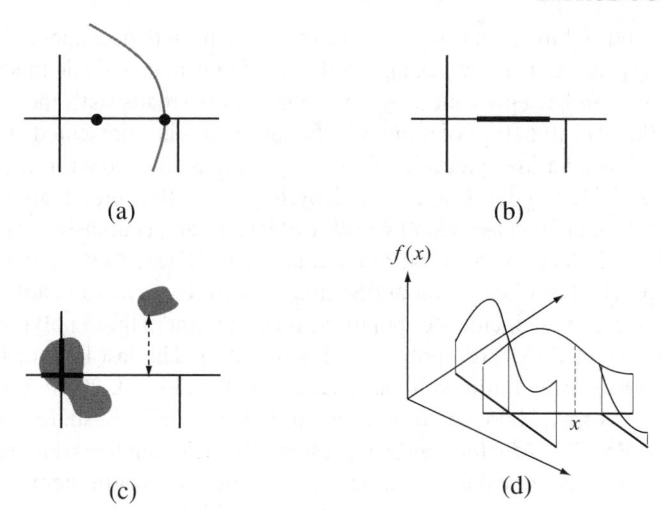

Figure 1.10 Influential factors classified according to geometrical forms: (a) points on a network, (b) line segments on a network, (c) line segments on a network included in a polygon and the distance from a polygon to the nearest point on a network, (d) a field function with a network as its domain.

literature that many kinds of animals can be victims of roadkills, including deer (e.g., Bellis and Graves, 1971), moose (Seiler, 2005), raccoon dogs (Saeki and Macdonald, 2004), wild boars (Malo, Suarez, and Diez, 2004), panthers (Forman and Alexander, 1998), otters (Philcox, Grogan, and Macdonald, 1999), rodents, voles, hedgehogs, rats, moles, shrews, and beech martens (Orlowski and Nowak, 2006), frogs and snakes (Forman and Alexander, 1998), and butterflies and burnets (Munguira and Thomas, 1992).

Statistical methods used in roadkill analysis are almost the same as those for traffic accident analysis. Differences appear in the factors influencing roadkills, although some are common. As in Section 1.2.2, the factors (entities) may be categorized by four types according to their geometrical forms. The first type of influential factor (entity) includes buildings alongside roads (Bashore, Tzilkowski, and Bellis, 1985; Malo, Suarez, and Diez, 2004), warning devices (Reed, Beck, and Woodard, 1982), swareflex reflectors (Schafer and Penland, 1985; Reeve and Anderson, 1993), underpasses (Ward, 1982; Foster and Humphrey, 1995; Malo, Suarez, and Diez, 2004), overpasses (Forman and Alexander, 1998), river crossings (Philcox, Grogan, and Macdonald, 1999), and crossroads or intersections (Malo, Suarez, and Diez, 2004; Seiler, 2005). These factors are represented by points on roads (Figure 1.10a).

The second type of influential factor (entity) includes bridges (Hubbard, Danielson, and Schmitz, 2000), fences (McKnight, 1969; Ward, 1982; Bashore, Tzilkowski, and Bellis, 1985; Philcox, Grogan, and Macdonald, 1999; Malo, Suarez, and Diez, 2004; Seiler, 2005), hedges (Malo, Suarez, and Diez, 2004), watercourses (Philcox, Grogan, and Macdonald, 1999), median strips (Bellis and Graves, 1971), gullies (Finder, Roseberry, and Woolf, 1999), the number of lanes (Hubbard,

Danielson, and Schmitz, 2000), road width (Forman and Alexander, 1998) and speed limit (Bashore, Tzilkowski, and Bellis, 1985; Forman and Alexander, 1998). These factors can be represented by line segments on roads with their attributes (Figure 1.10b). The third type of influential factor (entity) is represented by polygons characterized by land-use categories, for example, areas covered with woods (Bellis and Graves, 1971; Puglisi, Lindzey, and Bellis, 1974; Bashore, Tzilkowski, and Bellis, 1985; Finder, Roseberry, and Woolf, 1999; Hubbard, Danielson, and Schmitz, 2000; Saeki and Macdonald, 2004; Malo, Suarez, and Diez, 2004; Seiler, 2005) or grass or crops (Hubbard, Danielson, and Schmitz, 2000). These factors can be analyzed by the distance from an incidence spot to the nearest point in those polygons or road segments intersected by those polygons (Figure 1.9c). The last type of influential factor (entity) includes traffic volumes (Saeki and Macdonald, 2004; Seiler, 2005; Orlowski and Nowak, 2006) and in-line visibility (Case, 1978; Bashore, Tzilkowski, and Bellis, 1985). These factors can be represented by field functions defined on roads (Figure 1.10d). Geometrical formulations of these four types of influential factor are shown in Chapter 2 and computational methods are illustrated in Chapter 3.

1.2.4 Street crime

According to Sherman, Gartin, and Bürger (1989), 3.3% of street addresses and intersections in Minneapolis generated 50.4% of all dispatched calls for police service. These concentrated places are called *hot spots* (Brantingham and Brantingham, 1982; Sherman, Gartin, and Bürger, 1989; Sherman, 1995; Sherman and Weisburd, 1995; Bürger, Cohn, and Petrosino, 1995) or sometimes referred to as *mean streets* (Celik *et al.*, 2007) or *deviant streets* (Cohen, 1980). LaFree (1998) reported in the book subtitled *Street Crime and the Decline of Social Institutions in America* that in the early 1990s, nearly 25 000 Americans were being killed each year; in 1994, an estimated 620 000 Americans were robbed, 102 000 were raped, and 1.1 million were assaulted. In addition to these crimes, street crime (in a broad sense) includes car thefts, threats by street gangs, illegal drug trades, and graffiti and vandalism of public properties on and alongside streets. Street crime is a major subject of network spatial analysis.

Einstadter and Henry (2006) wrote that a French lawyer and statistician, Guerry, and a Belgian mathematician and astronomer, Quetelet first noted the uneven distribution of crime. This tendency was intensively surveyed in the 1920s by the Chicago School (e.g., Park, Burgess, and McKenzie, 1925); these are often referred to as *ecological crime studies*, and most of the research has been performed since then. Until the 1980s, studies mainly employed ordinary statistics without considering geographical proximity among data units, but since the 1990s, spatial statistical methods have been introduced to consider the configuration of crime spots (a comprehensive review is provided by Anselin *et al.* (2000), and Anselin, Griffiths, and Tita (2008)). The use of those methods was accelerated in practice by GIS-based tools, such as CrimeStat (Levine, 2004, 2006) and Spatial and Temporal Analysis of Crime (STAC) (Block, 1995). However, most methods and tools assume that crimes occur on the continuum of a plane (not restricted to streets) except for Shiode (2011). This assumption is, as noticed by Levine (2004), hard to accept for street crimes. We show proper statistical methods for analyzing crimes on streets in the following chapters.

The statistical methods for street crimes are common to those for traffic accidents and roadkills, but factors affecting street crimes are different from those studies. We classify those influential factors (entities) according to the types defined in Section 1.2.2. The first type (Figure 1.10a) includes street lights (MacDonald and Gifford, 1989; Perkins, Meeks, and Taylor, 1992; Perkins *et al.*, 1993; Painter, 1996; Loukaitou-Sideris, 1999), trees, security/watch signs (Perkins *et al.*, 1993), liquor shops (Block and Block, 1995; Buslik and Maltz, 1998; Loukaitou-Sideris, 1999), taverns (Roncek and Maier, 1991), bars (Sherman, Gartin, and Bürger, 1989), convenience stores (Duffala, 1976; Brantingham and Brantingham, 1982), elevated stations (Block, 1998), abandoned dilapidated housing (Perkins *et al.*, 1993; Buslik and Maltz, 1998), schools (Buslik and Maltz, 1998; Block, 1998), bus stops, adult bookstores and pawn shops (Loukaitou-Sideris, 1999), and graffiti and litter (Perkins *et al.*, 1993; Loukaitou-Sideris, 1999). These factors are represented by points. The second type (Figure 1.10b) includes arterial streets (Eck, 1998), wide streets and unlit streets (Cohen, 1980), which are represented by line segments. The third type (Figure 1.10d) includes pedestrian flows (Nasar and Fisher, 1993), represented by a filed function across a network. Geometrical treatments of those factors are shown in Chapter 2.

1.2.5 Events on river networks and coastlines

Networks are not limited to road networks. Rivers and coastlines also form networks. Roline (1988) examined the effects of heavy metal pollution of the upper Arkansas River on the distribution of aquatic macroinvertebrates (see also Soares *et al.* (1999) and Elbaz-Poulichet *et al.* (1999)). Anderson, Brown, and Rappleye (1968) surveyed water quality and plant distributions along the Upper Patuxent River, Maryland (Milovanovic, 2007; Perona, Bonilla, and Mateo, 1999). Garcia (1999) investigated the spawning distribution of fall Chinook salmon in the Snake River (Heise *et al.*, 2004). Hayes, Leathwick, and Hanchet (1989) studied fish distribution patterns with environmental factors in the Mokau River, New Zealand (also see Nicola and Almodovar (2004)). Fustec, Cormier, and Lode (2003) surveyed the beaver lodge locations in the upstream Loire River. Cabrero *et al.* (1997) studied the geographical distribution of B chromosomes in the grasshopper *Eyprepocnemis plorans* along rivers in the Sugura River Basin. Moss *et al.* (1989) studied the phytoplankton distribution in a temperate flood-plain lake and river system. Richter *et al.* (1998) made a spatial assessment of hydrologic alteration within a river network in the Upper Colorado Basin. Paul and Pillai (1986) investigated the distribution of radium in the Periar River from Karala to Moolampalli (El-Gamal, Nasr, and El-Taher, 2007). Gregory, Davis, and Tooth (1993) studied the spatial distribution of coarse woody debris dams in the Lymington Basin (also see Evans, Hungr, and Clague (2001)). In practice, tools for analyzing events along rivers are useful. Ries *et al.*

(2008) provided a software package, called *StreamStats*, for analyzing stream networks consisting of rivers, water bodies and constructed channels.

It should be noted that most of these studies employed ordinary statistics without explicitly considering spatial proximity among data units. A few exceptions are, however, found in the literature. Dacey (1960b) applied the reflexive nearest-neighbor method proposed by Clark (1956) (see Chapter 5) to the spacing of cities along rivers in the central lowland USA. Lewis (1977) applied the run test (Wald and Wolfowiz, 1940) to the distribution of biochemical oxygen demand along the River Trent. Cressie and Majure (1997) interpolate nitrate concentration on streams in the upper North Bask watershed (also see the studies in Section 9.2).

In addition to rivers, water lines include coastlines. Griggs and McCrory (1975) surveyed waste-water discharges along the Pacific Coast, California. Sivertsen (1997) investigated the distribution of kelp beds along the Norwegian coast. O'Driscoll (1998) examined the distribution of sea birds along a coastline. Storlazzi and Field (2000) studied sediment distribution along a rocky, embayed coast in the Carmel Bay, California. Vila *et al.* (2001) analyzed the distribution of algal blooms along the Catalan Coast.

1.2.6 Other events on networks

Studies of events on road networks are not restricted to the traffic accidents, roadkills and street crimes reviewed above. We next refer to other events that occur on road networks, together with events on other types of network.

Spooner *et al.* (2004) applied the network K function method (see Chapter 6) to *Acacia* populations on a road network in Lockhart Shire, Australia. Suzuki (1987), and Furth and Rahbee (2000) analyzed bus stop spacing along bus routes. Hillier (1984, 1996) analyzed straight-line visibility on roads using the concept of 'space syntax.' Using this concept, many researchers (e.g., Penn *et al.*, 1998; Jiang and Claramunt, 2002; Penn, 2003; Porta *et al.*, 2006a,b) have examined the structures of street networks in cities.

In a broad sense, road networks include passageways in complex facilities and corridors in buildings, and events on these networks are also within the scope of network spatial analysis. After the September 2001 terrorist attacks at the World Trade Center in New York City, Kwan and Lee (2005) drew attention to sites of accidents in corridor networks in high-rise buildings. Castle and Longley (2005) inspected interrupted routes at London's King's Cross Station (see also Pu and Zlatanova (2005)).

In addition, we find many kinds of network events that occur on networks other than road and river networks. Deckers *et al.* (2005) investigated the effect of landscape structure on the invasive spread of black cherry in the network of hedgerows in Flanders using the K function method (see Chapter 6) on a network. Maheu-Giroux, and Blois (2006) conducted landscape-scale analysis of *Phragmiets australis* invasion patterns in two preurban areas in Quebec, focusing on the interaction between the network of wetlands and the adjacent land uses using the *K* function method. Pipelines also form a network. Tucciarelli, Criminisi, and Termini (1999) analyzed reduction of leakages in a water pipe network by means of optimal valve regulation and Oliveira *et al.* (2008) studied breakages in water pipelines in Monroeville. Cevik and Topal (2003) surveyed damaged places on natural gas pipelines (also see Hwang *et al.* (2004)).

1.2.7 Events alongside networks

As illustrated in Figure 1.1, network events consist of on-network events and alongside-network events. The network events in the preceding subsections are all on-network events. However, in the real world, alongside-network events are also common, particularly location events in urbanized areas, i.e., various kinds of facilities located side by side alongside streets. In the literature, these location events are conventionally examined using planar spatial analysis, but the results may lead, as noted in Figure 1.4, to potentially false conclusions. We suggest reexamining these types of study with network spatial methods.

Of course, some studies do pay attention to alongside-network events. For example, Hodder and Orton (1976) investigated the distribution of late Iron Age coins alongside Roman roads, while Stark and Young (1981) applied a distance-based method to the distribution of archaeological sites alongside a trail in the Cabeza de la Vaca Arroyo in Mexico. Similarly, Li *et al.* (1990) surveyed the occurrence of abnormal hemoglobin in the populations of cities alongside the Silk Road in northwest China, and Xie *et al.* (2007) studied the distribution of ancient cities alongside the midsection of the Silk Road's He–Xi Corridor.

In related work, several studies more explicitly examine alongside-network events using network spatial methods. For instance, Okabe, Yomono, and Kitamura (1995) studied the distribution of liquor shops and car dealers in relation to railway stations in Nishinomiya, Japan using the nearest-neighbor distance method defined on a network (see Chapter 5). Using the same method, Sevtsuk (2010) considered the distribution of retail and food establishments in Cambridge and Somerville, Massachusetts. Moreover, because the service areas of facilities are sensitive to network patterns, the configuration of facilities is examined using Voronoi diagrams defined on a network (see Chapter 4). For example, Furuta, Uchida, and Suzuki (2005) estimated the service areas of ambulance stations in Seto, Japan, using the k-th nearest point Voronoi diagram defined on a network (Section 4.2.3 in Chapter 4).

Similarly, Okabe *et al.* (2008) determined the catchment areas of parking lots in Kyoto using the Voronoi diagram on a network. The *K* function method defined on a network (see Chapter 6) is also useful for analyzing facilities in urbanized areas. Using this particular network spatial method, Myint (2008) examined the distributions of banks, fast-food restaurants, schools, and churches in Norman City, Oklahoma. Besides these network spatial methods, Shiode (2008) examined the distribution of commercial facilities in Shibuya ward, Tokyo, using the cell-count method based on equal-size subnetworks (see Section 9.1.3 in Chapter 9), while Shiode and Shiode (2009b) found clumps of restaurants in Shibuya ward by using the clumping method defined on a network (see Section 8.2 in Chapter 8).

Finally, Borruso (2008) estimated the density of banks and insurance companies in Trieste, Italy, applying a kernel density estimation method to a network embedded in a plane (see Section 9.2 in Chapter 9).

1.3 Outline of the book

As the above review indicates, a wide range of subjects can be studied by network spatial analysis. This book intends to provide statistical and computational methods for such studies, with the expectation that the methods provide more suitable approaches than the existing methods. We outline those methods in this last subsection by referring to questions solved by the methods.

1.3.1 Structure of chapters

The book consists of twelve chapters including this introductory chapter (Chapter 1), and the structure of these chapters is illustrated in Figure 1.11.

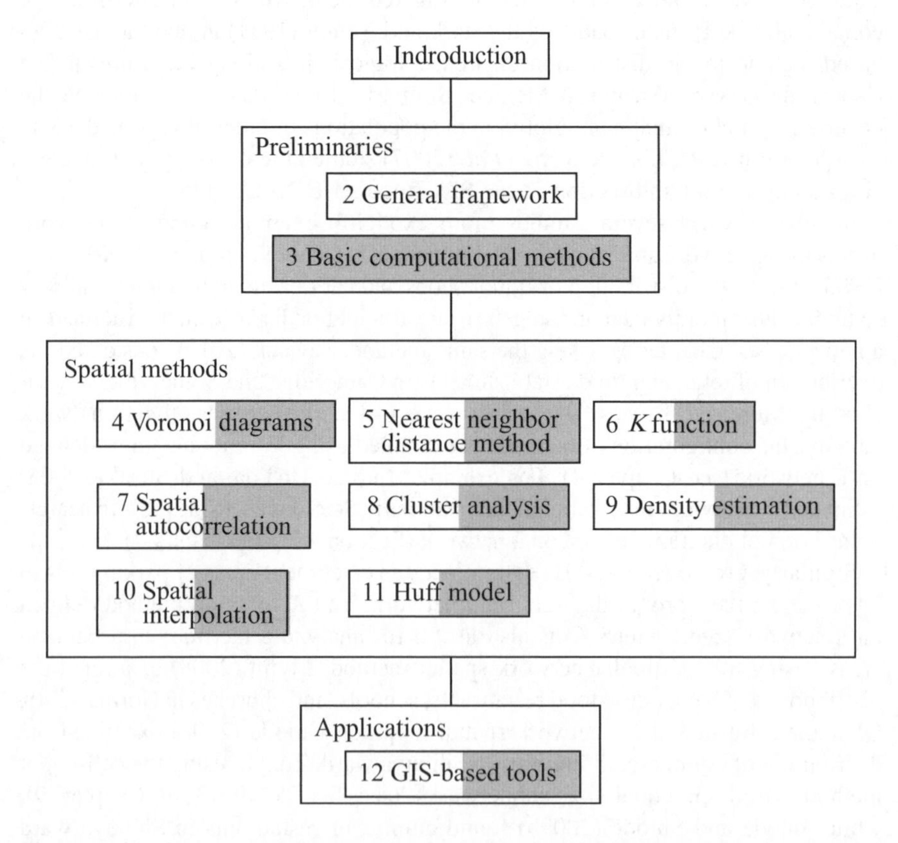

Figure 1.11 Structure of chapters (the shaded sections indicate computational parts).

The eleven remaining chapters constitute three parts: preliminaries, network spatial methods and applications.

The first part, consisting of Chapters 2 and 3 is concerned with preliminaries. Chapter 2 introduces fundamental concepts for network spatial analysis that are commonly used throughout this volume. In the chapter, we first show how conceptually to model real world networks and events on or alongside networks, and then illustrate how to numerically represent those conceptual models as data models. Third, we formulate a basic probabilistic model for events on a network, which is used as a basic null hypothesis throughout this book. In Chapter 3, we describe the elements of computational methods that are commonly used for carrying out the statistical calculations formulated in Chapters 4–11. We first show data structures for spatial analysis on one-layer and multiple-layer networks. Second, we illustrate geometric computational methods in a general way and then specify them for networks. Basic algorithms are illustrated with examples.

The main body of this book is the second part, consisting of Chapters 4–11. This part shows statistical methods for network spatial analysis and their efficient computational methods. The contents are described in the next subsection.

The last part, Chapter 12, shows tools for network spatial analysis and their applications to actual data. In practice, implementing the computational methods in each chapter in operational software requires much time and cost, particularly for those who are not skilled in programming. To simplify this task, the last chapter, Chapter 12, introduces GIS-based tools for network spatial analysis, in particular, a free software package, called *SANET* (Spatial Analysis along NETworks; Okabe, Okunuki, and Shiode, 2006a, 2006b; Okabe and Satoh, 2009). The book ends with references and an index.

1.3.2 Questions solved by network spatial methods

The second part of the book shows eight network spatial methods. We explain each method in terms of the questions it can solve.

Chapter 4 explains a family of *network Voronoi diagrams*, with which we can answer the question:

Q1′: For a given set of n generators, which may be points (e.g., fast-food shops), line segments (e.g., arterial streets), or circuits enclosing polygons (e.g., parks) on a network, how can we tessellate the network into n subnetworks associated with the n generators in such a way that the generator nearest from every point in a subnetwork is the generator assigned to the subnetwork?

A specific example of this general question is question Q1 in Section 1.1.1 (the catchment areas of parking lots). The resulting set of subnetworks is referred to as the *network Voronoi diagram*. The diagram can include inward and outward network Voronoi diagrams, weighted network Voronoi diagrams and k-th nearest point network Voronoi diagrams.

Chapter 5 formulates the *network nearest-neighbor distance method*, which gives answers to the questions:

Q2′: Given a set of points on a network, how can we test whether or not the shortest-path distance from each point to the next nearest point is significantly short (or long)?

Q3′: Given two sets of points (of types A and B) on a network, how can we test whether the shortest-path distance from each point of type A to the nearest point of type B is significantly short (or long)?

Specific examples are shown in questions Q2 (boutiques located side by side) and Q3 (burglaries occurring around stations) in Section 1.1.1.

Chapter 6 explains the *network* K *function method*. This method is similar to the above nearest-neighbor distance method, but different in that the method considers not only the nearest point but also points further away. This method is an alternative method to answer questions Q2 (boutiques located side by side) and Q3 (burglaries occurring around stations) in Section 1.1.1. More generally, the method can answer the questions:

Q2″: Given a set of points on a network, how can we test whether or not the number of points within a shortest-path distance from each point is significantly many (or few)?

Q3″: Given two sets of points (type A points and type B points) on a network, how can we test whether the number of type A points within a shortest-path distance from each point of type B is significantly many (or few)?

Chapter 7 discusses the *network spatial autocorrelation*, which can answer question Q4 (land value similarity along streets) in Section 1.1.1. This specific question can be generalized as:

Q4′: Given a set of attribute values of spatial units on a network (which may be represented by points, line segments, subnetworks) with the degrees of closeness between those spatial units (which may be categorical or numerical), are attribute values similar if their spatial units are close each other?

Chapter 8 illustrates *network spatial cluster analysis*, which gives an answer, for example, to question Q5 (clusters of fashionable boutiques) in Section 1.1.1, or more generally to the question:

Q5′: For a given set of points on a network, how can we find sets of clusters within which points are close to each other but between which the points are apart?

Chapter 9 shows the *network kernel density estimation method*. As in the example of question Q6 (detection of 'black spots' and 'hot spots') in Section 1.1.1, this method is useful for answering the question:

Q6′: For a given set of points on a network, how can we estimate the density of points along the network and detect high-density areas on the network?

Chapter 10 discusses *spatial interpolation* on a network, which examines, for instance, question Q7 (interpolating NO_x densities) in Section 1.1.1. More generally:

Q7′: Given known attribute values at a finite number of sample points on a network, how can we interpolate or predict an unknown attribute value at an arbitrary point on the network?

Chapter 11 formulates the *network Huff model*, a network version of the Huff model often used in marketing. The model can answer question Q8 (the choice probability of a fast-food shop), or the question:

Q8′: Given facilities located alongside a network and suppose that a user accesses the facilities through the network, how can we estimate the probability of the user choosing a facility among alternative facilities?

In each chapter, statistical and computational methods for answering these questions are described in detail.

1.3.3 How to study this book

This book can be read in several ways according to the reader's interest in network events. It is recommended that readers who wishes to understand network spatial analysis fully should read Chapters 1, 2 and 3 in that order; next Chapters 4–11 (the order depends on the reader's preference) and finally Chapter 12. It is recommended that readers who want to understand one of the network spatial methods shown in Chapters 4–11 should read Chapters 1, 2 and 3 in that order, and then proceed to the chapter containing the method that the reader wants to understand. The reader who wants to analyze data of network events with a specific network spatial method can first see whether a tool for that spatial method is provided in Chapter 12; if so, read the chapter containing that method in Chapters 4–11. As indicated by the subtitle of this book, *Statistical and Computational Methods*, the book shows each method in a separate subsection in every chapter. The reader who is mainly interested in statistical methods can read Chapters 1, 2 and the subsections on statistical methods in each chapter; the reader who is mainly interested in computational methods can read Chapters 1–3, the subsections on computational methods in each chapter and Chapter 12. However readers approach the book, we believe that it provides readers with theoretically as well as practically useful methods for network spatial analysis.

2

Modeling spatial events on and alongside networks

As shown in the review of the preceding chapter, there are various kinds of network events in the real world. To analyze them in a general manner, this chapter sets up a general framework and defines technical terms consistently used throughout this volume. General frameworks and technical terms for spatial analysis are found in many textbooks of geographic information science (Laurini and Thompson, 1992, Chapter 3; Johnston, 1998, Chapter 2; Burrough and McDonnell, 1998, Chapter 2; Bernhardsen, 2002, Chapter 3; Lo and Yeung, 2002, Chapter 3; O'Sullivan and Unwin, 2003, Chapter 2; Rigaux, Scholl, and Voisard, 2002, Chapter 2; Haining, 2003, Chapter 2; Worboys and Duckham, 2004, Chapter 3; Longley *et al.*, 2005, Chapter 2; and De Smith, Goodchild, and Longley, 2007, Chapter 2). However, when we compare technical terms in the different frameworks, there seems to be no standard nomenclature; even worse, a term in one framework can have a different meaning to the same term in another framework. To avoid such confusion, this chapter presents a consistent general framework together with technical terms for network spatial analysis.

The chapter has four sections. We first discuss how to conceptualize 'things' in the real world, and show two types of conceptual model: the *object-based model* and the *field-based model*. We next specify these general models to deal with network-like 'things' and 'things' on and alongside networks. Last, we introduce statistical concepts for the specified conceptual models.

Spatial Analysis along Networks: Statistical and Computational Methods, First Edition.
Atsuyuki Okabe and Kokichi Sugihara.

2.1 Modeling the real world

We consider that the real world is composed of 'things' (Laurini and Thompson, 1992; Raper, 1999; Haining, 2003). Here, the word 'things' has a broad implication, and includes, for instance, not only houses, streets, and parks, but also heat islands, crime occurrences, and traffic accidents. In the literature of geographical information science, the things are referred to as *entities* (Laurini and Thompson, 1992; Bernhardsen, 2002; Haining, 2003), *phenomena* (Lo and Yeung, 2002; O'Sullivan and Unwin, 2003; Johnston, 1998), or *events* (Burrough and McDonnell, 1998). To perform spatial analysis, the first step is to conceptualize a collection of entities forming the real world as a well-defined abstract model. We call the resulting model a *conceptual model*. Two types of conceptual model are frequently used in geographical information science (Couclelis, 1992; Goodchild, 1992; Worboys, 1995; Burrough and McDonnell, 1998): the *object-based model* and the *field-based model* (Figure 2.1), which are described in the following two subsections.

2.1.1 Object-based model

The *object-based model* assumes that entities are represented by discrete, distinct, and well-demarcated abstract 'things,' termed *objects*. The objects are described in terms of geometric elements representing the abstract forms of the entities together with the abstract characteristics of the entities (Figure 2.2). The *geometric elements* are points, lines, areas, and solids. An object may be represented by only one geometric element (e.g., a point) or by a set of many types of geometric element (e.g., points, lines, and areas). The abstract characteristics of entities are referred to as *attributes*, and the attributes are classified into *spatial attributes* and *nonspatial attributes* (Laurini and Thompson, 1992) (Figure 2.2).

In spatial analysis, we often deal with a set of areas or solids (each forming an object) that are mutually exclusive except at their boundaries and collectively exhaustive for a given space. Such a set forms a (*spatial*) *tessellation* for a given space (Rigaux, Scholl, and Voisard (2002) refer to it as a *tessellation mode*; for general discussion on spatial tessellations, see Okabe *et al.* (2000)). An example in

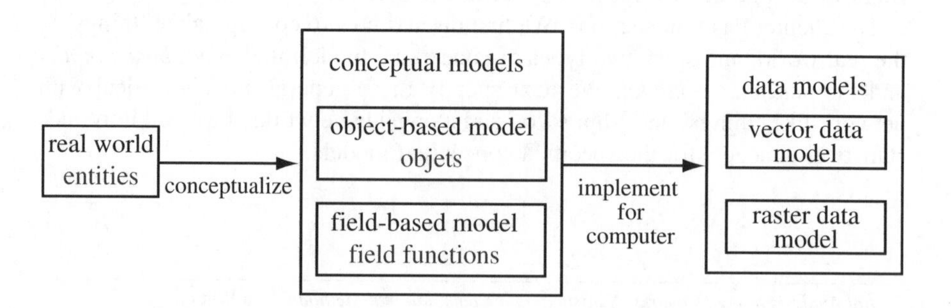

Figure 2.1 Modeling the real world.

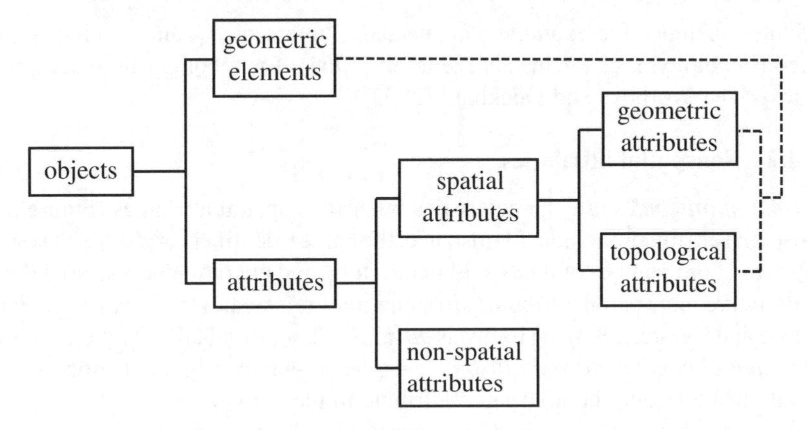

Figure 2.2 Attributes of objects in the object-based model.

the real world is a set of administrative districts (e.g., the 47 prefectures covering the whole of Japan).

2.1.1.1 Spatial attributes

Spatial attributes include *geometric attributes* and *topological attributes* (Figure 2.2). The *geometric attributes* are quantitative characteristics derived from the geometric elements forming an object, such as the length of a street or the area of a park. The *topological attributes* are qualitative characteristics derived from the geometric elements forming an object. In mathematical terms, *topological attributes* are characteristics that are preserved by topological transformation of the space in which geometric elements are embedded. Intuitively speaking, the *topological transformation* implies the deformation of a rubber sheet (representing a plane on which geometric elements are drawn) by stretching or shrinking without tearing or folding. For instance, a topological transformation is the transformation which transforms the network indicated by the black line segments on the shaded area in Figure 2.3a into the network on the shaded area in Figure 2.3b. Topological

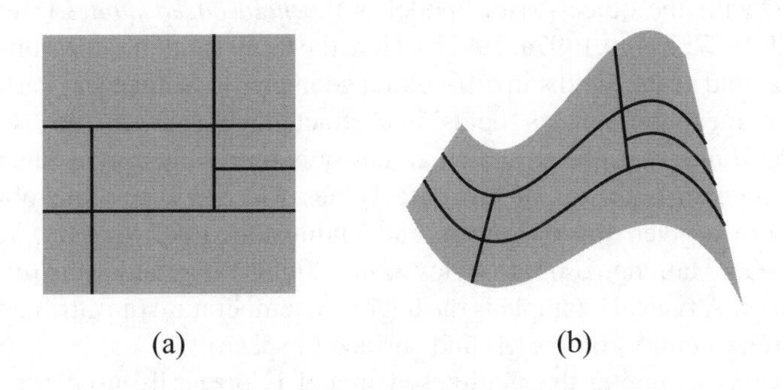

Figure 2.3 A topological transformation.

attributes include, for example, 'connected,' 'intersected,' and 'included' (for general topology, see, for instance, Pervin (1964); for topology in geographical contexts, see Worboys and Duckham (2004)).

2.1.1.2 Nonspatial attributes

Nonspatial attributes are characteristics other than spatial attributes (Figure 2.2). Nonspatial attributes include, for instance, the name (identifier) of a store, the owner of the store, the number of items sold in the store, and operating hours. Note that in the literature, nonspatial attributes are sometimes referred to as *thematic attributes* (Kresse and Fadaie, 2004) or simply as *attributes* (e.g., Bernhardsen (2002) uses the terms *spatial component* and *attribute component*, which might correspond to the spatial attributes and the nonspatial attribute in the above).

The attributes can or cannot be measured; if they can (the conditions for measurability are discussed in Krantz *et al.* (1971)), measured values are referred to as *attribute values*. The attribute values can be classified by scale type: *nominal scale* (e.g., Hilton Hotel), *ordinal scale* (one to five stars), *interval scale* (room temperature), and *ratio scale* (floor area) (Unwin, 1981, Chapter 2; Laurini and Thompson, 1992, Section 3.2; O'Sullivan and Unwin, 2003, Section 1.3; Longley *et al.*, 2005, Box 3.3). These scales are also applicable to geometric attributes if they can be measured.

It should be noted that the object-based model could be related to the *object-oriented model* in computer science. The object-oriented model has been introduced in geographical information science (e.g., Gahegan and Roberts, 1988; Oosterom and Van den Bos, 1989; Egenhofer and Frank, 1992; Worboys, 1994; Lo and Yeung, 2002; Kresse and Fadaie, 2004; Worboys and Duckham, 2004), but we do not adopt the specific terms used in the object-oriented model. Note also that the term *feature* is sometimes used in geographical information systems (GIS), but we do not use it here, because it can imply either an object or an entity as defined above (Johnston, 1998, Chapter 2).

2.1.2 Field-based model

Contrasted with the object-based model is the *field-based model* (Sachs, 1973; Goodchild, 1987; Kemp,1997a, 1997b). Here the term 'field' has the same meaning as in 'scalar and vector fields' in differential geometry (Millman and Parker, 1977), but does not mean the same as 'fields' in abstract algebra (Dean, 1966, Chapter 4). A *field-based model* is one in which an entity is represented by a function of an attribute value or a vector of attribute values for every possible point in the continuum of a given space (Laurini and Thompson, 1992; Rigaux, Scholl, and Voisard, 2002; Haining, 2003; Longley *et al.*, 2005). We refer to this function as a *field function*. A typical example is the degree in temperature (an attribute value) of atmosphere (an entity) on the ground surface (a space).

Besides this example, the field-based model is primarily used for elevation, precipitation, barometric pressure, ocean salinity, soil wetness, slope of terrain, and so on. The use of the field-based model is not restricted to those examples in which the attribute values have interval and ratio scales. The attribute values in a field-based model may take nominal and ordinal scales. For example, we can represent houses by a binary function defined on every point over a plane, where the attribute value takes 'house' if a point is included in a house; 'not house' if a point is not included in a house.

In a similar manner, we can represent land uses in a region by a multiple-value function defined on every point in the region where the attribute value can take one of several categorical values, say forests, lakes, rice fields, or wheat fields. Alternatively, land uses in a region can be represented in terms of a tessellation (in the object-based model), where a given space (the region) consists of objects (land parcels) and each object is associated with a categorical attribute value (a land-use category). As these examples indicate, many entities can be represented by both an object-based model and a field-based model, and the most appropriate one is used in a particular study (Bernhardsen, 2002; O'Sullivan and Unwin, 2003).

2.1.3 Vector data model

To perform spatial analysis in practice, the above conceptual models are not yet tractable to work with a computer. Therefore, we must describe the object-based model or the field-based model in terms of numerical values (Figure 2.1). This model is called a *data model* (Goodchild, 1992; Burrough and McDonnell, 1998; Bernhardsen, 2002). In the literature, two data models are frequently used: the *vector data model* and the *raster data model*. We describe these two models in this subsection and the next subsection.

By definition, a data model for an object-based model is decomposed into data models for spatial attributes and nonspatial attributes (Figure 2.2). The former data model is further decomposed into a data model for geometric attributes and one for topological attributes.

The data model for geometric attributes can be described in terms of the coordinates of the finite number of points forming geometric elements, or specifically the coordinates of points, those of the endpoints of line segments, and those of vertices forming polygons and polyhedrons. A curved line is approximated by a sequence of connected small straight line segments and a curved surface of a solid by connected small polygons. Because points, lines, polygons, and polyhedrons can be described in terms of vectors (Okabe *et al.*, 2000, Section 1.3), this model is referred to as a *vector data model*. The development of the vector data model has a long history, which is described by Cook (1998).

The data model for topological attributes is formulated in terms of topological relationships between geometric elements, and this model will be discussed in detail in Section 2.2.1.2 in this chapter and Chapter 3. The data model for nonspatial attributes is usually formulated using a table-like database, called a *relational database*, such as those used in Microsoft Excel and Access (Date, 2003, Chapter 4; DeMers, 2000, Chapter 4).

2.1.4 Raster data model

Contrasted with the vector data model is the *raster data model*, which is described in this subsection. As defined in Section 2.1.2, a field function is defined for every possible point on the continuum of a space, and the number of such points is infinite. In practice, however, a computer cannot deal with an infinite number of values. Therefore, in a data model for a field-based model, a field function is defined for a finite number of sample points. If necessary, attribute values at other than sample points are spatially interpolated using those at the sample points. This method is termed *spatial interpolation*, and is discussed in detail in Chapter 10 of this volume.

Several methods for sampling attribute values are proposed in the literature (e.g., Longley *et al.*, 2005, Section 3.6.3). The first method is to capture an attribute value at each of a lattice of regularly spaced points (Figure 2.4a). A grid lattice is most frequently used and the data based on a grid lattice (usually a fine grid lattice) are termed *raster data* and a lattice point is referred to as a *pixel*. The second method is to capture an attribute value at each point in the configuration of irregularly spaced points (Figure 2.4b).

The third and fourth methods assume that a given space is tessellated. The third method is to capture a single representative value (e.g., the mean, median, or maximum) of the attribute values over a regularly tessellated cell of the tessellation (Figure 2.4c). The fourth method is similar to the third method except that the shape of each cell is irregular (Figure 2.4d). The data captured by the third and fourth methods are referred to as *tessellation-based data* or *tessellation-mode data* (Rigaux, Scholl, and Voisard, 2002). When the irregularly spaced sample points are dense and the size of cells is small, the data captured by the second, third, and fourth methods could also be called *raster data* in a broad sense. The origin of the raster-based data model is old and its history is reviewed by Faust (1998).

To implement conceptual models, we use a vector data model and a raster data model (Figure 2.1). In principle, both models can be used for coding the data of an object-based model and those of a field-based model, but in practice there is a strong association between a raster data model and a field-based model, and between a vector data model and an object-based model (Couclelis, 1992; Worboys, 1995; Cova and Goodchild, 2002; Longley *et al.*, 2005).

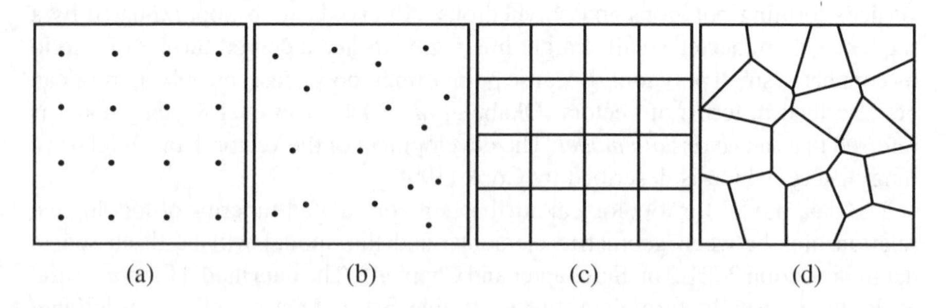

Figure 2.4 Raster data: (a) regularly spaced points, (b) irregularly spaced points, (c) regularly tessellated cells, (d) irregularly tessellated cells.

2.2 Modeling networks

Having identified frameworks for general spatial analysis, we now specify them for network spatial analysis. We conceptualize *network-like entities*, such as streets, rivers, and pipelines, using the object-based model (Section 2.1.1) and the field-based model (Section 2.1.2), which are described in the following two subsections.

2.2.1 Object-based model for networks

We first describe a network-like entity in terms of geometric elements defined in the object-based model.

2.2.1.1 Geometric networks

We represent a network-like entity by a set of lines embedded in two- or three-dimensional *Cartesian space,* i.e., Euclidean space with orthogonal axes. We assume that the lines satisfy the following conditions:

(i) The number of lines is finite.

(ii) The length of every line is finite; consequently the lines are line segments.

(iii) Every line segment is a straight line segment.

(iv) Line segments are connected only at the endpoints of the line segments.

(v) All line segments are *pairwise connected* in the sense that for any pair of points on any line segments, there is at least one path in the network connecting the two points.

We refer to the line segments satisfying the above conditions as *links* and their endpoints as *nodes* and denote the set of nodes and the set of links by $V = \{v_1, \ldots, v_{n_V}\}$ and $L = \{l_1, \ldots, l_{n_L}\}$, respectively. The pair of sets (V, L) is called a *geometric network* or simply a *network* and denoted by $N = (V, L)$ or N. The reader might expect N to denote nodes, but we use V, because N has already been used for a network (V stands for *vertex*, an alternative term for a node).

There are several points to note on the above conditions. The conditions (i) and (ii) imply that this book does not deal with networks that extend over an unbounded Cartesian space, such as a lattice network or a network generated by random lines (Miles, 1964; Bartlett, 1967; Santalo and Yanez, 1972). It follows from the conditions (i), (iii), and (iv) that the number of nodes is finite. We refer to the networks in which the numbers of nodes and links are finite and the lengths of links are finite as *finite networks* or *bounded networks*. This book deals only with finite networks. The condition (iii) might sound restrictive, but in practice, a curved line segment can be approximated by a connected sequence of small straight line segments in accordance with the required precision level.

A geometric network plays two roles in this volume. The first role is to show the geometrical and topological attributes of a network object (Figure 2.2). The second role, which is inherent in network spatial analysis, is to act as a 'container,' i.e., an underlying space for spatial analysis on a network, just as the Cartesian space acts as an underlying space for spatial analysis on a plane. In this case, we call the geometric network a *network space*. The network space may be denoted by $N = (V, L)$ or by the union, $\tilde{L}$, of $l_1, \ldots, l_{n_L}$, written as $\tilde{L} = \bigcup_{i=1}^{n_L} l_i$, the set of points constituting N. The former notation is used when we clarify the graph structure of a network (to be explained below), the latter when we indicate a point on the links (which include nodes) of a network. Note that a network space does not imply a space consisting of networks (Riviere and Schmitt, 2007) like a function space in mathematics (Pervin, 1964, Chapter 9).

2.2.1.2 Graph for a geometric network

A geometric network implicitly indicates the topological attributes of a network. The topological attributes of a geometric network can be made explicit by extracting its *graph* structure. We first abstractly state what a graph is (textbooks of graphs are, e.g., Busacker and Saaty (1965), Harary (1969), Chartrand (1984)).

Let V be a nonempty set, and let L be a set of pairs of elements of V (note that in general, L and V are not those used in the above). An example is illustrated in Figure 2.5a. Then, the pair $G = (V, L)$ is called an *abstract graph* or simply *graph*, and the elements of V and those of L are called *vertices* and *arcs*, respectively. If both V and L are finite sets, $G = (V, L)$ is called a *finite graph*. Conventionally, we can visualize an abstract graph in Cartesian space by considering vertices as distinct points and by connecting pairs of vertices by line segments (including curves) if associated pairs belong to L. Such a graph is termed a *geometric graph*. An example is depicted in Figure 2.5b.

Every geometric graph can be realized in three-dimensional Cartesian space without its line segments meeting except at their endpoints (Busacker and Saaty, 1965, Theorem 1.1). However, it may or may not be possible to represent a graph in two-dimensional Cartesian space. If a graph can be represented in two-dimensional

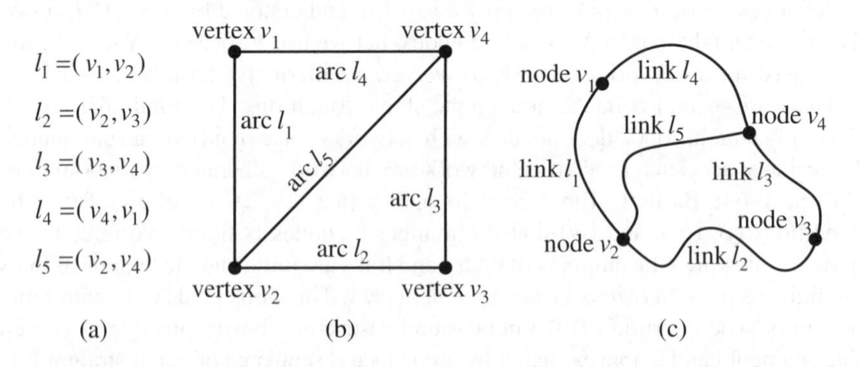

Figure 2.5 An abstract graph (a), a geometric graph (b), and a network (c).

Cartesian space without its line segments crossing except at endpoints, we call it a *planar graph*; otherwise, a *nonplanar graph*.

The endpoints of L can be considered as ordered pairs of vertices or as unordered pairs of vertices. If they are ordered pairs, the associated graph is called a *directed graph*, and its arcs are visualized as line segments with arrows. If the elements of L are unordered pairs of vertices, on the other hand, the associated graph is called an *undirected graph*, and its arcs are visualized by simple line segments.

For a given geometric network N (e.g., Figure 2.5c), we regard the set of nodes V as that of vertices V, and the set of links L as that of arcs L. Then, we obtain the graph $G = (V, L)$ (Figure 2.5a or b). This graph is interpreted as an abstract representation of the network such that we are concerned only with the pairs of nodes that are connected by links. In other words, the graph of the network N (Figure 2.5b) represents the qualitative (topological) aspect of the network (Figure 2.5c), neglecting the quantitative (geometrical) aspects such as the locations of the nodes and the lengths of the links (Section 2.1.1.1 or Figure 2.2).

2.2.2 Field-based model for networks

The field-based model for a network is defined by a field function with a domain given by the continuum of a network space, $\tilde{L}$ (Cova and Goodchild, 2002; Xu and Sui, 2007). In principle, the attribute may take ratio, interval, ordinal, or nominal scales. In practice, the first two scales are most frequently used. Examples are: the altitude of a road (a ratio scale) as shown in Figure 2.6, and the ground temperature of a road (an interval scale). The latter two scales could be used, for instance, for the name of each street (a nominal scale) and the rank of each road (an ordinal scale), but these are usually treated by the object-based model.

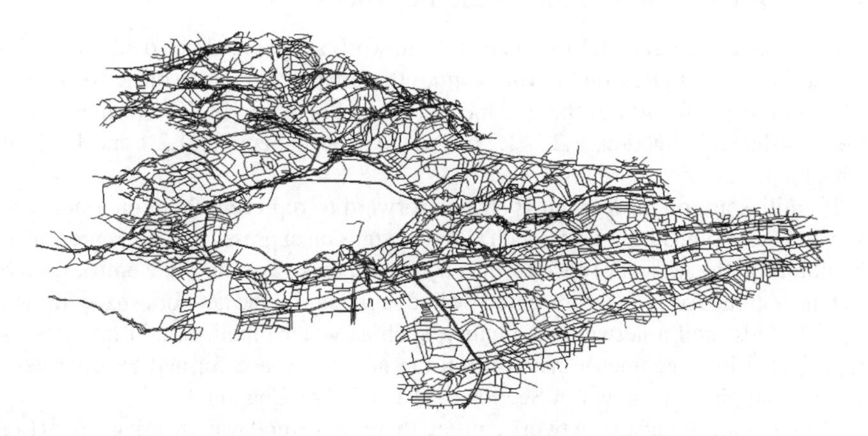

Figure 2.6 A field-based model representing the altitude of roads in Shibuya ward, Tokyo (the data provided by PASCO Corp.).

2.2.3 Data models for networks

The most commonly used data model for networks is the vector data model (Figure 2.1). This formulation is fully developed in Chapter 3. The raster data model is rarely used for networks in the literature except for Xu and Sui (2007). However, as shown in the preceding subsection, the field-based model is effective in describing attributes varying over a network; consequently, the raster data model for networks is indispensable for network spatial analysis.

In theory, raster data for the field-based model of a network could be constructed by extending the four sampling methods on a plane in Section 2.1.4 (Figure 2.4) to those on a network. In practice, however, such an extension is not always straightforward. Grid lattice points may correspond to equally spaced sample points along links, but equal spacing is difficult except for very special cases, because the length of each link varies and it is impossible to find a common divisor for all the links. Randomly sampled points are easy to generate, using the method in Section 3.4.4 of Chapter 3. Regularly shaped cells in the third method may correspond to equal-length links or more generally equal-length subnetworks. We refer to subnetworks forming a tessellation as *network cells* or *cells* in short. Constructing equal-length network cells is not a simple task and will be discussed in Chapter 9.

2.3 Modeling entities on network space

Network spatial analysis deals with entities on and alongside a network-like entity. Therefore, we must conceptualize those entities on a network space. As formulated in Section 2.1, we can conceptualize the entities with the object-based model or the field-based model (Figure 2.1).

2.3.1 Objects on and alongside networks

As illustrated in Figure 1.1 of Chapter 1, network spatial analysis deals with two broad classes of entities: *on-network entities* that are placed exactly on networks and *alongside-network entities* that are located alongside networks. The former examples are shown in Sections 1.2.2–1.2.6, and the latter in Sections 1.2.1 and 1.2.7 of Chapter 1.

If entities are on network, it is straightforward to represent them on a network space. Thus, point-like entities, such as traffic signs on expressways, are represented by points on the network space formed by the expressways; line-like entities, such as fences along roads are represented by line segments on the network space formed by the roads; and a network-like entity, such as water pipelines laid in roads, is represented by a geometric network on the network space formed by the roads. Other examples are shown in Sections 1.2.1–1.2.7 of Chapter 1.

If entities are alongside-network entities, there are several ways to assign them to a network space. To illustrate them, imagine, for instance, stores standing alongside a street in a city. Every store has an entrance or a gate adjacent to a street. Therefore, a

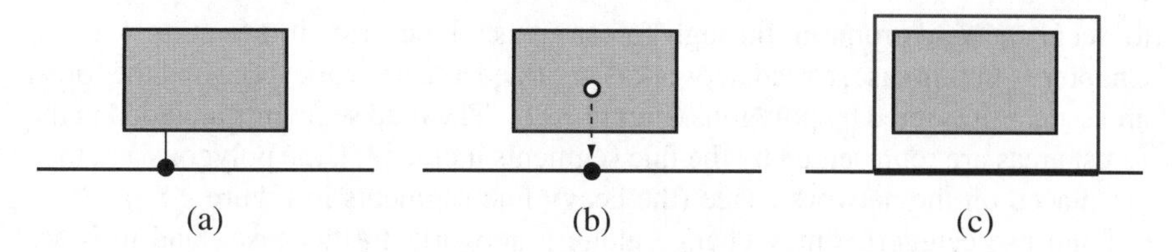

Figure 2.7 Stores represented by (a) the entrance of a store (the black circle), (b) the nearest point (the black circle) from the center (the white circle) of a store, and (c) the boundary line segment or frontage interval (the bold line segment) shared with the site of a store and the facing street.

store is naturally represented by its access point, which is represented by a point on a network space (the black circle in Figure 2.7a). In practice, data for access points are not always available. In such a case, we first compute the center of the polygon representing a store (the white circle in Figure 2.7b), and next, find the nearest point on a street from the center (the black circle). Last, we regard the nearest point as the access point of the store, and represent it by a point on its network space. Alternatively, we can represent a store by the boundary line segment shared with the site of the store and the facing streets (the heavy line segment in Figure 2.7c). Computational methods for finding the nearest point and the shared boundary line segment (frontage interval) are shown in Chapter 3 and access points obtained by a GIS-based tool is illustrated in Figure 12.2 of Chapter 12.

In network spatial analysis, intersection between entities is an important factor affecting spatial processes. For instance, suppose that we want to analyze the accidents resulting from slipping on wet ground on mountain trails in relation to creek crossing points. To achieve this analysis, we regard a trail network as a network space and indicate the intersections between the trail network (the black line segment in Figure 2.8a) and the creek network (the gray line segments) by points (the black circles) on that network space. Then a computation task is to find intersections between the two networks, and a method for this is shown in Chapter 3.

A network-like entity forming a network space may intersect not only with another network-like entity as in the above example but also with polygon-like entities, and such intersections may be an important factor for network spatial analysis. For instance, suppose that we want to test the hypothesis that roadkills tend

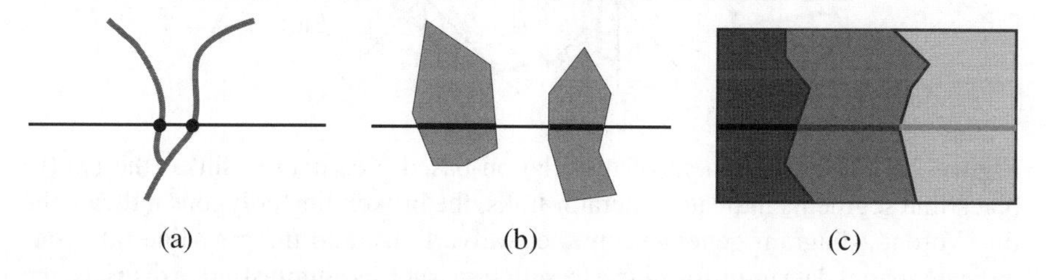

Figure 2.8 A line segment of a network intersecting (a) a network, (b) a polygons, and (c) polygons forming a tessellation.

to occur on roads running through forest areas, as described in Section 1.2.3 of Chapter 1. In this case, a road network is regarded as a network space and the forest areas are represented by polygons (Figure 2.8b). The road segments included in the forest areas are represented by the line segments included in the polygons and they are placed on the network space (the heavy line segments in Figure 2.8b).

Land use categories may change along a network. In this case, land uses are described in terms of a tessellation consisting of land parcels with their land use categories (Figure 2.8c). This tessellation divides a road network into road segments, and the collection of these road segments form a tessellation of the road network (Figure 2.8c). A computational method for constructing this tessellation is shown in Chapter 3.

In marketing, demand estimation is a crucial task and it is estimated from population data. The data units of population are usually polygons forming a tessellation (e.g., administrative districts). In reality, consumers access stores through a street network and hence population data with respect to street segments are useful. At present, however, such data are rarely available, and we must convert population data with respect to polygons into data with respect to links or network cells. A method for this conversion uses Voronoi diagrams for lines (Okabe *et al.*, 2000, Section 3.5) in the following manner.

We construct the Voronoi diagram for links $l_1, \ldots, l_{n_L}$ (the twelve black line segments in Figure 2.9; the computational methods are shown in Okabe *et al.* (2000, Section 3.5)). The resulting region associated with the line l_j is denoted by $V(l_i)$. We call the set of resulting Voronoi regions, $\mathcal{V}(L) = \{V(l_1), \ldots, V(l_{n_L})\}$, the *line Voronoi diagram* (the broken line polygons in Figure 2.9) generated by $l_1, \ldots, l_{n_L}$.

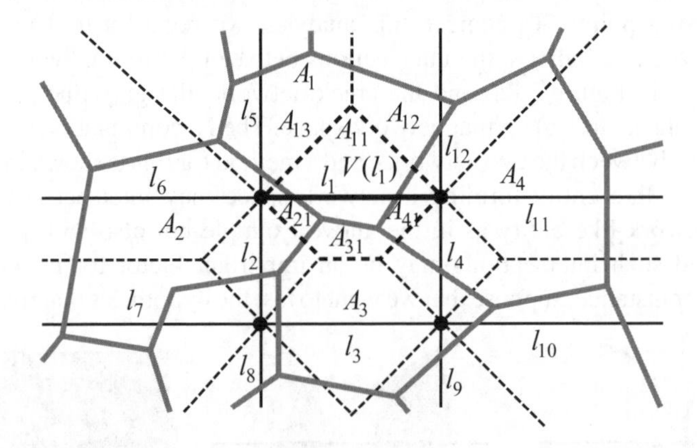

Figure 2.9 The assignment of tessellation-based area data to links (the twelve black line segments indicate generator links, the broken line polygons indicate the line Voronoi diagram generated by the twelve links, and the gray line polygons indicate spatial data units forming a tessellation, such as administrative districts; the bold black line segment l_1 is an example to which the populations of the districts that overlap the Voronoi region $V(l_1)$ are assigned).

Let A_i be the i-th administrative district (the gray line polygons in Figure 2.9) with population P_i, $i = 1, \ldots, n_A$. We want to assign $P_1, \ldots, P_{n_A}$ to $l_1, \ldots, l_{n_L}$. First, we superimpose $V(l_1), \ldots, V(l_{n_L})$ on $A_1, \ldots, A_{n_A}$. As a result, A_i is divided into subdistricts $A_{i1}, \ldots, A_{in_i}$ (e.g., A_{11}, A_{12}, A_{13} in Figure 2.9). Second, we assign the population in proportion to the area $|A_{ik}|$ of A_{ik}, i.e., population $P_i \times (|A_{ik}|/|A_i|)$ is assigned to the subdistrict A_{ik}. Third, we aggregate the population $P_i \times (|A_{ik}|/|A_i|)$ with respect to A_{ik} that is included in $V(l_j)$, i.e. the population $\sum P_i \times (|A_{ik}|/|A_i|)$ is assigned to the link l_j (e.g., in Figure 2.9, $P_1 \times (|A_{11}|/|A_1|) + P_2 \times (|A_{21}|/|A_2|) + P_3 \times (|A_{31}|/|A_3|) + P_4 \times (|A_{41}|/|A_4|)$ is assigned to l_1).

This assignment is very useful not only for population assignment but also for any attribute values of polygonal cells forming a tessellation on a plane to network cells.

2.3.2 Field functions on network space

We can obtain field functions on a network space from two sources: from attribute values of the network-like entity that forms a network space; and from those of entities other than a network-like entity. The former case has been discussed in Section 2.2.2. An example is the altitude along a road network as in Figure 2.6. In the latter case, field functions are constructed from the attribute values of point-like, line-like, and polygon-like entities on or alongside a network-like entity of concern. For instance, suppose that we want to test the hypothesis that street burglaries at night tend to occur inversely proportionally to the luminosity at points on streets (Section 1.2.4 in Chapter 1). In this case, we construct the field function of luminosity at every point on the streets in terms of the distances from the point to nearby streetlights (point-like entities). Another hypothesis is that frog-kills on roads tend to occur on roads near ponds. In this case, we construct the field function of nearness to ponds at every point on the roads in terms of the distances from the point to nearby ponds (polygon-like entities).

2.4 Stochastic processes on network space

Having established the general framework in the preceding sections, we now introduce statistical concepts to that framework. In statistical contexts, 'things' in the real world are often referred to as *events* rather than entities. Diggle (1983) refers to points being placed at locations as *events*. More generally, we refer to events that can be specified in a space as *spatial events* or *location events*. Examples of spatial events are car accidents on roads, purse snatches on pedestrian ways and fast-food shops standing alongside streets. Spatial events may occur deterministically or probabilistically. In the latter case, we call them *probabilistic spatial events* or *stochastic spatial events*.

To perform statistical spatial analysis, we formulate a collection of stochastic spatial events that occur according to some underlying random mechanism as a conceptual model. We call the resulting model a *probabilistic spatial model* or *stochastic spatial model* (Diggle (1983), Cressie (1991), Illian *et al.* (2008); note that their concepts are narrower than ours). Following the general framework in Section 2.1, we formulate stochastic spatial models with the object-based model and the field-based model, which are discussed in the following two subsections, respectively.

2.4.1 Object-based model for stochastic spatial events on network space

In the context of the object-based model, as can be seen in Figure 2.2, stochastic factors may exist in spatial attributes (including geometric and topological attributes) and nonspatial attributes. In the literature, two approaches are often used: stochastic spatial attributes with deterministic nonspatial attributes, and stochastic nonspatial attributes with deterministic spatial attributes. An example of the former approach is that the locations of fast-food shops are stochastically generated but their prices of foods are fixed. An example of the latter approach is that the prices of foods are stochastically generated but the locations of the fast-food shops are fixed. In spatial statistics, the former approach is referred to as *stochastic spatial processes* (Getis and Boots, 1978; Cox and Isham, 1980; Ripley, 1981, 1988; Tautu, 1986). Here, 'process' does not imply a dynamic evolution over time (Stoyan, 2006). The models that deal with spatial processes and time processes at the same time are referred to as *spatiotemporal models* or *spatiotemporal processes* (Cressie and Wikle, 2011). To avoid this confusion, Stoyan (2006) proposes the term *random fields* (which should be distinguished from 'field' as in the field-based model), but most textbooks conventionally adopt the term *stochastic spatial processes*. We follow this convention.

On a network space, possible geometric elements are points and lines. Stochastic spatial processes of points and those of lines are termed *stochastic point processes* and *stochastic line processes*, respectively. In spatial statistics, major attention has been paid to stochastic point processes on a plane (Matern, 1960; Bartlett, 1975; Ripley, 1981; Diggle, 1983; Møller and Waagepetersen, 2002; Baddeley *et al.*, 2006), but not on a network except for very simple networks, i.e., a line (Daley and Vere-Jones, 2003, Chapter 3; Lieshout, 2000). Similarly, most stochastic line processes are defined on a plane (Bartlett, 1967; Davidson, 1974; Stoyan, Kendall, and Mecke, 1995), but not on a network, except by Roach (1968, Chapter 2) who examined stochastic processes of line segments on a line.

2.4.2 Binomial point processes on network space

It follows from the finite network assumption in Section 2.2.1.1 that stochastic point processes take place on a bounded network. One of the most fundamental bounded spatial processes is the binomial point process, which is defined in this subsection.

We consider a finite network space, $\tilde{L} = \bigcup_{i=1}^{n_L} l_i$ embedded in two-dimensional Cartesian space with the orthogonal x_1 and x_2 axes, as shown in Figure 2.10 (to make the illustration simple, we assume a two-dimensional Cartesian space but the following formulation can be extended to the three-dimensional Cartesian space

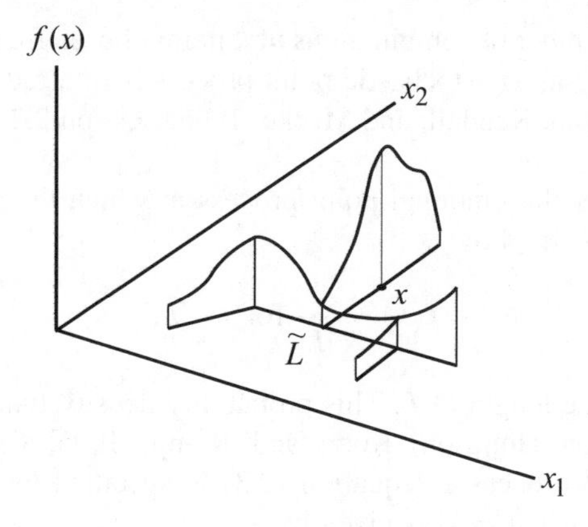

Figure 2.10 A probability density function $f(x)$ on $\tilde{L}$.

with slight modifications). An arbitrary point x on $\tilde{L}$ is represented by the coordinates $x = (x_1, x_2)$, and denoted by $x \in \tilde{L}$, where the binary relation $\in$ as in $a \in A$ means that a is an element of a set A. We consider a real-valued function of x, denoted by $f(x)$, with domain $\tilde{L}$ (Figure 2.10). We assume that the function $f(x)$ is (Riemann) integrable on $\tilde{L}$ (Bartle, 1964, Chapter 4; Bartle, 1966), and denote the integral of $f(x)$ over $\tilde{L}$ by $\int_{x\in\tilde{L}} f(x)\mathrm{d}x$, where d$x$ is the integration operator symbolically indicating an infinitesimal line segment on $\tilde{L}$ (it does not imply dx_1dx_2). If the function $f(x)$ satisfies the following conditions:

$$0 \le f(x) < \infty \text{ for } x \in \tilde{L}, \tag{2.1}$$

$$\int_{x\in\tilde{L}} f(x)\mathrm{d}x = 1, \tag{2.2}$$

the function $f(x)$ is termed a *probability density function* on $\tilde{L}$.

We now consider a stochastic point process in which a finite number, n, of points are independently and identically generated according to a probability density function $f(x)$. *Independently* implies that the n points do not interact; *identically* implies that the n points are all generated by the same probability density function $f(x)$. Let L_s be an arbitrary subset of $\tilde{L}$ (i.e., $L_s \subseteq \tilde{L}$), and denote the number of points placed on L_s by $n(L_s)$. In the context of the above stochastic point processes, the number $n(L_s)$ is a random variable, and this random variable follows a binomial distribution with parameters n and $\int_{x\in\tilde{L}} f(x)\mathrm{d}x$ (for general properties of the binomial distribution, see Johnson, Kotz, and Kemp (1992, Chapter 3)). Stated explicitly, the probability of $n(L_s)$ being equal to k, denoted by $\Pr[n(L_s) = k]$, is given by:

$$\Pr[n(L_s) = k] = {}_nC_k \left(\int_{x\in L_s} f(x)\mathrm{d}x \right)^k \left(1 - \int_{x\in L_s} f(x)\mathrm{d}x \right)^{n-k}, \tag{2.3}$$

where ${}_nC_k$ is the number of combinations of k items chosen from n items. Because of this property, the above stochastic point process is referred to as the *binomial point process* (Stoyan, Kendall, and Mecke, 1995, Section 2.2; Illian *et al.*, 2008, Section 2.2).

We next consider the binomial point process in which the probability density function $f(x)$ is specified by:

$$f(x) = \frac{1}{|\tilde{L}|} \text{ for } x \in \tilde{L}, \tag{2.4}$$

where $|\tilde{L}|$ means the length of $\tilde{L}$. This probability density function is termed the *uniform distribution* (Johnson, Kotz, and Kemp, 1995, Chapter 26). When the binomial point process of equation (2.3) is specified by equation (2.4), or mathematically, $\Pr[n(L_s) = k]$ is given by:

$$\Pr[n(L_s) = k] = {}_nC_k \left(\frac{|L_s|}{|\tilde{L}|}\right)^k \left(1 - \left(\frac{|L|_s}{|\tilde{L}|}\right)\right)^{n-k}, \tag{2.5}$$

the process is termed the *homogeneous binomial point process*. A binomial point process with the probability density function not given by equation (2.4) is referred to as an *inhomogeneous binomial point process*.

When n points are generated on a network space $\tilde{L}$ according to the homogeneous binomial point process, the probability of a point being placed on an arbitrary subnetwork L_s of $\tilde{L}$ depends only on the length of L_s; it depends neither on the shape of L_s nor the location of L_s in $\tilde{L}$. For example, in Figure 2.11, the probability of a point being placed on L_s and that of a point being placed in L_{Si} are the same for $i = 1, \ldots, 6$ because $|L_S| = |L_{Si}|$ holds for $i = 1, \ldots, 6$.

The randomness characterized by the homogeneous binomial point process is termed the *complete spatial randomness* and it is abbreviated to *CSR* (Diggle,

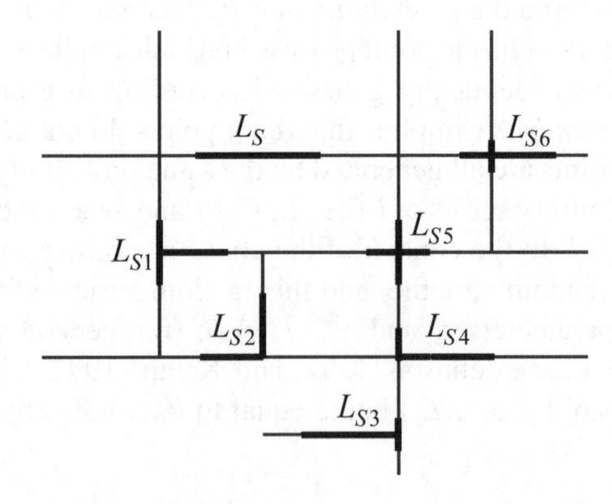

Figure 2.11 In the homogeneous point process, the probability of a point being placed on L_s is equal to that on L_{Si} for $i = 1, \ldots, 6$ (the bold line segments) because $|L_S| = |L_{Si}|$ holds for $i = 1, \ldots, 6$.

2003, Section 1.4). Note that Diggle (2003) defines CSR in terms of the Poisson point process, but in this book we define it in terms of the homogeneous binomial point process, because, as noted above, we always deal with finite networks. A slight difference exists between CSR as defined by the Poisson point process and that defined by the binomial point process. In the latter case, the covariance between the random number of points being generated in a small line segment and that in another small line segment is close to zero but not perfectly zero (Wilks, 1962, Section 6.3).

In statistical spatial analysis, one of the most fundamental hypotheses is that points are generated according to the homogeneous binomial point process, and this hypothesis is referred to as the *CSR hypothesis*. When the CSR hypothesis is rejected, we next examine an inhomogeneous binomial point process, which is specified by a nonuniform probability density function. For example, when we want to test the hypothesis that car accidents occur in proportion to traffic volume (Section 1.2.2 in Chapter 1), we derive the probability density function from a traffic density function (a field function) over a road network (Okabe and Satoh, 2005).

2.4.3 Edge effects

We theoretically assume that spatial processes occur on a bounded network space and no events occur outside the network (recall Section 2.2.1.1). In practice, this theoretical assumption may or may not be acceptable. Bounded networks used in empirical studies may be *naturally bounded networks* or *artificially bounded networks*. An example of the former is a road network in an island (e.g., the bold black and gray line segments in the shaded area in Figure 2.12); that of the latter is a street network in an urbanized area (the black bold line segments in Figure 2.12) obtained by discarding streets outside the area (the gray line segments in Figure 2.12). The edges of a naturally bounded network, referred to as *natural edges*, are real, but those of an artificially bounded network, referred to as *artificial*

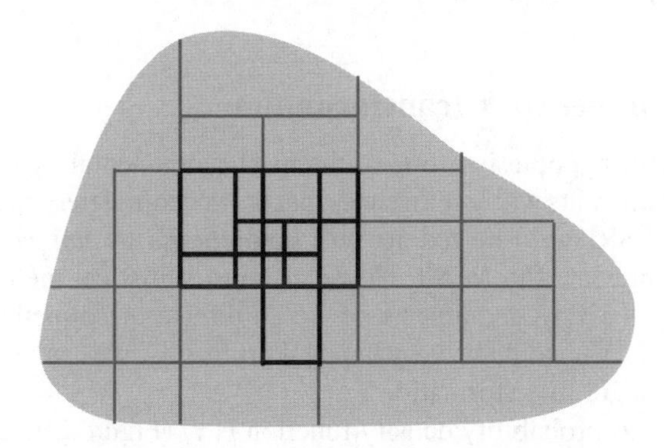

Figure 2.12 A naturally bounded network (the gray and black line segments) in an island (the shaded area) and an artificially bounded network (the black line segments).

edges, are fictive. If a bounded network is naturally bounded, the theoretical assumption of no events outside the bounded network is acceptable. However, if a network is artificially bounded, the assumption is not acceptable. In addition, we often have no data about the events outside an artificially bounded network. As a result, we inevitably ignore spatial processes outside an artificially bounded network. This ignorance results in a bias in understanding spatial processes on an artificially bounded network, because a spatial process outside the network (i.e., in the gray network in Figure 2.12) may affect the interior of the network (the bold black network). This resulting bias is termed *edge effect* or the *edge bias*.

The edge effects on a plane are discussed in depth in the literature (Boots and Getis (1988, Section 3.3); a review by Yamada (2009)), but they are rather different from those on a network. Most theoretical spatial processes on a plane assume an unbounded plane (e.g., the Poisson point process). However, networks used in empirical studies are always bounded. Therefore, edge effects inevitably appear. Many methods for correcting the edge effect, termed *edge correction* (Ripley, 1988), are proposed in the literature, such as toroidal mapping (Dacey, 1975; Ingram, 1978; Ripley, 1979; Griffith and Amrhein, 1983; a review by Yamada (2009)), but note that those methods are not applicable to network spatial analysis.

We should make two remarks on the edge effect on a network. The spatial methods to be shown in the following chapters all take the boundedness of networks into account. Therefore, if we apply those methods to a naturally bounded network, we do not have edge effects; if we apply them to an artificially bounded network, we do have edge effects. In practice, we may determine an artificially bounded network in such a way that the number of artificial edges is small. For example, if we choose a road network in a city surrounded by a rural area, the road network has a small number of artificial edges. As a result, the edge effect or bias is weakened. Therefore, the edge effect is often not so serious in network spatial analysis as in planar spatial analysis. Yet, edge effects may not be neglected. A theoretical study was initiated by Morita and Okunuki (2006), which is referred to in Section 6.3 of Chapter 6.

2.4.4 Uniform network transformation

Although the CSR hypothesis is one of the most fundamental hypotheses, we are more interested in events with occurrences that are not completely spatially random. In contrast to CSR, we referred to this randomness as *not complete spatial randomness*, abbreviated to *NCSR*. However, most statistical methods are developed for testing the CSR hypothesis. If we could use these methods for testing NCSR hypotheses, we would be very happy. Fortunately, we can use them through the following 'magic' transformation.

For a continuous probability density function $f(x)$ (Figure 2.13a), consider the transformation given by the *cumulative distribution function* of $f(x)$ (Figure 2.13b), or mathematically:

$$y = F(x) = \int_{-\infty}^{x} f(x)\mathrm{d}x, \tag{2.6}$$

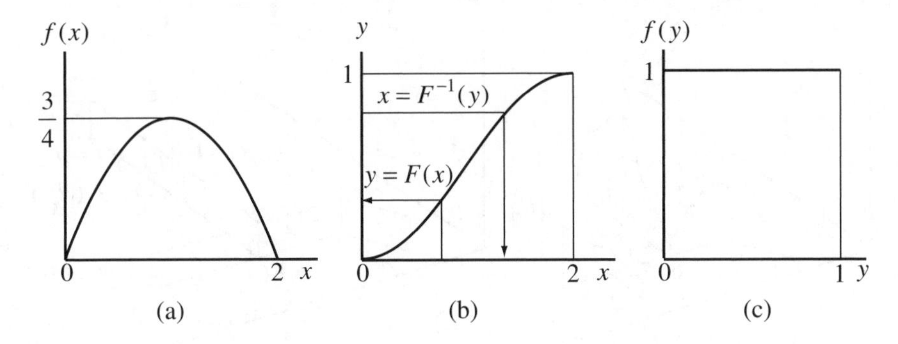

Figure 2.13 A probability integral transformation, for example, (a) a probability density function, $-\frac{3}{4}(x-1)^2+\frac{3}{4}$, (b) a probability integral transformation, $y=F(x)=-\frac{1}{4}x^3+\frac{3}{4}x^2$, and (c) a uniform density function, $f(y)=1$.

and let $x=F^{-1}(y)$ be the inverse function of $F(x)$ (Figure 2.13b). Then, we have the following theorem.

Theorem 2.1: probability integral transformation

If a variable x is randomly generated according to a probability density function $f(x)$, then the transformed variable $y=F(x)$ follows the uniform distribution with domain $0 \leq y \leq 1$. Conversely, if a random variable y is generated by the uniform distribution with domain $0 \leq y \leq 1$, then the transformed variable $x=F^{-1}(y)$ follows the probability density function $f(x)$.

This transformation is termed the *probability integral transformation*. The original idea was found in Fisher (1932, Section 21.1), although he used the term *pivotal function* (a review is provided by Pearson (1938)). The proof is shown in Freund (1971, Theorem 4.2).

This transformation provides a very powerful tool for generating random numbers for various probability density functions. Random numbers in a statistics table are usually from the uniform distribution (Figure 2.13c), but we often want to have random numbers generated by a nonuniform distribution $f(x)$ (Figure 2.13a). In such a case, we first obtain random numbers y from a table of random numbers from the uniform distribution (Figure 2.13c); next, we transform the numbers by the inverse function, $x=F^{-1}(y)$ (Figure 2.13b). Then, the resulting numbers x are regarded as random numbers generated by the nonuniform distribution $f(x)$ (Figure 2.13a). Conversely, if we want to have random numbers of the uniform distribution (Figure 2.13c), we first generate random numbers x according to the nonuniform distribution $f(x)$ (Figure 2.13a); next, we transform the numbers by the function $y=F(x)$ (Figure 2.13b). Then, the resulting numbers y follow the uniform distribution (Figure 2.13c).

It is straightforward to apply this transformation to a nonuniform density function on a line (the simplest network space), because random numbers generated by a

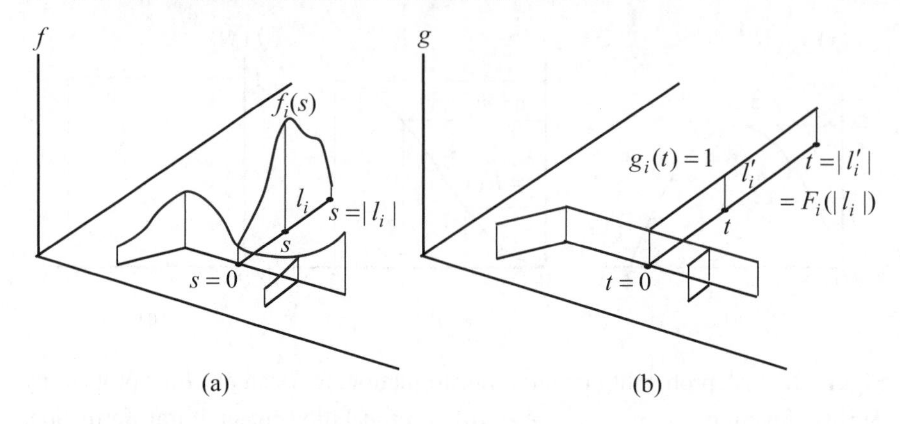

Figure 2.14 A uniform network transformation: (a) a nonuniform function, (b) its transformed uniform function.

probability density function can be regarded as points on a line. Therefore, points generated by an inhomogeneous binomial point process on a line can be transformed to points generated by the homogeneous binomial point process on a line through the probability integral transformation.

The extension of the probability integral transformation from the simplest network space (a line) to a general network space is not difficult. Because points are independently generated in the binomial point process, we can decompose a probability density function $f(x)$ on a network space $\tilde{L} = \bigcup_{i=1}^{n_L} l_i$ into probability density functions on links, $l_1, \ldots, l_{n_L}$. Let $f_i(s)$ be part of the probability density function of $f(x)$, where the domain of $f_i(s)$ is link l_i, s is the parametric distance along l_i from an end node of l_i ($s = 0$) to a point on l_i, and $f_i(s)$ is the density at s (note that the value of s at the other end is $s = |l_i|$, the length of l_i) (Figure 2.14). Then, we can define a transformation similar to the probability integral transformation for each link l_i as:

$$t = \int_0^s f_i(u)\mathrm{d}u = F_i(s). \tag{2.7}$$

We now transform from the network space $N(V, L)$ to a new network space $N'(V', L')$ in such a way that l_i corresponds to l'_i, where $|l'_i| = F_i(|l_i|)$; the node at $s = 0$ in V corresponds to the node at $t = 0$ in V'; the node at $s = |l_i|$ in V corresponds to the node at $t = |l'_i|$ in V'; and a point at s on the link l_i is mapped to the point at t on the link l'_i by $t = F_i(s)$ (Figure 2.14). We call this transformation the *uniform network transformation* (Okabe and Satoh, 2005), because any nonuniform distribution on $\tilde{L}$ is transformed into the uniform distribution on $\tilde{L}' = \bigcup_{i=1}^{n_L} l'_i$, i.e., $g_i(t) = 1,\ 0 \le t \le |l'_i|$ for all i. Actual applications of the uniform network transformation will be shown in the subsequent chapters.

3

Basic computational methods for network spatial analysis

In Chapter 2, we presented two conceptual models, namely the *object-oriented model* and the *field-based model*, for representing *network events*, i.e., events that occur on and alongside networks. We then described these models in terms of numerical values, known as *data models*, for which we introduced two types, the *vector data model* and the *raster data model*. Using these models, in subsequent chapters we formulate statistical methods for *network spatial analysis*, i.e., spatial analysis for examining network events. Unfortunately, the statistical methods require demanding geometric computation. Therefore, in practice, we can perform network spatial analysis only when those computational methods are feasible. In the following nine chapters, we illustrate practical computational methods for this purpose. To assist this purpose, this chapter focuses on the basic computational methods common to the individual computational methods presented in subsequent chapters.

Spatial datasets for network spatial analysis are usually very large, and hence computational methods should be efficient with respect to both computation time and required memory. In many cases, however, there must be a trade-off between time and memory. For example, we can more quickly find the shortest path between two points on a network by performing additional precomputation in the preprocessing stage, but the memory requirements for storing the precomputed data are correspondingly greater. Hence, in this chapter we emphasize the balance between time and memory, and present well-balanced computational methods.

Spatial Analysis along Networks: Statistical and Computational Methods, First Edition.
Atsuyuki Okabe and Kokichi Sugihara.

The chapter consists of four sections. The first two sections are devoted to the geometric modeling of networks, with Section 3.1 concerning planar networks and Section 3.2 addressing nonplanar networks. The final two sections are devoted to basic computational methods with the geometric models. In Section 3.3, we survey basic methods for computing geometric objects and for testing geometric properties, and in Section 3.4 we review the basic computational methods on networks including shortest paths and minimum spanning trees.

3.1 Data structures for one-layer networks

3.1.1 Planar networks

In spatial analysis, a network is often drawn in a plane, just like a road map. Two types of drawn network should be distinguished: 'planar' and 'nonplanar.' For example, crossings of city roads are usually 'planar,' as shown in Figure 3.1a, in the sense that one can choose any branch at a road crossing to go out, independent of the branch through which one comes in. In this case the crossing is modeled in the network in such a way that, as shown in Figure 3.1b, a node is placed at the crossing and four links representing the four branches are connected to this node. We call this type of crossing *planar* and say that the associated network structure is *embedded* in the plane.

As contrasted to Figure 3.1a, the second type crossing is illustrated in Figure 3.1c. City roads sometimes have a three-dimensional crossing consisting of an overpass and an underpass in such a way that one cannot move from one path to the other at the crossing. In this case, as shown in Figure 3.1d, the associated network is modeled with two links crossing at a point other than a node. We call this crossing *nonplanar* and say that the associated network structure is *not embedded* in the plane. In the real world, we can find many other nonplanar structures: for example, a public transportation network consisting of foreground railroads and subways, a road network consisting of ordinary roads at the ground level and expressways constructed at a higher level to avoid crossing, and a network of customer paths in a shopping mall with two or more floors (Figure 1.8 in Chapter 1). In general, we call the network a *planar* network if it is embedded in a plane without mutual crossing except at nodes; otherwise, a *nonplanar* network. Sometimes a planar network is alternatively called

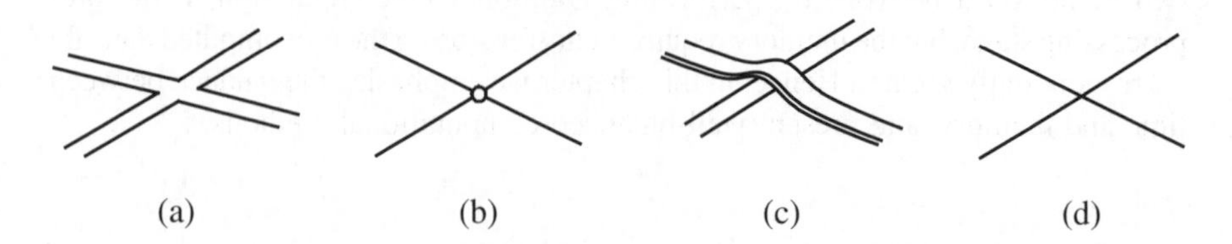

Figure 3.1 Crossings: (a) a planar crossing, (b) the geometric representation of (a), (c) a nonplanar crossing, (d) the geometric representation of (c).

a *one-layer* network. In this section, we first present data structures for planar networks and then extend them to nonplanar networks in the next section, Section 3.2.

3.1.2 Winged-edge data structures

In general, as shown in Figure 2.2 of Chapter 2, a planar network, $N = (V, L)$ consisting of a node set V and a link set L, is described in terms of topological data and geometric data. The topological data are represented by the graph structure consisting of nodes and links (Section 2.2.1.2), while the geometric data include coordinates of the nodes in V, lengths of the links in L and so forth.

A layman's method for representing a network is to make use of the list of nodes and the list of links. The node list consists of the x and y coordinates of the nodes, while the link list consists of the start and end nodes of the links. In concept, this data structure is simple and easy to understand. However, in practice, this is not a good representation, because we cannot retrieve local substructures efficiently. For example, when we want to enumerate all links incident to a given node, we have to scan all links in the link list to find such links that their start nodes or end nodes coincide with the given node.

There are two basic requirements for the representation of N in a computer: *conciseness* and *accessibility*. Since an actual network is often very large, we want to represent it as concisely as possible; otherwise, it may exceed memory limits. On the other hand, we want to retrieve necessary data as quickly as possible, and hence the representation should be rich enough to travel through the network smoothly. Therefore, we have to consider a practical balance between the speed in computation and the amount of memory used. To realize this balance, the topological representation of a network is crucial. We show one of the standard representations of the network topology in the following.

Since network N is embedded in a plane, the links partition the plane into a finite number of connected regions. Let R be the set of all such regions. We represent the topological information of network N as the incidence relation among V, L and R. First, for each node $v \in V$, as shown in Figure 3.2a, we assign an arbitrarily chosen link incident to v, and denote it by node-to-link[v]:

node-to-link[v]: a link incident to v.

This datum represents a relation between a node v and one of the incident links, or this datum point from v to the link. Thus we call such data *pointers*. Note that each node v has one pointer, node-to-link[v].

Second, for each region $r \in R$, as shown in Figure 3.2b, we assign an arbitrarily chosen link on the boundary of R, and denote it by region-to-link[r]; that is, r has the following pointer:

region-to-link[r]: a link on the boundary of r.

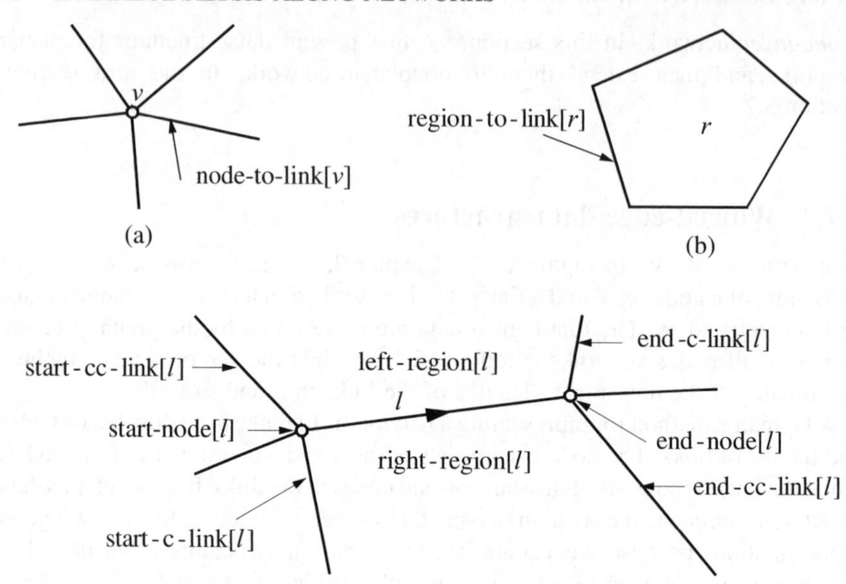

Figure 3.2 Ten pointers in the winged-edge data structure: (a) node-to-link[v], (b) region-to-link[r], (c) start-node[l], end-node[l], right-region[l], left-region[l], start-c-link[l], start-cc-link[l], end-c-link[l], end-cc-link[l].

Third, for each link $l \in L$, as shown in Figure 3.2c, we arbitrarily choose and fix the direction of l, and then assign the following eight pointers:

start-node[l]: start node of l,
end-node[l]: end node of l,
right-region[l]: right region of l,
left-region[l]: left region of l,
start-c-link[l]: clockwise next link at the start node of l,
start-cc-link[l]: counterclockwise next link at the start node of l,
end-c-link[l]: clockwise next link at the end node of l,
end-cc-link[l]: counterclockwise next link at the end node of l.

Consequently, we generate and store ten kinds of pointers: one pointer from a node to a link, one pointer from a region to a link and eight pointers from a link to geometric elements around it; the topological information is mainly represented around links.

Note that the directions of the links fixed here are for a representational purpose. They do not necessarily correspond to the direction of the original links. The links of a road network, no matter whether they are one-way or bidirectional, are represented by a directed link, and these directions are arbitrary. Therefore if we want to distinguish between one-way streets and bidirectional streets, we assign the directional attributes to the links. This way of representing the topological structure of a network is called a *winged-edge data structure*. This term comes from the analogy of the structures of links around a link l to the spread wings of a bird, and it was coined by Baumgart (1972, 1974) in the context of representing geometric data in computer vision (Ballard and Brown, 1982).

3.1.3 Efficient access and enumeration of local information

The winged-edge data structure allows us to retrieve local information efficiently. We present some examples of the procedures for this retrieving in the form of 'algorithms.' To concisely describe algorithms, we use several conventions. First, we often use the left-arrow notation:

$$A \leftarrow B, \tag{3.1}$$

This notation represents the instruction that the value of B is substituted in A. Hence, after this step is executed, the variable A is set to have the value of B. Another typical notation is a sentence-like form such as:

if A then B else C endif,

where 'A' represents a proposition, and 'B' and 'C' represent instructions. This notation implies: check A, and execute B if A is true; otherwise execute C. The last word, 'endif,' represents the end of this instruction. The part 'else C' can be removed if there is no instruction in the case where A is not true. We also use other notations, but most of them can be understood easily according to the ordinary meanings of the associated English words. These notations are used throughout the book.

In the above terms, we describe two basic algorithms. The first algorithm lists all the links incident to $v \in V$. We first describe it formally and then explain it with an example in Figure 3.3.

Algorithm 3.1 (Enumeration of the links around v)

$l_0 \leftarrow$ node-to-link[v]
$l \leftarrow l_0$
label 1:
  report l
  if (v = start-node[l])
    then $l \leftarrow$ start-c-link[l]
    else $l \leftarrow$ end-c-link[l]
  endif
  if ($l = l_0$) then stop endif
go to label 1

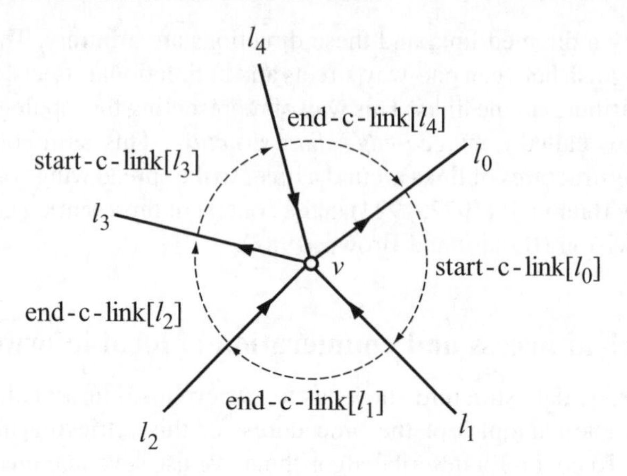

Figure 3.3 Enumeration of the links around a node, v.

This algorithm, following the pointer 'node-to-link,' first visits the link l_0 that is incident to v, and then visits the other links one by one clockwise around v. At link l, the algorithm tests whether v is the start node of l. If that is the case, as with the links l_0 and l_3 in Figure 3.3, the algorithm follows the pointer 'start-c-link' to reach the next link. Otherwise, as with the links l_1, l_2 and l_4 in Figure 3.3, it follows the other pointer 'end-c-link.' Each time the algorithm goes to the next link, it checks whether it reaches the start link at line 9 and if it does, it terminates the processing; otherwise it goes to 'label 1' and repeats the enumeration.

Note that the enumeration of links around a node can be done locally in the sense that we need not visit other links. Hence, this algorithm runs in time proportional to the number of links incident to a node. Thus, the enumeration time does not depend on the size of the whole network; it is very quick no matter how large the whole network is. This demonstrates an example of the power of the pointers of the winged-edge data structure.

The second algorithm lists, for a given region $r \in R$, all the links on the boundary of r. We illustrated it with an example in Figure 3.4.

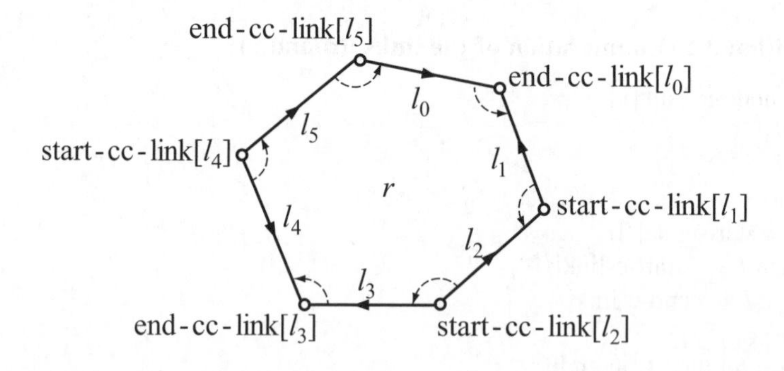

Figure 3.4 Enumeration of the links on the boundary of a region, r.

Algorithm 3.2 (Enumeration of the links around r)

$l_0 \leftarrow$ region-to-link[r]
$l \leftarrow l_0$
label 1:
  report l
  if (r = right-region[l])
    then $l \leftarrow$ end-cc-link[l]
    else $l \leftarrow$ start-cc-link[l]
  endif
  if ($l = l_0$) then stop endif
go to label 1

In this algorithm, we first visit a link l_0 according to the pointer 'region-to-link,' and then visit the other links in clockwise order around r. At link l, we check whether region r is to the right of l or not. If it is, as with links l_0, l_3 and l_5 in Figure 3.4, we move to the counterclockwise next link at the end node according to 'end-cc-link.' If it is not, as with links l_1, l_2 and l_4, then we move to the counterclockwise next link at the start node according to 'start-cc-link.' We repeat this procedure until we reach the start link l_0 again.

The above two algorithms are just examples of local enumeration. Other sets, such as the set of all adjacent nodes of a given node and the set of all adjacent regions of a given region can be enumerated efficiently in a similar manner. Thus, the winged-edge data structure is a powerful way of representing a one-layer network. Note that in computational geometry, there are also other data structures for planar networks; they include the quad-edge data structure (Guibas and Stolfi, 1985) and the half-edge data structure (Yamaguchi, 2002).

3.1.4 Attribute data representation

The attribute data of a network N are assigned to the nodes, the links or the regions of N. Nodes and links are assigned values, for example, a node is assigned x and y coordinates and height above the sea level; a link is assigned length, cost and capacity, and a region is assigned area and population.

Sometimes numerical data are distributed on links; for example, the height and width of a road are given. They can be represented as a function defined on the associated link. To define the function explicitly, we uniquely specify a point by a distance t from the start node of l to the current point measured along l (the length is d). We call the value of t the *along-link distance* of the point from the start node, and denote this point by $p(t)$ for $0 \leq t \leq d$, where $p(0)$ and $p(d)$ indicate the start and end nodes of l, respectively. Let $f(t)$ be the value (such as the width of a road) at the point $p(t)$. Then, $f(t)$ is a function defined on a link l, termed the *field function* (Section 2.3.2 in Chapter 2). Sometimes values are given at only discrete points $p_1 = p(t_1)$, $p_2 = p(t_2), \ldots, p_k = p(t_k)$ of the link l. In that case, $f(t)$ is replaced with a sequence of values at discrete points as $f(t_1), f(t_2), \ldots, f(t_k)$ (the raster data model for a field function in Section 2.2.3 of Chapter 2).

An event specified by a point on a link l is represented by the location of the event point, i.e., the name of the link and the along-link distance of the event point from the start node of the link.

3.1.5 Local modifications of a network

Sometimes we want to make a slight modification to a network in use. For example, inserting new nodes and deleting existing nodes. The insertion of new nodes may result from overlaying two networks. These modifications can be done easily if the network is represented by the winged-edge data structure.

3.1.5.1 Inserting new nodes

Suppose that we want to insert a new node v' into l connecting node v_1 and node v_2, as shown in Figure 3.5a. To achieve this insertion, we first change the end node of l to v', and then generate a new link l' with start node v' and end node v_2, as shown in Figure 3.5b. We next assign the values of other pointers associated with l', and change the values of the pointers associated with the geometric elements around l' accordingly. As is seen in this example, this modification changes only a local part of the winged-edge data structure.

3.1.5.2 New nodes resulting from overlying two networks

The insertion of new nodes plays a fundamental role when we want to overlay two networks. Given two networks N_1 and N_2 (for example, a national road network and a local road network) in the same area, we want to merge them into one network by overlaying.

First, consider the case (Figure 3.6a) in which the two networks are positioned such that the nodes are mutually distinct with no node of one network falling on a link belonging to the other network. Then, the merge can be done by merging all pairs of intersecting links one by one. To describe it explicitly, let l_1 be a link in N_1, and l_2 a

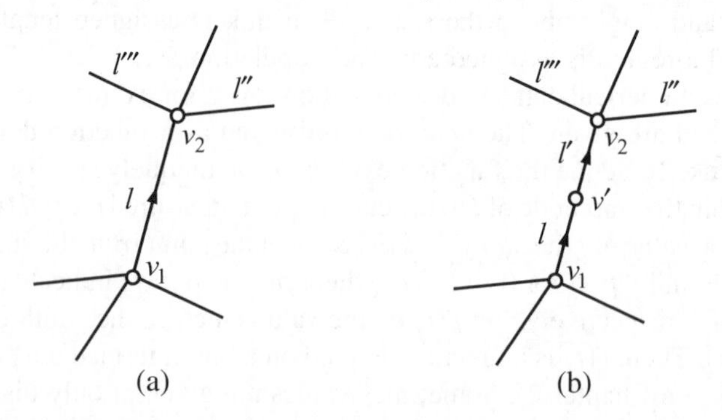

Figure 3.5 Insertion and deletion of a degree-2 node, v'.

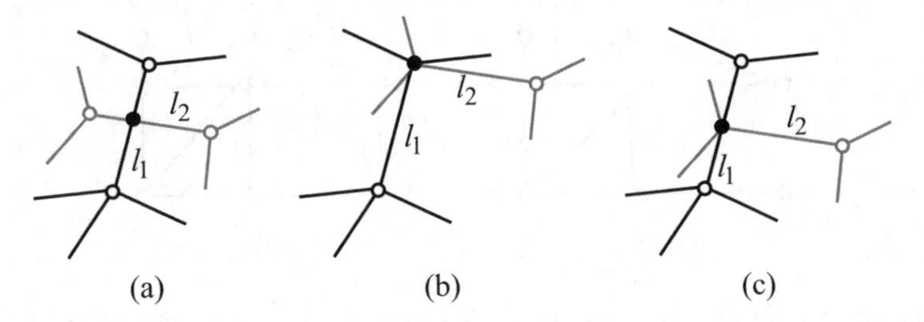

Figure 3.6 New nodes (the black circles) resulting from overlying two networks (the black line segments and the gray line segments): (a) a new node is positioned at the intersection of two links of the two networks, (b) a node of one network is positioned at a node of the other network, (c) a node of one network is positioned at a point on a link of the other network.

link in N_2 that intersects with l_1. First, we compute the point of intersection. Next, we generate a new node at the point of intersection (the black circle in Figure 3.6a). Thirdly, we execute the node insertion procedure for both l_1 and l_2 by dividing l_1 into l_1 and l'_1, and l_2 into l_2 and l'_2. Finally, we determine the clockwise order of the four links l_1, l'_1, l_2, l'_2 around the new node and reconnect them by changing the pointers accordingly. Applying this merge procedure to all pairs of intersecting links, we can construct a new network that provides the overlay of the two networks.

Next consider the case in which a node of one network is positioned exactly at a node (Figure 3.6b) or on a link (Figure 3.6c) of the other network. In the former case we need not generate any new node, while in the latter case we need to generate a new node only on one link. The procedure for reconnecting the links around the node is almost the same as above. Another way to deal with this case is to employ a technique called *symbolic perturbation* (for the precise definition, see Edelsbrunner and Mucke, 1988; Yap, 1988). Intuitively speaking, in symbolic perturbation, we introduce symbols that correspond to infinitesimally small positive numbers, and perturb the coordinates of the nodes by adding polynomials with those symbols in such a way that exceptions such as node-line alignments never arise. As the result of this, we can deal with the networks as if there were no accidental alignments, and thus can apply the general case algorithm.

3.1.5.3 Deleting existing nodes

In order to simplify a network, sometimes we want to remove a node with exactly two links and thus merge the two links into one (Douglas and Peucker, 1973). This modification is the opposite operation of inserting a new node, that is, the modification from Figure 3.5b to Figure 3.5a. To show this procedure, suppose that link l connects v_1 to v' and link l' connects v' to v_2, and that the node v' is incident to only l and l' (i.e., v' is a degree-2 node). We delete v' from V, delete l and l' from L, and insert into L a new link, say l (different from l in Figure 3.5b), which connects v_1 to v_2. In addition, the values of some of the other pointers need to be adjusted

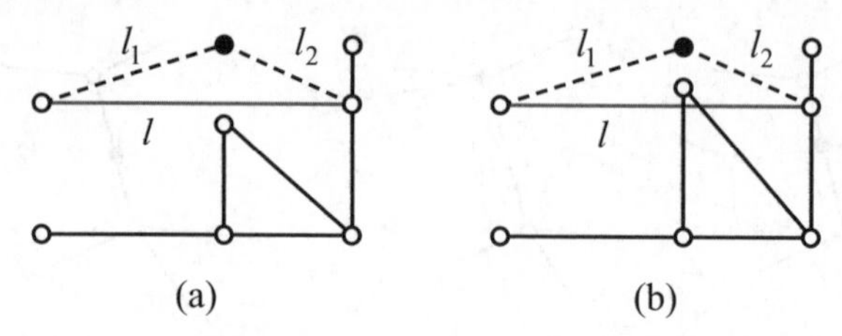

Figure 3.7 New links (the gray line segments) by deleting a degree-2 node (the black circle): (a) a new link not crossing the other links (the black line segments), (b) a new link crossing the other links.

accordingly, although we omit the details. This procedure decreases the sizes of L and V by one. Hence, if the network contains many degree-2 nodes, we can decrease the sizes of L and V greatly and thus can simplify the network. However, we must be careful because the resulting network might have self-intersection. If the links are on a common line (Figure 3.5b), the replacement of those links with the new link does not change the appearance of the embedded network except for the intermediate node v_2 being removed. On the other hand, if the links l_1 and l_2 are not on a common line, these two lines are replaced by a single line segment (the gray line segment in Figure 3.7a), and hence the location of the new link l differs from the two links l_1 and l_2. Therefore, it might cross other links (Figure 3.7b). If we wish to represent each link as straight line segments, we should avoid removing nodes that cause links to cross. However, if the links are abstract objects and we do not care whether they are straight or not, then we can remove degree-2 nodes freely.

3.2 Data structures for nonplanar networks

3.2.1 Multiple-layer networks

Network structures are not necessarily planar. In some networks, links cross each other when they are drawn in a plane. For those networks, the winged-edge data structure cannot be applied directly because the concept of a region bounded by an alternating sequence of links and nodes cannot be defined clearly. For example, consider the network composed of six links $l_1, l_2, \ldots, l_6$ forming a cycle as shown in Figure 3.8, where the directions of the links are arbitrarily chosen and fixed. The links l_2 and l_5 cross each other (note that this cross is as in panel (d) in Figure 3.1, not as in panel (b)), and hence the left side of l_2 partly belongs to the left side of l_5 and partly belongs to the right side of l_5. Thus, an alternating sequence of the links and nodes cannot determine a bounded region. Therefore the data structure for planar networks cannot be applied directly to the representation of nonplanar networks. However, there is a class of nonplanar networks, called *multiple-layered networks*, which can be represented by a slightly modified version of the winged-edge data structure. We introduce this class in this subsection.

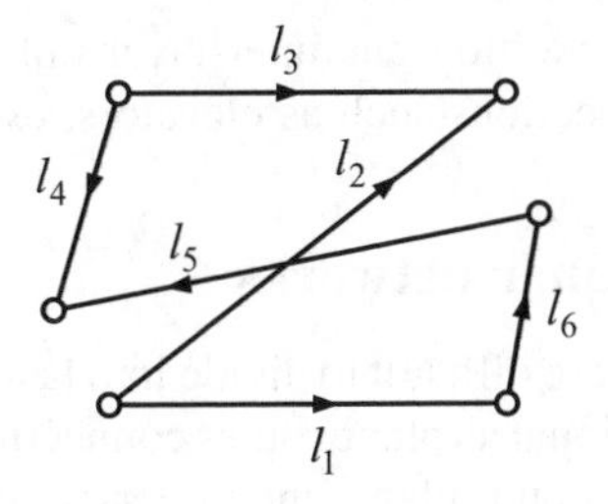

Figure 3.8 A nonplanar network (a cycle) that cannot determine a bounded region.

A multiple-layer network consists of two or more planar networks, called *layers*, and inter-layer links connecting two different layers. A typical example is a shopping mall with two or more floors. As shown in Figure 3.9, passages on each floor form a single-layer network, and equipment connecting different floors such as elevators, escalators and staircases can be regarded as links connecting floors. Elevators and staircases are bidirectional; that is, we can move from one layer to another in both directions. Hence, they are represented by undirected inter-layer links. On the other hand, escalators are one-way; that is, each escalator is only for going either up or down. These are represented by directed links from the start points of the escalators to their endpoints. Another example of a multiple-layer network is a road/subway network, where the network is composed of two layers, one for an ordinary road network on the ground and the other for a subway network under the ground; the two layers are connected at subway stations. In this manner, we can represent each layer in a standard manner such as winged-edge data structures, and generate additional links for inter-layer connections. In the data structure, each node has an additional flag showing whether it is connected by inter-layer links, and the terminal nodes of the inter-layer links contain additional

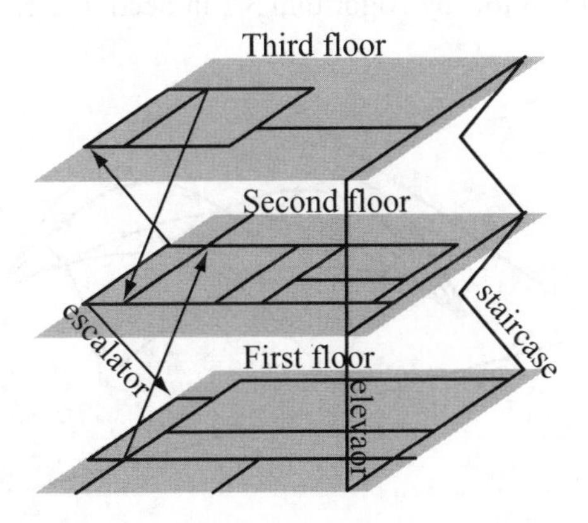

Figure 3.9 Layers and inter-layer links.

pointers to those links. In addition, the inter-layer links should contain attributes showing the kinds of connections such as elevators, escalators and staircases.

3.2.2 General nonplanar networks

Many nonplanar networks are difficult to divide into layers in a natural manner. For example, consider international airplane routes connecting airports (Figure 3.10). If we draw the routes by arcs on a plane, they intersect each other at many points. However, it is difficult to divide the routes into layers. Actually, each airplane should keep to one of the prespecified discrete heights depending on the direction of flight in order to avoid collision, but even within each height, the airplane routes intersect, and hence the height is different from the 'layer' in which a planar network is embedded. Consequently, we cannot apply the winged-edge data structure to general nonplanar networks.

In order to represent general nonplanar networks, we cannot use 'regions,' and hence the incidence relations between links and nodes should be represented directly in the following manner. At each node $v \in V$, we assign an arbitrary cyclic order to the links incident to v, and regard this cyclic order as *virtual clockwise order* ('virtual' in the sense that it does not necessarily correspond to physical clockwise). For each link l, we virtually use the pointers 'start-c-link' and 'end-c-link' to represent the next links around the start node and the end node, respectively, and also the pointers 'start-cc-link' and 'end-cc-link' to represent the next links in the reverse order around the start node and the end node, respectively. We also use the virtual pointers 'start-node,' 'end-node' and 'node-to-link' in the same manner as in the winged-edge data structure. In other words, we delete from the winged-edge data structure the three pointers 'right-region,' 'left-region' and 'region-to-link,' and use the remaining seven pointers (i.e., 'start-node,' 'end-node,' 'start-c-link,' 'end-c-link,' 'start-cc-link,' 'end-cc-link' and 'node-to-link') to represent nonplanar networks. This data structure enables us to list neighboring elements efficiently in the same manner as the winged-edge data structure. For example, for node $v \in V$, we can list all the links incident to v by Algorithm 3.1 in Section 3.1.3.

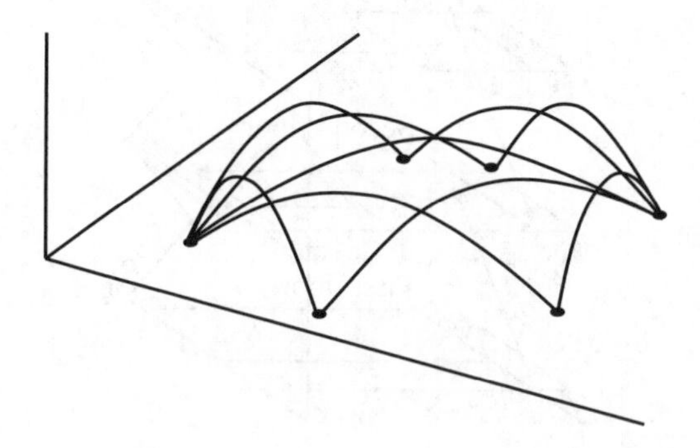

Figure 3.10 Airplane routes.

3.3 Basic geometric computations

Assuming the data structures of planar and nonplanar networks in the preceding section, in this section, we briefly review the methods for representing basic geometric objects such as line segments and polygons, those for computing basic geometric values such as the area of a polygon, and those for geometric tests such as the point-polygon inclusion test. For more details, refer to standard textbooks on computational geometry, such as Preparata and Shamos (1985), Edelsbrunner (1987) and Berg *et al.* (2008).

3.3.1 Computational methods for line segments

A line segment can be represented by the two endpoints. Let v_1 and v_2 be the two endpoints of a line segment in a plane. Then the line segment connecting them is represented by the pair (v_1, v_2) of points, and each point is represented by its coordinates. We denote the coordinates of a node v_i by (x_i, y_i) for any positive integer i throughout this book. An (x, y) coordinate system is said to be a *counterclockwise coordinate system* if the counterclockwise rotation of the x axis by 90 degrees results in the y axis. Throughout this book we assume that the (x, y) coordinate system is counterclockwise unless otherwise mentioned.

3.3.1.1 Right-turn test

Suppose that we are given three points v_1, v_2 and v_3, and that we visit these points one after another in this order. We want to judge whether we turn to the right at v_2. For this purpose we can use the function:

$$F(v_1, v_2, v) = \begin{vmatrix} 1 & x_1 & y_1 \\ 1 & x_2 & y_2 \\ 1 & x & y \end{vmatrix} = (y_1 - y_2)(x - x_1) - (x_1 - x_2)(y - y_1), \quad (3.2)$$

where (x, y) represents the coordinates of an arbitrary point v. The equation $F(v_1, v_2, v) = 0$ represents the line passing through v_1 and v_2. We can confirm this as follows. Substituting x_1 and y_1 in x and y, respectively in equation (3.2), we obtain $F(v_1, v_2, v_1) = 0$, which implies that the line passes through v_1. Similarly, substituting x_2 and y_2 in x and y, respectively in equation (3.2), we obtain $F(v_1, v_2, v_2) = 0$, which implies that the line passes through v_2. Therefore, this line passes through both v_1 and v_2.

The function $F(v_1, v_2, v)$ is continuous in v, and it becomes 0 when v is on the line. Therefore, $F(v_1, v_2, v)$ is positive if v is on one side of the line, and is negative on the other side. Using this property, we can judge whether we turn to the right at v_2 from the sign of $F(v_1, v_2, v_3)$. The sign depends on the orientation of the coordinate system. Because we adopt the counterclockwise coordinate system, $F(v_1, v_2, v_3) > 0$ implies a left turn and $F(v_1, v_2, v_3) < 0$ implies a right turn.

3.3.1.2 Intersection test for two line segments

Suppose that we are given two line segments l and l', l connects v_1 and v_2, and l' connects v_3 and v_4. Then v_3 and v_4 are on mutually opposite sides of the line l if and only if:

$$\begin{vmatrix} 1 & x_1 & y_1 \\ 1 & x_2 & y_2 \\ 1 & x_3 & y_3 \end{vmatrix} \cdot \begin{vmatrix} 1 & x_1 & y_1 \\ 1 & x_2 & y_2 \\ 1 & x_4 & y_4 \end{vmatrix} < 0. \tag{3.3}$$

Similarly, v_1 and v_2 are on mutually opposite sides of the line l' if and only if:

$$\begin{vmatrix} 1 & x_3 & y_3 \\ 1 & x_4 & y_4 \\ 1 & x_1 & y_1 \end{vmatrix} \cdot \begin{vmatrix} 1 & x_3 & y_3 \\ 1 & x_4 & y_4 \\ 1 & x_2 & y_2 \end{vmatrix} < 0. \tag{3.4}$$

Therefore, l and l' intersect at their interior point if and only if both in equations (3.3) and (3.4) are satisfied.

3.3.1.3 Enumeration of line segment intersections

Given a finite set of line segments, we want to enumerate all the intersection points. An efficient algorithm called the *plane sweep method* was proposed by Bentley and Ottmann (1979). In this method, as shown in Figure 3.11, the plane in which the line segments are placed is swept by a vertical sweepline l_s (the bold line segment) from left to right, with a list, denoted by A, that manages the intersection between the line segments and the sweepline in the following manner. As the sweepline moves from left to right, three kinds of events happen. The first kind of an event is that the sweepline hits the left endpoint of a line segment (e.g., at $x_1, x_2, x_3, x_4, x_{10}, x_{11}$ in Figure 3.11). In this case, the new line segment is inserted in the list A. The second kind of an event is that the sweepline hits the right endpoint of a line segment (e.g., at $x_6, x_7, x_8, x_9, x_{12}, x_{13}$ in Figure 3.11). In this case, the line segment is deleted from the list A. The third kind of an event is that the sweepline hits a point of intersection of two line segments (e.g., at x_5 in Figure 3.11). In this case, the two line segments

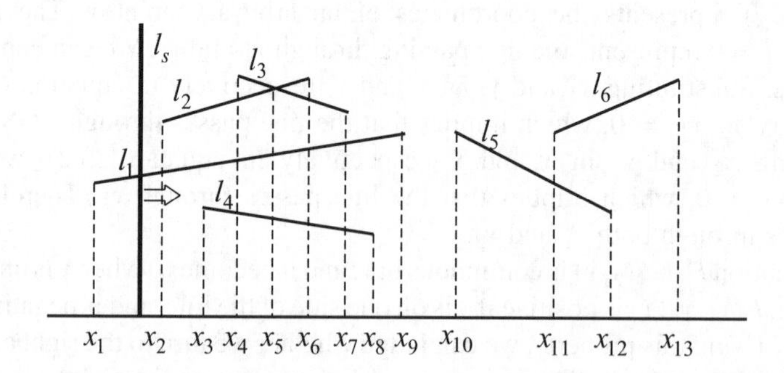

Figure 3.11 Plane sweep method for the enumeration of intersections between line segments.

are interchanged in the list A. As the sweepline l_s moves, the list changes, for instance, in Figure 3.11: $A(x_1) = (l_1)$, $A(x_2) = (l_1, l_2)$, $A(x_3) = (l_4, l_1, l_2)$, $A(x_4) = (l_4, l_1, l_2, l_3)$, $A(x_5) = (l_4, l_1, l_3, l_2)$, $A(x_6) = (l_4, l_1, l_3)$, $A(x_7) = (l_4, l_1)$, $A(x_8) = (l_1)$, $A(x_9) = (\emptyset)$, $A(x_{10}) = (l_5)$, $A(x_{11}) = (l_5, l_6)$, $A(x_{12}) = (l_6)$, $A(x_{13}) = (\emptyset)$, where $A(x_i)$ is the list A when the sweepline l_s is at x_i. For pairs of line segments that become neighbors in the list A, we check whether they intersect, and report the intersection if they do. In the case shown in Figure 3.11, l_2 and l_4 never become neighbors of each other in the list A, and nor do l_3 and l_4, and consequently we can skip the intersection check for these pairs of line segments. Moreover, as l_1 and l_5 in Figure 3.11, two line segments whose projections on the x axis do not overlap (such a pair will never intersect) do not appear in the list A simultaneously, and hence we can similarly skip the intersection check. Thus we can expect to save computational cost if the number of intersections is small. This method can enumerate all the intersections efficiently. In order to evaluate the efficiency of this algorithm, we need a new concept called *time complexity*, which is defined in the next subsection.

3.3.2 Time complexity as a measure of efficiency

Because a network dataset is usually very large, computational methods should be efficient. In this subsection, we show how to measure the efficiency of a computational algorithm. Let $f(n)$ and $g(n)$ be two positive-value functions of n. If there exists a constant C satisfying:

$$\frac{f(n)}{g(n)} \leq C$$

for any n, we write:

$$f(n) = \mathrm{O}(g(n))$$

and say that $f(n)$ is of *order* $g(n)$. The fact that $f(n) = \mathrm{O}(g(n))$ implies that the rate of increase of $f(n)$ is no faster than that of $g(n)$, and hence $g(n)$ is regarded as the upper bound on the growth of $f(n)$. In particular, if $f(n)$ represents the number of steps necessary for executing an algorithm, we choose a simple function $g(n)$ satisfying $f(n) = \mathrm{O}(g(n))$, and regard $g(n)$ as a measure for evaluating the efficiency of the algorithm; we call this the *time complexity* of the algorithm. Specifically, if $f(n)$ represents the average number of steps, the associated $\mathrm{O}(g(n))$ is called the *average time complexity*, while if $f(n)$ represents the worst-case number of steps, $\mathrm{O}(f(n))$ is called the *worst-case time complexity* (Aho, Hopcroft and Ullman, 1974; Cormen *et al.*, 2001).

Using this measure of time complexity, we can see that the plane sweep method can enumerate all the intersections in $\mathrm{O}((n + k)\log n)$ time steps, where n is the number of line segments and k is the number of intersections (Bentley and Ottmann, 1979). Chazelle and Edelsbrunner (1992) improved the time complexity to $\mathrm{O}(n \log n + k)$ by constructing a different algorithm, and Balaban (1995) further improved the algorithm by reducing required memory size.

3.3.3 Computational methods for polygons

Let Π be a polygon whose vertices are $p_1, p_2, \ldots, p_n$, located counterclockwise on the boundary of Π. We represent this polygon by $\Pi = (p_1, p_2, \ldots, p_n)$. Because there is freedom in the choice of the start vertex of this list, this polygon is equivalent to $\Pi = (p_i, p_{i+1}, \ldots, p_n, p_1, \ldots, p_{i-1})$ for any i $(1 \leq i \leq n)$.

3.3.3.1 Area of a polygon

The area A of polygon $\Pi = (p_1, p_2, \ldots, p_n)$ can be computed by:

$$A = \frac{1}{2}\sum_{i=1}^{n}(x_i y_{i+1} - x_{i+1} y_i), \tag{3.5}$$

where (x_i, y_i) is the coordinates of x_i (x_{n+1} and y_{n+1} should be read as x_1 and y_1, respectively). This formula can be regarded as the summation of the following signed areas of the triangles forming the polygon Π. First, note that:

$$A_i = \frac{1}{2}(x_i y_{i+1} - x_{i+1} y_i)$$

represents the signed area of the triangle formed by the vertices p_i, p_{i+1} and the origin of the coordinate system (Figure 3.12a), where $A_i > 0$ if the origin is to the left of the directed line from p_i to p_{i+1}, $A_i < 0$ if the origin is to the right, and $A_i = 0$ if the origin is on the line passing through p_i and P_{i+1} (Section 3.3.1.1). For example, for the polygon in Figure 3.12a, the signed area A_1 for the triangle $\mathrm{O}p_1p_2$ is positive and similarly the areas A_2, A_3, A_4 are positive (Figure 3.12b), while the area A_5 for the triangle $\mathrm{O}p_5p_6$ and the area A_6 for the triangle $\mathrm{O}p_6p_1$ are negative (Figure 3.12c). Therefore, the summation of these signed areas cancels the part of the triangles outside Π and consequently results in the net area of Π.

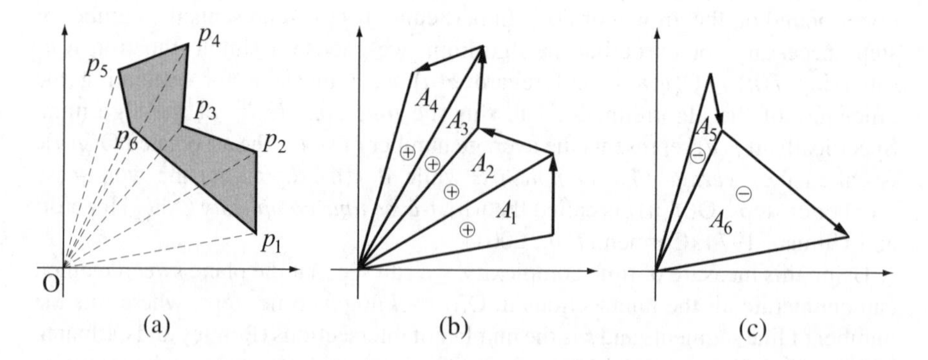

Figure 3.12 Computation of the area of a polygon: (a) a polygon, (b) positively signed areas, (c) negatively signed areas.

3.3.3.2 Center of gravity of a polygon

The center of gravity of a triangle, (x_g, y_g), with vertices v_1, v_2 and v_3 can be computed by:

$$(x_g, y_g) = \left(\frac{x_1 + x_2 + x_3}{3}, \frac{y_1 + y_2 + y_3}{3}\right). \tag{3.6}$$

To compute the center of gravity, (x_g, y_g), of a polygon Π, we first partition the polygon into triangles, $T, i = 1, \ldots, n_i$, next compute the centers of gravity, $(x_{gi}, y_{gi}), i = 1, \ldots n$, of these triangles, and finally compute the weighted average of these centers of gravity with the areas $|T|, i = 1, \ldots, n_i$ chosen as the weights, i.e.:

$$(x_g, y_g) = \left(\sum_{i=1}^{n} \frac{|T_i|}{\sum_{j=1}^{n}|T_j|} x_{gi}, \sum_{i=1}^{n} \frac{|T_i|}{\sum_{j=1}^{n}|T_j|} y_{gi}\right). \tag{3.7}$$

3.3.3.3 Inclusion test of a point with respect to a polygon

To test whether a point p is included in a polygon Π, we first generate a half line starting at p in an arbitrary direction, and next count the number of intersections between the half line and the boundary of Π. Then, we can judge that p is in Π if the number of intersections is odd as in Figure 3.13a, while we judge p is outside Π if the number of intersections is even as in Figure 3.13b.

Note that this procedure is not perfect because we often come across exceptional cases. An example of an exception is shown in Figure 3.13c, where the half line emanating from p includes a complete link of Π. Exceptional cases in geometric configurations are said to be *degenerate*. In general, there are many different degenerate situations, and it is usually difficult to give a complete algorithm that can cope with all the degenerate cases. However, as mentioned in Section 3.1.5, there is a powerful technique, called *symbolic perturbation*, which can remove degenerate cases. With this method, we can construct algorithms without worrying about exceptional branches for degenerate cases. For details, see Edelsbrunner and Mucke (1988) and Yap (1988).

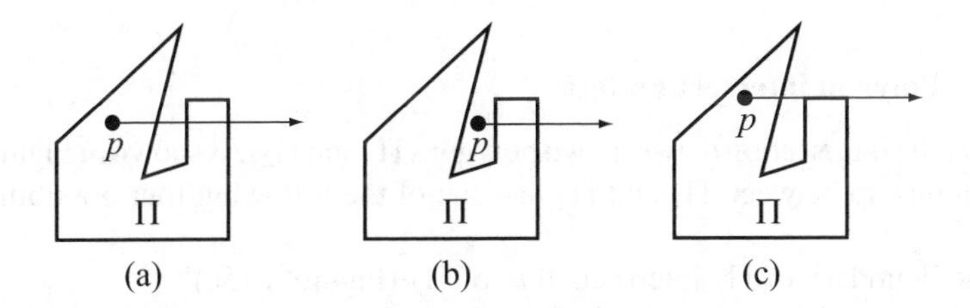

Figure 3.13 Inclusion test of a point with respect to a polygon: (a) a point (the black circle) included in a polygon, (b) a point outside a polygon, (c) a degenerate case.

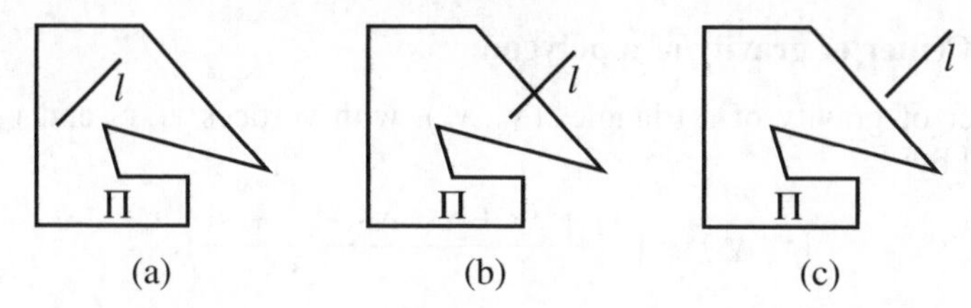

Figure 3.14 Possible relations between a polygon and a line segment: (a) a line segment l included in a polygon, (b) a line segment intersected a polygon, (c) a line segment outside a polygon.

3.3.3.4 Polygon-line intersection

Consider the intersection between a polygon Π and a line segment l (e.g., Figure 3.14). The relative relationship between Π and l has one of the following three possibilities:

(i) l is completely in Π as in Figure 3.14a,
(ii) l intersects with the boundary of Π as in Figure 3.14b, or
(iii) l is completely outside Π as in Figure 3.14c.

We can identify these possibilities by the following algorithm.

Algorithm 3.3 (Point-line intersection)

Input: polygon Π and line segment l.
Output: one of the possibilities (i), (ii), (iii) of the intersection.
Procedure:
1. Test whether l intersects one of the edges of Π. If it intersects, report (ii).
2. Otherwise, test whether one endpoint of l is included in Π. If it is, report (i).
3. Otherwise, report (iii).

Note that in Steps 1 and 2, the intersection test (Section 3.3.1.2) and the inclusion test (Section 3.3.3.3) are used, respectively.

3.3.3.5 Polygon intersection test

Consider the intersection between two polygons Π_1 and Π_2. As shown in Figure 3.15, the relationship between Π_1 and Π_2 has one of the following four possibilities:

(i) the boundary of Π_1 intersects that of Π_2 (Figure 3.15a),
(ii) Π_1 is completely included in Π_2 (Figure 3.15b),
(iii) Π_2 is completely included in Π_1 (Figure 3.15c), or
(iv) Π_1 and Π_2 do not overlap at all (Figure 3.15d).

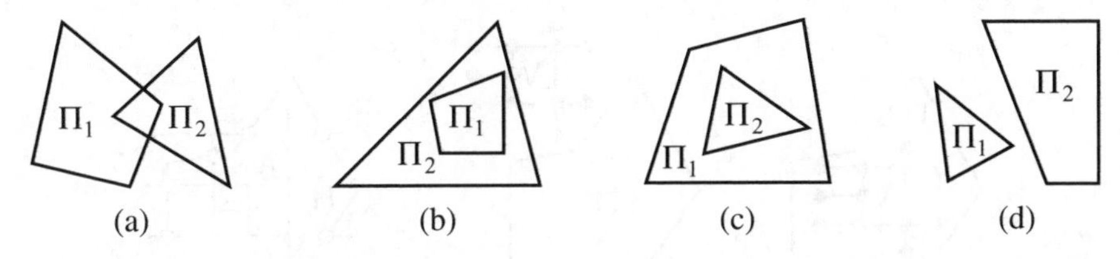

Figure 3.15 Possible relations between two polygons: (a) intersected polygons, (b) polygon Π_1 completely included in polygon Π_2, (c) polygon Π_2 completely included in polygon Π_1, (d) nonintersected polygons.

We can identify these possibilities by the next algorithm.

Algorithm 3.4 (Polygon intersection)

Input: two polygons Π_1 and Π_2.
Output: one of the possibilities (i), (ii), (iii), (iv) of the intersection.
Procedure:

1. Test whether an edge of Π_1 and an edge of Π_2 intersect for all pairs of edges. If there exists a pair that intersects, report (i).
2. Otherwise, choose a vertex of Π_1 and test whether the vertex is in Π_2.
3. If it is, report (ii).
4. Otherwise, choose a vertex of Π_2, and test whether the vertex is in Π_1. If it is, report (iii).
5. Otherwise, report (iv).

Again the intersection test and the inclusion test are used in this algorithm.

3.3.3.6 Extraction of a subnetwork inside a polygon

We consider the problem of extracting the part of a network that is included in a given polygon, as shown in Figure 3.16. Let N be a planar connected network and Π be a polygon, and we want to extract the subnetwork of N that is included in Π. This problem can be solved in the following manner.

First we overlay N and Π and find all the points of intersections between the links of N and the edges of Π by the plane sweep method (Section 3.3.1.3). Then we have two possible cases, Case 1: no points of intersection are found (Figure 3.16a and b), or Case 2: one or more points of intersections are found (Figure 3.16c).

Case 1. Suppose that we find no point of intersection between N and Π. Then, we arbitrarily choose one node of N (the black circles in Figure 3.16a and b) and execute the inclusion test of the node with respect to Π (Section 3.3.3.3). If it is included, we report that the whole network N is included in Π (Figure 3.16a). Otherwise, we report that no part of N is included in Π (Figure 3.16b).

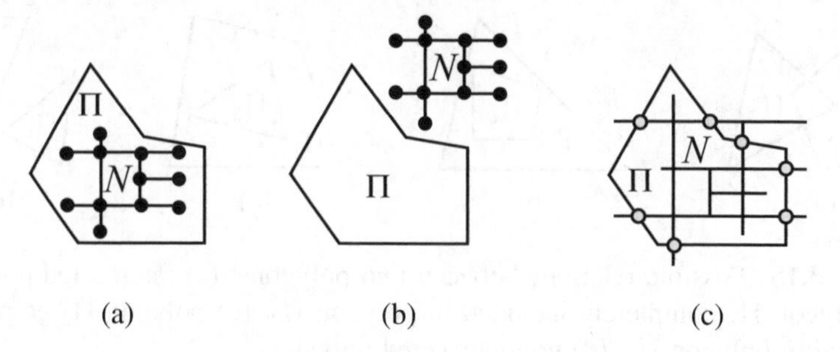

Figure 3.16 Extraction of a subnetwork inside a polygon: (a) a network included in a polygon (Case 1), (b) a network outside a polygon (Case 1), (c) a network intersected with the boundary line segments of a polygon (Case 2). The white circles are intersection points.

Case 2. Suppose that we find one or more points of intersections between N and the boundary edges of Π (the white circles in Figure 3.16c). In this case, we trace the subnetwork from the intersection points toward the inside of Π in order to enumerate all the nodes and links in Π. Note that in this case, the subnetwork included in Π is not necessarily connected although the whole network is connected. Therefore, we have to trace the network until all the connected components are found. Because every connected component of the subnetwork in Π is incident to at least one point of intersection between N and Π, we can exhaust the subnetwork by tracing the network from all the points of intersections.

On the basis of this observation, we can list all the links and the nodes in Π in the following way. First, we generate new nodes at the points of intersections between N and Π (the white circles in Figure 3.16c), and remove the links outside Π by changing the pointers. Then we choose an arbitrary point of intersection and apply the single-source shortest path algorithm (Section 3.4.1) from that point. Then we report the nodes that have finite distances from the start point and the links incident to those nodes. If there remain points of intersection not yet enumerated, we choose one and apply the single-source shortest path algorithm again. We repeat this procedure until all the points of intersections have been reported.

3.3.3.7 Set-theoretic computations

We can use Algorithm 3.4 to achieve set-theoretic operations, specifically, *intersection*, *union* and *subtraction* between two polygons (Figure 3.15). First, consider the *intersection* of two polygons. If the output of the above algorithm is (i) (Figure 3.15a), we collect the parts of the boundaries of the polygons that are included in the other polygons. If the output is (ii), (iii) or (iv) (Figure 3.15b–d), the intersection is Π_1, Π_2 or empty, respectively.

Second, consider the *union* of the two polygons. If the output of the above algorithm is (i), we collect the parts of the boundaries of the polygons that are outside of the other polygons. If the output is (ii), (iii) or (iv), the union is Π_2, Π_1 or the collection of Π_1 and Π_2, respectively.

Third, consider the *subtraction* of Π_2 from $\Pi_1$or the *complement* of Π_1 with respect to Π_2. If the output of the above algorithm is (i), we collect the part of the boundary of Π_1 outside of Π_2 and the part of the boundary of Π_1 included in Π_2. The resulting collection of boundary parts constitutes the boundary of the subtraction. If the output is (ii), (iii) or (iv), the subtraction is empty, Π_1 having the hole Π_2, and Π_1 itself, respectively. The subtraction of Π_1 from Π_2 can be computed by interchanging the roles of the two polygons.

3.3.3.8 Nearest point on the edges of a polygon from a point in the polygon

Given a polygon $\Pi = (p_1, p_2, \ldots, p_n)$ and a point p_0 within it (for example, Figure 3.17), we want to find the point, say q, on the boundary of Π nearest to p_0. This problem arises when we want to find the access point on a street to enter a house, where Π corresponds to the region bounded by city roads that contains the house, and p_0 to the entrance to the house.

We first compute the minimum distance from p_0 to the i-th edge e_i connecting p_i and p_{i+1} (the coordinates of p_i is (x_i, y_i)). The line l_i passing through p_i and p_{i+1} is written by equation (3.2), i.e., $(x_i - x_{i+1})(y - y_i) - (y_i - y_{i+1})(x - x_i) = 0$. The distance $d(p_0, l_i)$ from p_0 to this line is written as:

$$d(p_0, l_i) = \frac{|(x_i - x_{i+1})(y_0 - y_i) - (y_i - y_{i+1})(x_0 - x_i)|}{\sqrt{(x_i - x_{i+1})^2 + (y_i - y_{i+1})^2}}, \qquad (3.8)$$

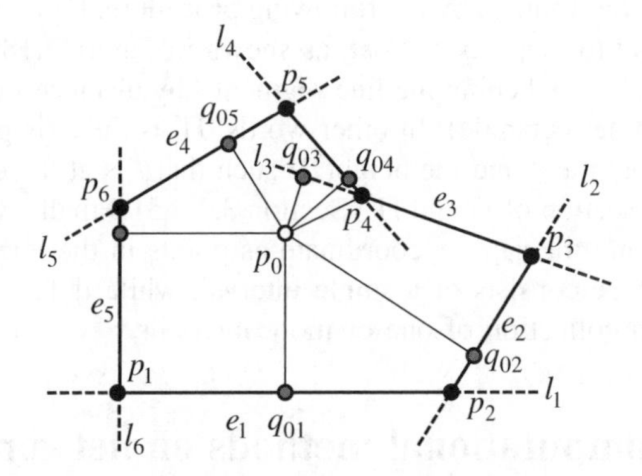

Figure 3.17 The nearest point on the edges of a polygon from a point p_0 in a polygon.

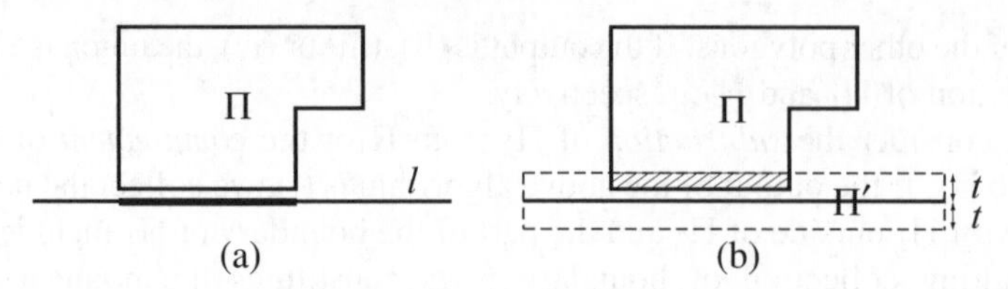

Figure 3.18 Frontage of a house on a street and its computation: (a) the frontage of a polygon Π (the bold line segment), (b) thickening a line segment l (the broken line rectangle).

and the intersection q_{0i} of the line perpendicular to l_i passing through p_0 is given by:

$$q_{0i} = (x_0 - d(p_0, l_i)(x_i - x_{i+1}), \quad y_0 - d(p_0, l_i)(y_i - y_{i+1})). \tag{3.9}$$

Therefore, if q_{0i} is between p_i and p_{i+1}, q_{0i} is the nearest point on the i-th edge from p_0. Otherwise, one of either p_i or p_{i+1} is the nearest point on the i-th edge to p_0 (e.g., l_3 in Figure 3.17). We compute the distance from p_0 to all the edges of Π and take the minimum. Then we can find the required point q.

3.3.3.9 Frontage interval

As shown in Figure 3.18a, let Π be a polygon representing the top view of a house, l a line segment representing a street, and t a positive real number. We want to find the interval on l such that, for any point p in the interval, if we move from p perpendicular to l, then we hit the house Π within the distance t. This interval is called the *frontage* of the house Π on the street l (the bold line segment in Figure 3.18a). Formally we define the frontage as:

$$\text{frontage}(\Pi, l, t) = \{p \in l | s(p, l, t) \cap \Pi \neq \emptyset\},$$

where $s(p, l, t)$ denotes the line segment with the length $2t$ and the midpoint p perpendicular to l.

We can obtain the frontage by the following procedure. First, we rotate Π and l so that l is parallel to the x axis. Next, as shown in Figure 3.18b, we generate a rectangle, say Π′, by thickening the line segment l by distance t on the both sides of l (the broken line rectangle). In other words, Π′ is the axis parallel rectangle with the same length as l and the height $2t$ such that l is at the center. Third, we compute the intersection of Π and Π′ (Section 3.3.3.5). Finally, we collect all the intervals of l having the same x coordinate as points in the intersection. If Π is convex, the frontage consists of a single interval, while if Π is not convex, the frontage can be a collection of one or more intervals.

3.4 Basic computational methods on networks

In this section, we present three basic computational methods on networks: single-source shortest paths (including a connectivity test, shortest-path trees, extended shortest-path trees, and all nodes within a prespecified distance), the shortest path between two nodes and the minimum spanning tree.

3.4.1 Single-source shortest paths

In network spatial analysis, the most frequently required task is to compute the shortest path from one node to other network nodes. More precisely, let $N = (V, L)$ be a network consisting of a node set V and a link set L and $c(l), l \in L$ be the cost we must pay to move from one endpoint of l to the other. A typical example of $c(l)$ is the length of l, and hence $c(l)$ is often referred to as the *length* of l. Sometimes, $c(l)$ is written as $c(v, v')$ when a link l connects two nodes v and v'. We then consider an alternating sequence of nodes and links, $(v_0, l_1, v_1, l_2, v_2, \ldots, l_k, v_k)$, such that l_i connects v_{i-1} and v_i for $i = 1, 2, \ldots, k$. This sequence is called a *path* from v_0 to v_k. The *length* of a path is defined as the sum of the lengths of the links on this path, or mathematically $\sum_{i=1}^{k} c(l_i)$. The *shortest-path distance* between v_0 and v_k is defined as the length of the shortest path between v_0 and v_k among possible paths between them. The shortest-path distances from v_0 to all other nodes $v_k \in V \backslash \{v_0\}$ can be computed by a well-known algorithm, the *Dijkstra method* (Dijkstra, 1959; Knuth, 1973; Aho, Hopcroft, and Ullman, 1974). We first describe this in algorithmic terms. We then explain the task of each step. Lastly, we provide a numerical example to enable the reader to understand the process more easily.

Algorithm 3.5 (Single-source shortest paths; single-source Dijkstra method)

Input: network $N = (V, L)$ with node set V and link set L, the length $c(l)$ for links $l \in L$, and start node $v_0 \in V$.

Output: shortest-path distance $D(v)$ from v_0 to v, and the immediate predecessor node $\text{pred}(v)$ of node v on the shortest path from v_0 to v, for all $v \in V$.

Procedure:

1. $D(v_0) \leftarrow 0$.
2. $D(v) \leftarrow \infty$ for all $v \in V \backslash \{v_0\}$.
3. $A \leftarrow \{v_0\}$.
4. Choose and delete the node v with the smallest $D(v)$ from A.
5. For each node v' adjacent to v, if
   $D(v') > D(v) + c(v, v')$,
   then
   $D(v') \leftarrow D(v) + c(v, v')$,
   $\text{pred}(v') \leftarrow v$
   and insert v' to A
   endif.
6. If A is empty, stop (at this point, the shortest-path distance from v_0 to v is represented by $D(v)$). Otherwise, go back to Step 4.

At any stage of this algorithm, $D(v)$ represents the length of the shortest path found from the source node v_0 to v. Consequently, in Step 1, $D(v_0) = 0$, and in

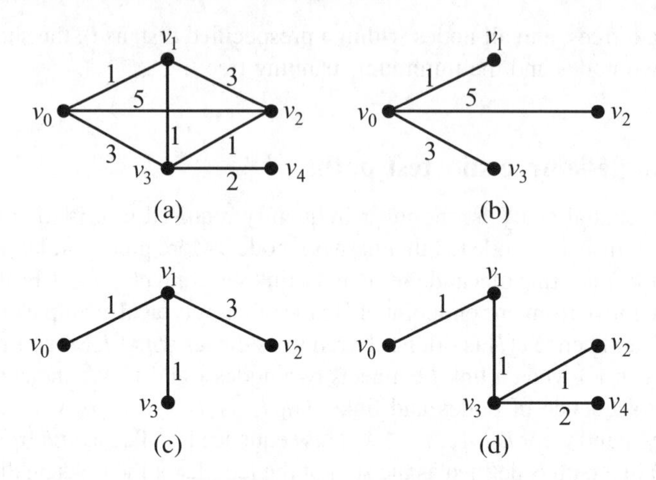

Figure 3.19 The repetition process of finding the shortest path from a node v_0 by the Dijkstra method: (a) a network, (b) the paths found in the first repetition of Step 4, (b) the paths found in the second repetition of Step 4, (c) the path found in the third repetition of Step 4 (Table 3.1).

Step 2, $D(v) = \infty$ for $v \neq v_0$, because we are initially at v_0 and have not found any route to other nodes. Each time a shorter path is found, $D(v)$ is updated. The storage A contains nodes v such that we have found a path (which is not necessarily the shortest) from v_0 to v but the links from v to its neighbors have not yet been examined; A is initialized as a set containing only v_0 in Step 3. In Step 4, the node, v, with the smallest $D(v)$ is chosen and deleted from A. At this stage $D(v)$ represents the length of the shortest path from v_0 to v. In Step 5, for each neighbor v' of v, we compare two paths from v_0 to v': the path from v_0 to v' that has been already found, and the new path composed of the shortest path from v_0 to v followed by the link (v, v'). If the new path is shorter, we update $D(v')$ and store the node v into pred(v'), i.e., the immediate predecessor node along the new path to v'. We repeat this procedure until A becomes empty.

We illustrate the repetition process of this algorithm with a simple example in Figure 3.19a, where $V = \{v_0, v_1, \ldots, v_4\}$ and the number placed at each link l represents the length $c(l)$ of the link. The algorithm applied to this network changes the values of $D(v)$ and pred(v) step by step, as shown in Table 3.1, where the leftmost column represents the number i of repetitions of Step 4. The second column represents the node v chosen from A at the beginning of each repetition of Step 4, and the rightmost column represents the contents of the storage A at the end of each repetition of Step 4. The other columns represent the values of $D(v')$ and pred(v') for nodes $v' \in V \backslash \{v_0\}$.

In the initialization stage (i.e., Steps 1, 2, and 3 of Algorithm 3.5), we set $D(v_0) = 0$, $D(v) = \infty$ for $v \in V \backslash \{v_0\}$, and put v_0 into A; no value is assigned to pred(v) at this stage. In Table 3.1, '–' indicates that the value is not assigned.

Table 3.1 Repetition process of Algorithm 3.5 for the example network in Figure 3.18a.

i-th repetition of Step 4	Node chosen from *A*	v_1		v_2		v_3		v_4		*A*
		D	pred	*D*	pred	*D*	pred	*D*	pred	
Initialization		∞	–	∞	–	∞	–	∞	–	v_0
1	v_0	1	v_0	5	v_0	3	v_0			v_1, v_2, v_3
2	v_1			4	v_1	2	v_1			v_2, v_3
3	v_3			3	v_3			4	v_3	v_2, v_4
4	v_2									v_4
5	v_4									$\emptyset$

In the first repetition of Step 4, $v = v_0$ is chosen from A, and the adjacent nodes $v' = v_1, v_2, v_3$ are checked one by one. For $v' = v_1$, the relation $D(v_1) = \infty > D(v_0) + c(v_0, v_1) = 0 + 1 = 1$ holds and hence we substitute $D(v_1) \leftarrow 1$ and $\text{pred}(v_1) \leftarrow v_0$, as shown in the row of $i = 1$ and the column for v_1 in Table 3.1. Similarly, for $v' = v_2$, the relation $D(v_2) = \infty > D(v_0) + c(v_0, v_2) = 0 + 5 = 5$ holds and hence we substitute $D(v_2) \leftarrow 5$, and $\text{pred}(v_2) \leftarrow v_0$; for $v' = v_3$, the relation $D(v_3) = \infty > D(v_0) + c(v_0, v_3) = 0 + 3 = 3$ holds and hence we substitute $D(v_3) \leftarrow 3$, and $\text{pred}(v_3) \leftarrow v_0$. Thus, at the end of the first repetition of Step 4, the paths and their lengths from v_0 to v_1, v_2, and v_3 are found, as shown in Figure 3.19b, and the storage A contains v_1, v_2, and v_3, as shown in the rightmost column of Table 3.1. Note that nothing is changed for v_4 because v_4 is not adjacent to v_0; the associated cells in Table 3.1 are left blank if $D(v')$ and $\text{pred}(v')$ are not changed.

In the second repetition of Step 4, at the beginning the storage A contains v_1, v_2, and v_3, among which v_1 has the smallest D value $D(v_1) = 1$. Hence we choose v_1 from A, and check the adjacent nodes $v' = v_0, v_2, v_3$ one by one. For $v' = v_0$, the relation $D(v_0) = 0 > D(v_1) + l(v_1, v_0) = 1 + 1 = 2$ does not hold and hence nothing is changed for v_0. For $v' = v_2$, the relation $D(v_2) = 5 > D(v_1) + l(v_1, v_2) = 1 + 3 = 4$ holds and hence we substitute $D(v_2) \leftarrow 4$ and $\text{pred}(v_2) \leftarrow v_1$. Similarly for $v' = v_3, D(v_3) = 3 > D(v_1) + l(v_1, v_3) = 1 + 1 = 2$ and hence we substitute $D(v_3) \leftarrow 2$ and $\text{pred}(v_3) \leftarrow v_1$. At this point, we have found the paths from v_0 to v_1, v_2, and v_3, as shown in Figure 3.19c.

In the third repetition of Step 4, we choose v_3 from $A = \{v_2, v_3\}$ because $D(v_3) = 2 < D(v_2) = 4$ at the end of the second repetition. We check the adjacent nodes $v' = v_0, v_1, v_2, v_4$ one by one. For $v' = v_0, D(v_0) = 0 < D(v_3) + l(v_3, v_0) = 2 + 3 = 5$ and hence we do nothing. For $v' = v_1, D(v_1) = 1 < D(v_3) + l(v_3, v_1) = 2 + 1 = 3$, and hence we do nothing. For $v' = v_2, D(v_2) = 4 > D(v_3) + l(v_3, v_2) = 2 + 1 = 3$, and hence we substitute $D(v_2) \leftarrow 3$ and $\text{pred}(v_2) \leftarrow v_3$. For $v' = v_4$, $D(v_4) = \infty > D(v_3) + c(v_3, v_4) = 2 + 2 = 4$, and hence we substitute $D(v_4) \leftarrow 4$ and $\text{pred}(v_4) \leftarrow v_3$. At this point, we have found the paths from v_0 to all the other nodes, as in Figure 3.19d.

In the fourth repetition of Step 4, v_2 is chosen, but nothing is changed because we have not found the shorter paths from v_0 to the adjacent nodes. In the fifth repetition of Step 4, v_4 is chosen, but again nothing is changed. Thus, the storage A becomes empty and the algorithm terminates. As the result, we obtain $D(v_0) = 0, D(v_1) = 1, D(v_2) = 3, D(v_3) = 2, D(v_4) = 4$.

The shortest paths can be retrieved from the values of 'pred.' For example, consider v_4. We obtained $\text{pred}(v_4) = v_3$. This implies that v_3 is the immediate predecessor node of v_4 on the shortest path from v_0 to v_4. For v_3, we obtained $\text{pred}(v_3) = v_1$, whereas for v_1, we obtained $\text{pred}(v_1) = v_0$. This implies that (v_4, v_3, v_1, v_0) is the reverse order of the nodes along the shortest path from v_0 to v_4. In this manner, we can retrieve the shortest paths from v_0 to all of the nodes.

To consider the time complexity of this algorithm, let $n_V = |V|$ (the number of elements in set V) and $n_L = |L|$. Steps 1, 2, and 3 can be done in O(1), $\text{O}(n_V)$, and O(1) time, respectively. Step 4 can be done in $\text{O}(n_V)$ time simply by scanning all of the nodes in A. Step 5 can also be done in $\text{O}(n_V)$ time because there are at most $\text{O}(n_V)$ adjacent nodes and the processing for each adjacent node v' can be done in constant time. Step 6 is done in O(1) time. Because Steps 4, 5, and 6 are repeated n_V times, the total time complexity of Algorithm 3.5 is $\text{O}(n_V^2)$. This is the time complexity of the Dijkstra algorithm in a general network.

However, in many cases the number of adjacent nodes to be checked in Step 5 is not as large as $\text{O}(n_V)$. In these cases, we can evaluate the time complexity of Algorithm 3.5 in a different way. Step 4 is repeated $\text{O}(n_V)$ times. At each repetition of Step 4, we are required to choose the node v with the smallest value of $D(v)$ from A. This can be done by the data structure called 'heap' in $\text{O}(\log n_V)$ time (for definition, see Section 3.4.1.6 or Aho, Hopcroft, and Ullman (1974)). Hence, n_V repetition of Step 4 in total requires $\text{O}(n_V \log n_V)$ time. At each repetition of Step 5, we check all the adjacent nodes v' of the node v. Hence, in all the repetitions of Step 5, every link is checked twice: once in one direction and once more in the opposite direction. Thus, the total number of checking edges is $2n_L = \text{O}(n_L)$. In each checking, $D(v')$ might be updated and v' is inserted into A. The insertion of a node to the heap requires $\text{O}(\log n_V)$ time and hence repetition of Step 5 requires $\text{O}(n_L \log n_V)$ time. Therefore, Algorithm 3.5 runs in $\text{O}(n_L \log n_V)$ time.

Algorithm 3.5 is used to solve many related problems. In the following subsections, we refer to five typical problems: the connectivity test, the shortest-path tree, the extended shortest-path tree, all nodes within a prespecified distance, and the center of a network.

3.4.1.1 Network connectivity test

The network connectivity test judges whether a given network N is connected or not. First, we choose an arbitrary node v_0 from N, and apply Algorithm 3.5. Next, we check whether there remains node v with $D(v) = \infty$. Then we can see that N is connected if and only if all the nodes have values $D(v)$ different from ∞. The time complexity of this method is $\text{O}(n_L \log n_V)$ for a network with n_V nodes and n_L links, because we solve the single-source shortest-path problem once.

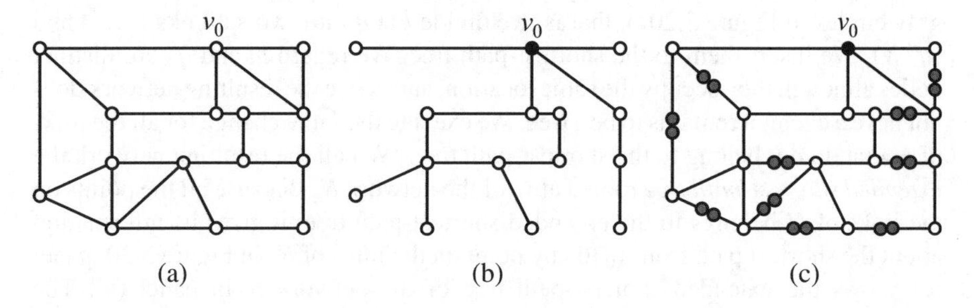

Figure 3.20 Computation for a shortest-path tree and its extended shortest-path tree: (a) a network, (b) the shortest-path tree rooted at the black node of the network shown in (a) (the lengths of the links are the lengths of the line-segments drawn in this figure), (c) the extended shortest-path tree of the shortest-path tree in (b) (the gray circles are breakpoints).

3.4.1.2 Shortest-path tree on a network

To construct a shortest-path tree, we first execute Algorithm 3.5 for a connected network N. We then obtain pred(v) for any node $v \in V$, which indicates the immediate predecessor of v in the shortest path from v_0 to v. Hence, we can trace back the shortest path from any node v to the start node v_0. Collecting the links connecting pred(v) to v for all $v \in V \backslash \{v_0\}$, we obtain the tree rooted at v_0. An example is illustrated in Figure 3.20, where panel (a) shows a network (its nodes are depicted by the white circles), and panel (b) shows the shortest-path tree rooted at v_0 (the black circle). This tree is termed the *shortest-path tree* of N rooted at node v_0. The shortest-path tree can be computed in $O(n_L \log n_V)$ time for a network with n_V nodes and n_L links.

3.4.1.3 Extended shortest-path tree on a network

The shortest-path tree rooted at a node v_0 of network N (Figure 3.20b) tells us the shortest path and its length from any node of N to v_0. However, it does not necessarily tell us the shortest path from a point on a link of N to v_0, because some of the links of N are not included in the shortest-path tree. In order to include this information, we extend the shortest-path tree to a larger tree that includes all the points on the links of the original network N.

To construct such a tree, let l be a link that belongs to N but does not belong to the shortest-path tree (i.e., the complement of the network in panel (b) with respect to the network in panel (a) in Figure 3.20); u and v be the terminal nodes of l; and $D(u)$ and $D(v)$ be the shortest-path distances from v_0 to u and v. Because the shortest path from v_0 to u and that to v do not include l, there is a point, say q, that satisfies $D(u) + d(u, q) = D(v) + d(v, q)$ (the gray circles in Figure 3.20c). In other words, there are two shortest paths from v_0 to q; one passes through u and the other passes through v. We call this point a *breakpoint*. We divide the link l at the breakpoint q into two pieces, and name this point q' for one piece and q'' for the other piece (the gray circles in Figure 3.20c); that is, we divide l at q into two sublinks (u, q') and (q'', v). We insert them to the shortest-path tree. We regard q' and q'' are distinct nodes although they occupy the same location, and hence the resulting network does not have a cycle; it remains to be a tree. We execute the same change for all the links of N that do not belong to the shortest-path tree. We call the resulting network the *extended shortest-path tree* rooted at v_0 of the network N. Because all the points on the links of N belongs to the extended shortest-path tree, it gives us information about the shortest path from v_0 to any point on the links of N. In Figure 3.20, panel (c) shows the extended shortest-path tree of the network N in panel (a). The extended shortest-path tree plays an important role in many contexts of network spatial analysis throughout this book.

3.4.1.4 All nodes within a prespecified distance

All the nodes within distance D_0 from $v_0 \in V$ can be enumerated by a procedure similar to Algorithm 3.5. The only change is to stop the procedure as soon as a node v with the distance $D(v)$ greater than D_0 is deleted from A. The set of nodes deleted from A earlier than v is the set of nodes within distance D_0 from v_0. This procedure can be executed in $O(n_L \log n_V)$ time for a network with n_V nodes and n_L links.

3.4.1.5 Center of a network

To define the center of a network $N = (V, L)$, we define *the radius with respect to a node* $v \in V$ as:

$$\text{Radius}(N, v) = \max_{w \in V} d_S(v, w),$$

where $d_S(v, w)$ is the shortest-path distance between v and w. The radius depends on the choice of v, and the smallest one:

$$\phi(N) = \min_{v \in V} \text{Radius}(N, v)$$

is called the *radius of the network* N. We term the node v that attains the radius $\phi(N) = \text{Radius}(N, v)$ the *center node* of N.

Radius(N, v) can be computed by Algorithm 3.5. Actually, the node that is deleted from the storage A lastly attains Radius(N, v). Therefore, Radius(N, v) can be computed in $O(n_L \log n_V)$ time. The radius $\phi(N)$ of the network N can be obtained by applying Algorithm 3.5 to all the nodes as the source, and hence can be computed in $O(n_V n_L \log n_V)$ time.

Note that the center node is not necessarily unique. For example, suppose that N is a cycle of n_V nodes in which all the links are of the same length. Then, all nodes attain the same radius and consequently, every node is the center node. However, such situations are exceptional; in many cases, the center node is unique and it can be a representative of the nodes.

3.4.1.6 Heap data structure

In Step 4 of Algorithm 3.5, we choose the element v in the set A that gives the minimum value of $D(v)$ for elements v in A. If we were to naively carry out this task by checking all the elements of A sequentially, we would require $O(|A|)$ time. As stated in Section 3.4.1, this task can be done more efficiently using the data structure known as a 'heap.' In this subsection, we review the basic concepts of this type of data structure.

A heap is typically described in terms of *items* $x, x', x'', \ldots$ and real numbers, $k(x), k(x'), k(x''), \ldots$, called *keys*, assigned to the associated items. In the context of Algorithm 3.5, nodes $v, v', v'', \ldots$ are the items and $D(v), D(v'), D(v''), \ldots$ are the corresponding keys. We then wish to design a data structure for which we can efficiently carry out the following two tasks:

(i) Insert new item x with key $k(x)$ into A.

(ii) Remove the element v in the set A that gives the minimum value of $D(v)$ for elements v in A.

This data structure is realized by a *rooted tree* H, i.e., a tree having a special node, called the *root*, denoted by r. As shown in Figure 3.21a, we can construct a rooted tree in such a way that we place the root at the top and the adjacent nodes at one lower level horizontally, and so on. A node of the tree H is then said to be the *level* k *node* if there are exactly k links between this node and r. More specifically, the root r itself is the *level* 0 *node*. Figure 3.21a depicts an example of a rooted tree with levels 0, 1, 2, 3 nodes. If a level k node connects to a level $k+1$ node by a link, we then refer to the former as the *parent* of the latter, and the latter as a *child* of the former.

The tree H is said to be *binary* if every node has at most two child nodes, called a *left child node* and a *right child node*. We can assign ordinal numbers to the nodes of a binary tree in the following recursive manner. The root node is numbered 1 and the left and right child nodes of the root node are numbered 2 and 3, respectively. In

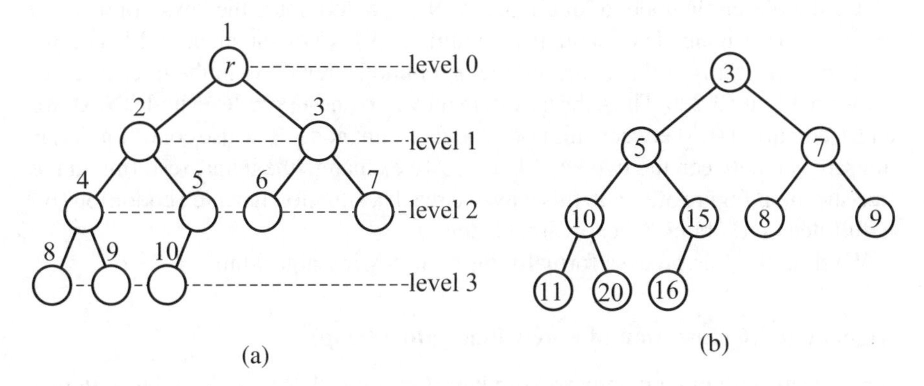

Figure 3.21 An example of a heap: (a) a complete binary tree, (b) a heap (the values circled are keys).

general, for a node with number k, the left child node is numbered $2k$, and the right child node is numbered $2k + 1$. The nodes in Figure 3.21a show the ordinal numbers assigned in this way. In what follows, we refer to the node with the ordinal number k as the k-th *node*. Note that the parent node of the k-th node is the $\lfloor k/2 \rfloor$-th node, where $\lfloor m \rfloor$ represents the largest integer not greater than m (e.g., the parent node of the ninth node in Figure 3.21a is $\lfloor 9/2 \rfloor = 4$), and the left and right child nodes of the k-th node are the $2k$-th node and the $(2k + 1)$-th node, respectively. A binary tree with n nodes is said to be *complete* if it comprises nodes with the numbers $1, 2, \ldots, n$ numbered in the above manner, and the tree is termed a *complete binary tree*. Figure 3.21a is a complete binary tree with 10 nodes.

The data structure characterized by the rooted tree H containing items x_i with key $k(x_i), i = 1, 2, \ldots, n$ is called a *heap* if the following conditions are fulfilled.

(i) H is a binary tree.

(ii) H is complete.

(iii) Each node of H contains exactly one item.

(iv) If item x_i is contained in a child node of a node containing item x_j, then $k(x_i) \geq k(x_j)$ holds.

The tree in Figure 3.21b provides an example of a heap, where the numbers placed in the nodes are the keys; we can see that for any link of this tree, the child node contains a larger key than its parent node.

We employ the heap to achieve two tasks: first, insertion of a new item, and second, deletion of the item with the minimum key. We now illustrate how efficiently the heap can achieve these tasks. First, consider the first task, i.e., the insertion of a new item x with the key $k(x)$ to the heap H. We first illustrate this task using an example and then by employing an algorithmic procedure. Suppose that we wish to insert an item with key 4 into the heap shown in Figure 3.21b. For this purpose, we first generate the 11th node into which we insert the new item, as shown by the double-circle node in Figure 3.22a. Next, we compare the keys of this node and the parent node. The inequality condition (iv) is not satisfied, and hence we exchange the items of this node and the parent node, resulting in the tree structure shown in Figure 3.22b. Thus, the new item moves to one-upper-level node. Next, we check condition (iv) between this node and its parent node. Because condition (iv) is not satisfied between the two keys 4 and 5, we exchange the items, resulting in the tree shown in Figure 3.22c. In this way, we reach a situation in which condition (iv) is fulfilled, and obtain a heap with 11 items.

We describe this process formally in the following algorithm.

Algorithm 3.6 (Insertion of a new item into a heap)

Input: heap containing n items x_i with keys $k(x_i), i = 1, 2, \cdots, n$, and new item x with key $k(x)$.
Output: heap containing $n + 1$ items $x_1, x_2, \ldots, x_n, x$.

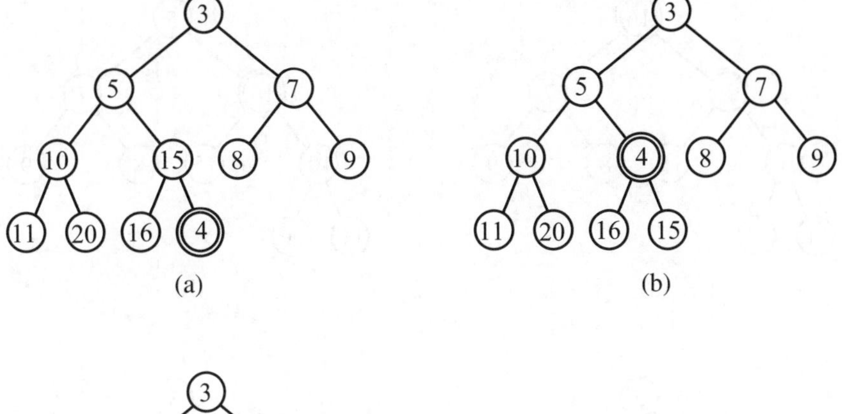

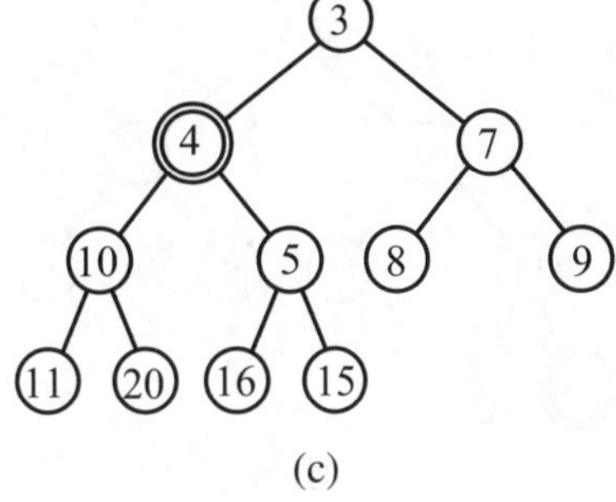

Figure 3.22 The process of inserting a new item into a heap: (a) the generation of the 11th node with key 4, (b) the first exchange, (c) the second exchange.

Procedure:

1. Generate the $n+1$st node and store x in it.
2. $i \leftarrow n+1$.
3. $j \leftarrow \lfloor i \rfloor$.

   (Comment: the jth node is the parent node of the i-th node.)
4. Let y be the item stored in the j-th node.
5. If $k(x) \geq k(y)$, stop. Otherwise exchange x and y, i.e., store x in the j-th node and store y in the i-th node.
6. If the j-th node is the root node, stop. Otherwise $i \leftarrow j$ and go to Step 3. □

We can accomplish each step in O(1) time. Steps 3, . . ., 6 are repeated O(log n) times, because the complete binary tree having n nodes has level log n, and each repetition moves x to one-upper-level node. Therefore, the time complexity of this algorithm is O(log n).

Next, we consider the second task, i.e., deletion of the item with the minimum key. We again illustrate this task with an example, followed by a general

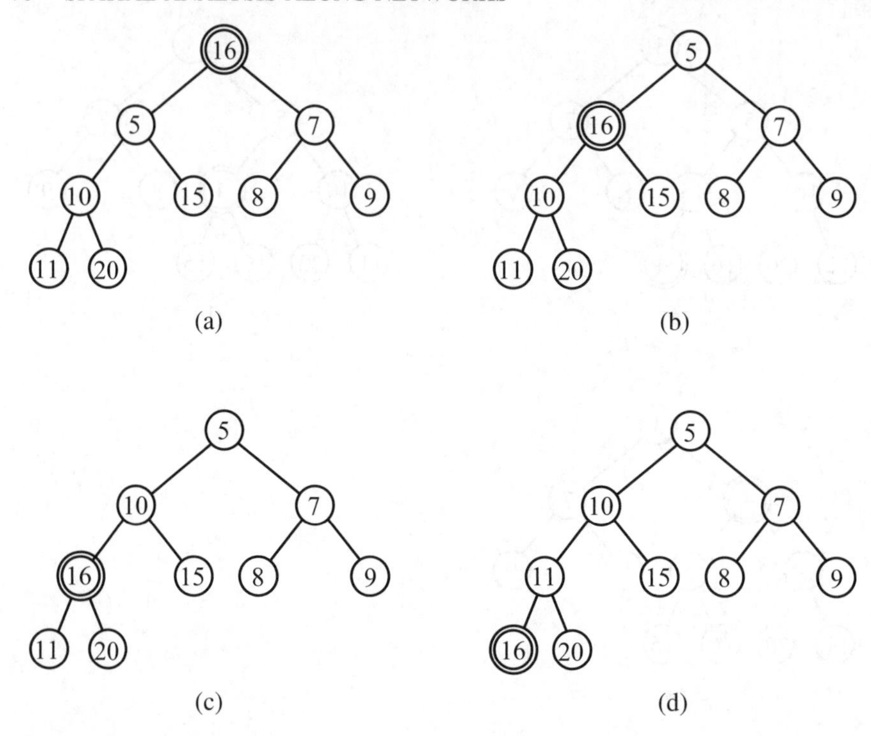

Figure 3.23 The process of deleting the item with the minimum key and reconstructing a heap with the remaining items: (a) the item with key 16 in Figure 3.21b placed at the root node, (b) the first exchange, (c) the second exchange, (d) the third exchange.

algorithm. Suppose that we are given the heap in Figure 3.21b and that we wish to delete the item with the minimum key (i.e., key 3 in the figure) placed at the root node. We first delete this item; the number of items then changes from 10 to nine (in general, n to $n-1$). As a result, we are supposed to construct the heap consisting of nine nodes for the nine items (in general, $n-1$). Recalling that the heap should be a complete binary tree – that is, the nodes are orderly arranged from the top down and from left to right at the same level – we delete the last node (the rightmost node at the lowest level, i.e., the node with 16). We then move the item on that node (i.e., the item with key 16 in Figure 3.21b) to the top node. Figure 3.23a depicts the resulting tree. We then check condition (iv). Because the key of the root node (i.e., key 16) is greater than its child nodes, we exchange 16 and the smaller child, i.e., key 5, resulting in the tree in Figure 3.23b. Similarly, key 16 is larger than its child nodes, i.e., 10 and 15, and hence we exchange 16 and the smaller key 10, resulting in the tree in Figure 3.23c. Key 16 is still greater than its left child, i.e., key 11, and so is also exchanged. Thus, we obtain the tree in Figure 3.23d, which is a heap.

We generally describe the above procedure as the following algorithm.

Algorithm 3.7 (Deletion of the item with the minimum key from a heap)

Input: heap containing n items x_i with keys $k(x_i), i = 1, 2, \cdots, n$.
Output: item with the minimum key, and a heap containing the remaining items.
Procedure:

1. Delete the item at the root node, and report it.
2. Move the item in the n-th node to the root node, and remove the n-th node.
3. $i \leftarrow 1$.
4. $j_1 \leftarrow 2i$ and $j_2 \leftarrow 2i + 1$.

   (Comment: the j_1-th node and the j_2-th node are the child nodes of the i-th node.)

5. If $j_1 \geq n$, stop. Otherwise, insert j using the following rule:

   5.1 $j \leftarrow j_1$, if $j_1 = n - 1$.
   5.2 $j \leftarrow j_1$, if the j_1-th node contains an item with a smaller key than the j_2-th node.
   5.3 $j \leftarrow j_2$ otherwise.

6. If the j-th node contains an item with a smaller key than the i-th node, then exchange the items in the i-th node and the j-th node.
7. $i \leftarrow j$ and go to Step 4. □

Steps 1, 2 and 3 are initialization, through which the item with the minimum key is removed and the item in the n-th node is moved to the root node. At this stage, the tree is complete, but not necessarily a heap, because condition (iv) is not guaranteed. Then in Steps 4, ..., 7, we exchange the item until condition (iv) is satisfied, and thus we obtain the heap. This algorithm also runs in $O(\log n)$ time, because each step can be done in $O(1)$ time, and Steps 4, ..., 7 are repeated at most $\log n$ times.

3.4.2 Shortest path between two nodes

In the last subsection we considered how to compute the shortest paths from one source node to all other nodes. In addition to this, we sometimes wish to compute the shortest path between any two nodes. For this purpose we can also employ Algorithm 3.5: that is, for two arbitrary nodes of a network, v_0 and v_1, we compute the distance from v_0 to v_1 using a similar procedure to Algorithm 3.5 except that we stop the procedure as soon as v_1 is deleted from A in Step 4. At that stage, $D(v_1)$ represents the distance between v_0 and v_1.

A modification may be effective when a network is fixed and the shortest-path queries between two nodes are put many times. To illustrate this, we depict Figure 3.24, where the area containing a network is partitioned into rectangles, known as *buckets*, by horizontal and vertical lines (a rectangular grid). For every

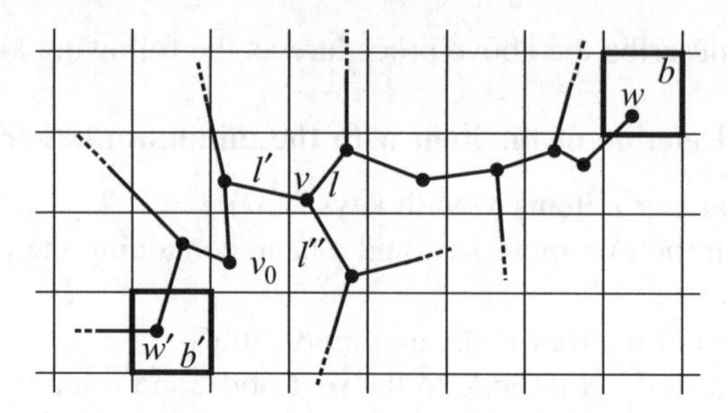

Figure 3.24 Buckets and preprocessing for the shortest path between two nodes.

node v, every bucket b, and every link l incident to v, we define a function $f(v, l, b)$, which shows whether a link l incident to v is included in the shortest paths from v to some node, say w, in bucket b. Specifically, for a link l, we set $f(v, l, b) = 1$ if there is at least one node w in bucket b such that l is the start link on the shortest path from v to w, otherwise $f(v, l, b) = 0$.

To illustrate how this function works, consider a node v with three links l, l', l'' and an arbitrary bucket b, as in Figure 3.24. If the link l is the first link on the shortest path from v to a node in bucket b, then we set $f(v, l, b) = 1$; if neither l' nor l'' appears on any shortest path from v to a node in bucket b, then we set $f(v, l', b) = 0$ and $f(v, l'', b) = 0$. We next consider another bucket b'. If l' is the first link on the shortest path from v to some node in b', then we set $f(v, l', b') = 1$; if no shortest path from v to a node in b' contains l or l'', then we set $f(v, l, b') = 0$ and $f(v, l'', b') = 0$. In this manner, we set $f(v, l, b) = 1$ or $f(v, l, b) = 0$ for every node v, every bucket b and every link incident to v. Once this function is obtained, we can find the shortest path from node v to a node in bucket b by checking only link l', such that $f(v, l, b) = 1$. We need not check every link l where $f(v, l, b) = 0$ as they are not included in the shortest path we wish to find. To utilize this property in Algorithm 3.5, we compute $f(v, l, b)$ in a preprocessing stage. Suppose that we want to find the shortest path from v_0 to v_1 and that v_1 belongs to b (for example, in Figure 3.24, v_0 and $v_1 = w$). Then, at each repetition of Step 4 of Algorithm 3.5, we check only those nodes v' adjacent to v such that $f(v, l, b) = 1$ where l is the link connecting v to v'. This modification can reduce computation time drastically (Wagner and Willhalm, 2003, 2007; Bauer *et al.*, 2010; Holzer, Shulz, and Wagner, 2005, 2008).

3.4.3 Minimum spanning tree on a network

The shortest tree connecting all the nodes is called the *minimum spanning tree* of N. The minimum spanning tree also plays one of main roles in network spatial analysis. It can be constructed by the next algorithm (Bentley, Weide, and Yao, 1980).

Algorithm 3.8 (Minimum spanning tree)

Input: connected network $N = (V, L)$.
Output: set of links belonging to the minimum spanning tree of N.
Procedure:

1. Make storage A empty.
2. Sort the links in L in the increasing order of the lengths, and let $(l_1, l_2, \ldots, l_n)$ be the resulting list.
3. For $i = 1, 2, \ldots, n$, do:
   If l_i connects distinct connected components of the graph induced by A, then insert l_i to A; else do nothing.
4. Report A and stop.

Figure 3.25 illustrates the process of constructing the minimum spanning tree for the network in panel (a) according to Algorithm 3.8. The numbers in panel (a) indicate the increasing order of the lengths of links forming the network, and the broken lines in panels (f), (h), (i) and (j) imply 'do nothing' in Step 3. The time complexity of this algorithm is $O(n_L \log n_L)$ (Aho, Hopcroft, and Ullman, 1974).

3.4.4 Monte Carlo simulation for generating random points on a network

We often need to generate points at random on a network. One method is first to separate links, then to connect them in a line segment, and finally locate a new point uniformly according to a generated random number.

Sometimes we want to generate points according to a nonuniform distribution. For each link $l \in L$, let $w(l)$ be the weight assigned to l. Suppose that the weight $w(l)$ is proportional to the probability density for a point to be generated in the unit length

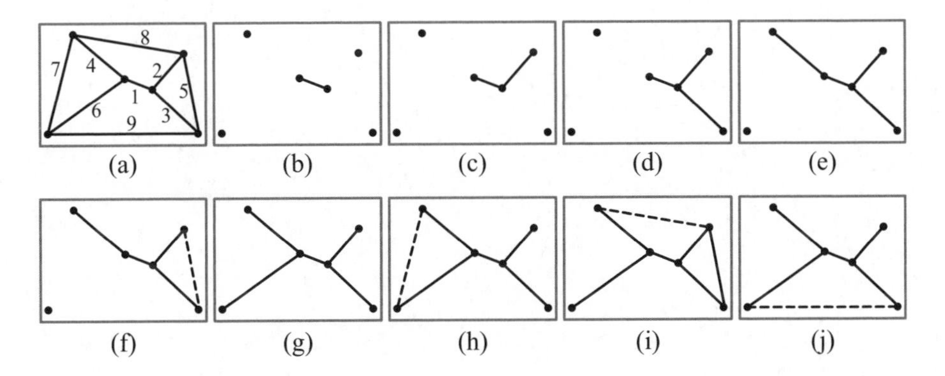

Figure 3.25 The process of constructing a minimum spanning tree: (a) a given network, (b), ..., (j) intermediate minimum spanning trees of the network in panel (a) (the broken line segments mean 'do nothing').

of this link. In this case, we first separate links, next change their lengths by multiplying $w(l)$ to the length of l for all $l \in L$, then connect them in a line, and finally locate a new point uniformly according to the random number generated. This method is conceptually similar to the uniform network transformation in Section 2.4.4 of Chapter 2.

Most of the computational methods introduced in this chapter are implemented in a software package, called *SANET* (*S*patial *A*nalysis along *NET*works; Okabe, Okunuki, and Shiode, 2006a, 2006b; Okabe, Satoh, and Sugihara, 2009), which is illustrated in the final chapter, Chapter 12.

4

Network Voronoi diagrams

This chapter presents a family of diagrams known as *network Voronoi diagram,* abbreviated to *N-VD* (Okabe *et al.*, 2000, Section 3.9). This diagram is directly and indirectly (i.e., background computing) utilized in many spatial analysis methods, as shown in subsequent chapters. In particular, this diagram is useful for solving, for example, question Q1 in Chapter 1, that is, Q1: How can we estimate the catchment areas of parking lots in a downtown area including one-way streets, assuming that drivers access their nearest parking lots? More generally,

Q1′: For a given set of n generators, which may be points (e.g., fast-food shops), line segments (e.g., arterial streets), or circuits enclosing polygons (e.g., parks) on a network, how can we tessellate the network into n subnetworks associated with the n generators in such a way that the nearest generator from every point in a subnetwork is the generator assigned to the subnetwork?

This chapter responds to this general question through formulating the tessellations in Q1′, together with developing their efficient computational methods.

The N-VD is an extension of the Voronoi diagram (VD) originally defined on the Cartesian plane or a higher-dimensional Cartesian space. The Voronoi diagram is named in honor of Georgy Voronoi (1868–1908), a mathematician specializing in analytic number theory (Medvedev and Syta, 1998). An early concept of the VD itself dates back to Rune Descartes in the seventeenth century (Mahoney, 1979). Okabe *et al.* (2000, Section 1.2) describes the history of the VD and its wide application in various fields. An extension of the VD on the Cartesian plane to a VD on a network is in Hakimi (1965) and Hakimi, Labbe, and Schmeichel (1992), who

Spatial Analysis along Networks: Statistical and Computational Methods, First Edition.
Atsuyuki Okabe and Kokichi Sugihara.

referred to it as the *Voronoi partition*, while Yomono (1993) and Okabe *et al.* (1995) provide an explicit definition of an N-VD. Okabe and Okunuki (2001) develop an application of an N-VD to market area estimation. Erwig (2000) define the *graph Voronoi diagram* (for graph, see Section 2.2.1.2 in Chapter 2). The computational methods for VDs on transportation networks (e.g., roads and railways) have been studied by Abellanes *et al.* (2003), Aichholzer, Aurenhammer, and Palop (2004), and Georke and Wolff (2005), among others. Okabe *et al.* (2000, Section 3.9) provides a comprehensive review of N-VDs. Okabe *et al.* (2008) provides generalizations of these N-VDs and this chapter proceeds along the same line.

The chapter itself consists of three sections. Section 4.1 formulates a standard type of the N-VD. Section 4.2 then generalizes this diagram in five ways: namely, the directed N-VD, the weighted N-VD, the k-th nearest point N-VD, the line N-VD (including the polygon N-VD), and the point-set N-VD. The final section, Section 4.3, develops efficient computational methods for constructing these N-VDs.

4.1 Ordinary network Voronoi diagram

In this section, we first formulate a standard type of VD on a network. We contrast this with a standard type of VD on the Cartesian plane, referred to as the *ordinary planar Voronoi diagram* or *ordinary P-VD*. We then illustrate its geometric properties, which will help elucidate the distinction between the N-VD defined in this section and the generalized N-VDs defined in the following section.

4.1.1 Planar versus network Voronoi diagrams

To illustrate the distinction between N-VDs and P-VDs, let p be an arbitrary point on a plane, $p_1, \ldots, p_n$ be points on the plane, $P = \{p_1, \ldots, p_n\}$, and $d_{\mathrm{E}}(p, p_i)$ be the Euclidean distance between p and p_i. In these terms, we define a set of points by:

$$V_{\mathrm{VD}}(p_i) = \{p | d_{\mathrm{E}}(p, p_i) \leq d_{\mathrm{E}}(p, p_j), j = 1, \ldots, n\}. \tag{4.1}$$

We respectively refer to the set $V_{\mathrm{VD}}(p_i)$ and the points $p_1, \ldots, p_n$ as the *Voronoi polygon* of p_i and the *generator points*. We refer to the resulting set, $\mathcal{V}(P) = \{V_{\mathrm{VD}}(p_1), \ldots, V_{\mathrm{VD}}(p_n)\}$, as the *ordinary P-VD* generated by P. Figure 4.1a depicts a simple example of a P-VD. Because the edges of the Voronoi polygons form a network (the line segments in Figure 4.1a), the literature sometimes refers to this as a *Voronoi network* (Medvedev and Syta, 1988; Bartkowiak and Mahan, 1995). Note that we should distinguish the Voronoi network from the N-VDs as defined. To avoid any confusion, we refer to the N-VDs as *network-constrained Voronoi diagrams* or *network-based Voronoi diagrams*, but for simplicity, we use N-VDs. Also, note that in this chapter, we provide only simple artificial examples of N-VDs, because the interested reader can easily observe the characteristics of these examples. Chapter 12 of this volume and Okabe *et al.* (2008) include many actual examples.

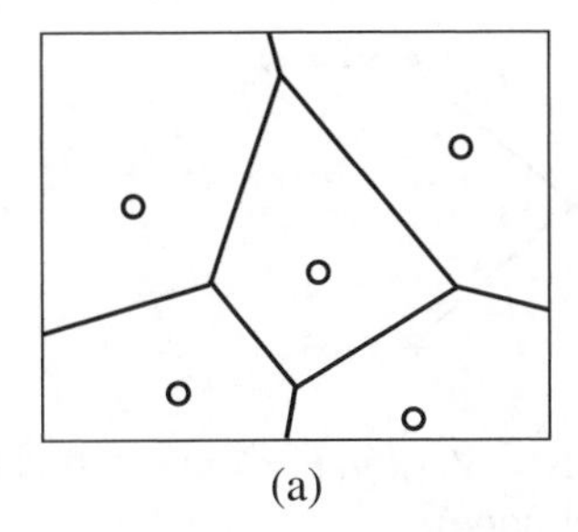

(a)

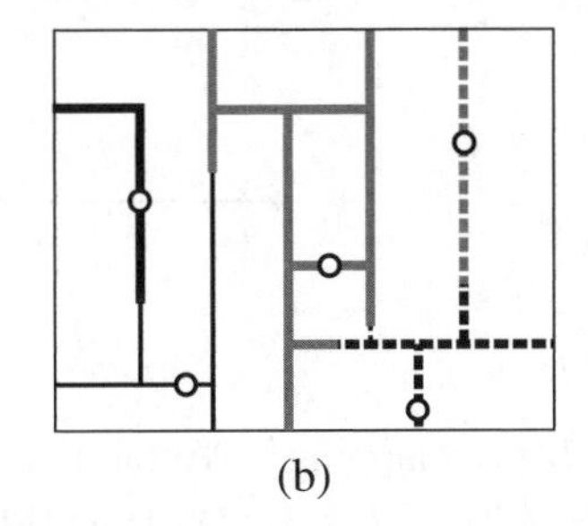

(b)

Figure 4.1 Voronoi diagrams: (a) an ordinary planar Voronoi diagram, (b) an ordinary network Voronoi diagram (the configuration of points in (a) and that in (b) are the same).

The ordinary P-VD includes many variants, characterized by three factors: space, generators, and distance. For instance, the characteristics of the ordinary P-VD are two-dimensional Cartesian space, generators as points, and Euclidean distance. Consequently, we can generalize the ordinary P-VD with respect to space, generators, and distance. Because there are many kinds of spaces, generators, and distances, the number of possible generalizations is large. In fact, the literature has already proposed a variety of generalizations since Voronoi (1908). Okabe, Boots, and Sugihara (1994) and Okabe *et al.* (2000, Chapter 5) reviews many of these generalizations. As an alternative, this chapter focuses on the generalizations on network space.

One of the most basic extensions from VDs on a plane to a network is obtained by replacing the two-dimensional Cartesian space and the Euclidean distance in equation (4.1) with a network space $N = (V, L)$ (composed of a set of nodes V and a set of links L) and the shortest-path distance, respectively. Mathematically, let $d_S(p, p_i)$ be the shortest-path distance between p and p_i, and $V_{VD}(p_i)$ be a set of points defined by:

$$V_{VD}(p_i) = \{p | d_S(p, p_i) \leq d_S(p, p_j), j = 1, \ldots, n\}. \tag{4.2}$$

Set $V_{VD}(p_i)$ is termed the *Voronoi subnetwork* of p_i or the *Voronoi cell* of p_i, while the set of resulting Voronoi cells, $\mathcal{V}(P) = \{V_{VD}(p_1), \ldots, V_{VD}(p_n)\}$, is the *ordinary network Voronoi diagram*, hereafter the *ordinary N-VD*. Figure 4.1b depicts an example of an ordinary N-VD, which corresponds to the ordinary P-VD in Figure 4.1a (the configuration of generators is identical in both panels). Figure 1.8 in Chapter 1 depicts the three-dimensional ordinary N-VD generated by toilets on passageways in a four-floor department store. Chapter 12 includes an application of the two-dimensional ordinary N-VD to railway stations on the street network in Shibuya ward in Tokyo (Figure 12.5).

4.1.2 Geometric properties of the ordinary network Voronoi diagram

Having defined the ordinary N-VD, we now wish to observe its geometric properties. It follows from the definition that every point on $\tilde{L} = \bigcup_{i=1}^{n_L} l_i$ (where

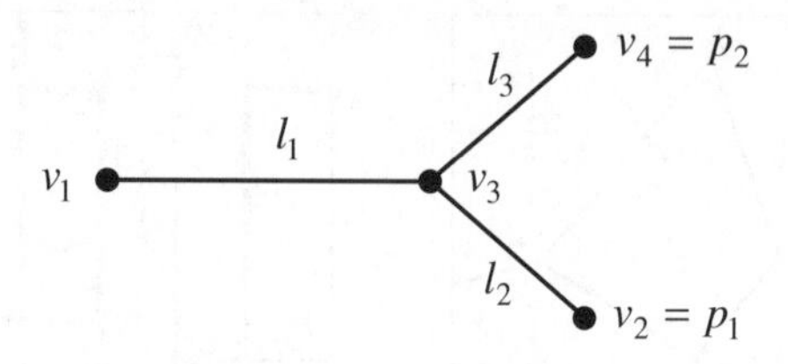

Figure 4.2 Overlapping Voronoi cells $V_{\mathrm{VD}}(p_1) = \{p|p \in l_1 \cup l_2\}$ and $V_{\mathrm{VD}}(p_2) = \{p|p \in l_1 \cup l_3\}$ (v_3 is a critical node).

$L = \{l_1, \ldots, l_{n_L}\}$) is assigned to at least one $V_{\mathrm{VD}}(p_i)$ of the n Voronoi cells, $V_{\mathrm{VD}}(p_1), \ldots, V_{\mathrm{VD}}(p_n)$. Therefore, the n Voronoi cells are collectively exhaustive for the entire network $\tilde{L}$. However, they may or may not be mutually exclusive. This property is distinct from that of the ordinary P-VD (i.e., Voronoi polygons are mutually exclusive, except at the boundary lines). An example of the nonexclusive case is shown in Figure 4.2, where the generator points are p_1 and p_2, and the lengths of the links l_2 and l_3 are equal. Voronoi cell $V_{\mathrm{VD}}(p_1)$ is then given by l_1 and l_2, and Voronoi cell $V_{\mathrm{VD}}(p_2)$ is given by l_1 and l_3. Points on l_1 are thus shared by Voronoi cells $V_{\mathrm{VD}}(p_1)$ and $V_{\mathrm{VD}}(p_2)$.

To exclude this special case, we consider a node v satisfying the following conditions: m links meet at v where $m \geq 3$; there are m' shortest paths, each of which connects v and a generator point in P; and there are no other generator points on the path, where m' satisfies $2 \leq m' < m$, and the lengths of the shortest paths are equal. In the example in Figure 4.2, $m = 3$, $m' = 2$ and the length of the shortest path between v_3 and p_1 and that between v_3 and p_2 (i.e., l_2 and l_3) are equal. We refer to this as a *critical node*. For example, node v_3 in Figure 4.2 is a critical node. In terms of critical nodes, we describe a property of the ordinary N-VD as:

Property 4.1

If a network $N = (V, L)$ does not include critical nodes, Voronoi cells $V_{\mathrm{VD}}(p_1), \ldots, V_{\mathrm{VD}}(p_n)$ generated by $P = \{p_1, \ldots, p_n\}$ are mutually exclusive, except at boundary points between Voronoi cells, and collectively exhaustive for $\tilde{L} = \bigcup_{i=1}^{n_L} l_i$ (where $L = \{l_1, \ldots, l_{n_L}\}$).

If a network has this property, we are unbothered by any special treatment in developing computational methods for constructing N-VDs. In the following theoretical discussion, we assume this property. Under this assumption, two Voronoi cells may share only boundary points or may not share any points. If they share, we would say that the Voronoi cells are *adjacent*; otherwise, *not adjacent*. Employing this concept, we state the following property:

Property 4.2

For a generator point p_i in $P = \{p_1, \ldots, p_n\}$, the generator point that achieves the minimum value among the shortest-path distances between generator p_i and the remaining generator points, i.e., $\min\{d_S(p_i, p_j), j \neq i, j = 1, \ldots, n\}$ is found in the generator points satisfying that Voronoi cells $V_{\mathrm{VD}}(p_j)$ are adjacent to Voronoi cell $V_{\mathrm{VD}}(p_i)$.

This property is very useful for solving the following problems, which correspond to the same-name problems defined on a plane (Shamos and Hoey, 1975; Bentley, Weide, and Yao, 1980).

The all-nearest neighbor problem: For a set of points $P = \{p_1, \ldots, p_n\}$ on a network, find the minimum value among the shortest-path distances between p_i and the remaining points in P, i.e., $\min\{d_S(p_i, p_j), j \neq i, j = 1, \ldots, n\}$ for every p_i in P.

The closest-pair problem: For a set of points $P = \{p_1, \ldots, p_n\}$ on a network, find the minimum value among the shortest-path distances between all possible pairs of points in P, i.e., $\min\{d_S(p_i, p_j), i \neq j, i = 1, \ldots, n, j = 1, \ldots, n\}$. This pair is the *closest pair*.

A simple method for solving these problems is to compare all of the shortest-path distances between p_i and the remaining points in P. However, we have a more efficient method. That is, because Property 4.2 holds, we compare only the shortest-path distances between p_i and the generator points of the Voronoi cells adjacent to $V_{\mathrm{VD}}(p_i)$.

The above problems deal with points in $P = \{p_1, \ldots, p_n\}$, whereas the following problem deals with an arbitrary point p on $\tilde{L}$ and a point p_i in P.

The nearest-search problem: For a set of points $P = \{p_1, \ldots, p_n\}$ and an arbitrary point p on a network $\tilde{L}$, find the point in P that attains the minimum value among the shortest-path distances between point p and the points in P, i.e., $\min\{d_S(p, p_i), i = 1, \ldots, n\}$.

We can readily solve this problem by use of an apparent property (almost equivalent to the definition) of the ordinary N-VD, or specifically:

Property 4.3

Generator point p_i in $P = \{p_1, \ldots, p_n\}$ generating $V_{\mathrm{VD}}(p_i)$ attains the minimum value among the shortest-path distances between a point p on $\tilde{L}$ and generator points $p_1, \ldots, p_n$, i.e., $\min\{d_S(p, p_i), i = 1, \ldots, n\}$, if and only if $V_{\mathrm{VD}}(p_i)$ contains p.

We employ this property in developing efficient computational methods for almost all network spatial methods in the following chapters in this volume.

4.2 Generalized network Voronoi diagrams

In this section, we generalize the ordinary N-VD in five ways: namely, the directed N-VD, the weighted N-VD, the k-th nearest point N-VD, the line N-VD (including the polygon N-VD), and the point-set N-VD.

4.2.1 Directed network Voronoi diagram

The ordinary N-VD assumes that network $N = (V, L)$ is a *nondirected network*, that is, one can move from one end node v_i to the other end node v_j of every link as well as from v_j to v_i. In the real world, however, especially in urbanized areas, street networks are *directed networks*, that is, the link set L includes at least one link on which one can move from one end node to the other end node, but not in reverse. In fact, for the street network in downtown Kyoto, about 30 percent of the total network length is directed (Okabe *et al.*, 2008). The difference in directed and nondirected brings about a large difference in the properties of N-VDs. To demonstrate these explicitly, consider a directed network, denoted by $\vec{N} = (V, \vec{L})$. We can then define the shortest-path distance between p and p_i on $\vec{N}$ in two ways: the shortest-path distance from p_i to p, denoted $\vec{d}_S(p_i, p)$, and that from p to p_i, denoted $\vec{d}_S(p, p_i)$. Referring to p_i, we term the former distance as the *outward (shortest-path) distance from p_i* and the latter as the *inward (shortest-path) distance to p_i*. It should be noted that on nondirected networks, the shortest-path distance is *symmetric*, i.e., equation $d_S(p_i, p) = d_S(p, p_i)$ always holds, while on directed networks, the shortest-path distance is *asymmetric*, i.e., equation $\vec{d}_S(p, p_i) = \vec{d}_S(p_i, p)$ does not always hold. This implies that we can define N-VDs in two ways for $\mathcal{V}(P) = \{V_{VD}(p_1), \ldots, V_{VD}(p_n)\}$, where:

$$V_{VD}(p_i) = \{p|\vec{d}_S(p_i, p) \leq \vec{d}_S(p_j, p), j = 1, \ldots, n\}, \tag{4.3}$$

$$V_{VD}(p_i) = \{p|\vec{d}_S(p, p_i) \leq \vec{d}_S(p, p_j), j = 1, \ldots, n\}. \tag{4.4}$$

We refer to the former as the *outward N-VD* and the latter as the *inward N-VD*. Together, these are *directed N-VDs*. We then alternatively refer to the ordinary N-VD as a *nondirected N-VD*. Figure 4.3 illustrates an outward and an inward N-VD, from which we note the difference between them. Okabe *et al.* (2008) explain the actual difference between a nondirected N-VD and a directed N-VD based on the Kyoto street network.

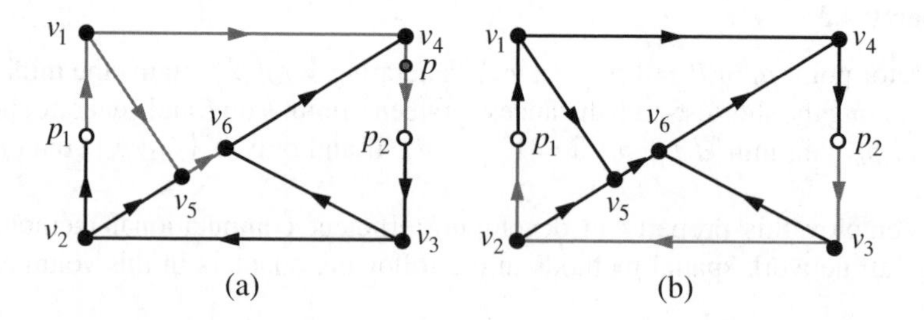

Figure 4.3 Directed network Voronoi diagrams (N-VDs): (a) an outward N-VD ($V_{VD}(p_1)$ is indicated by the gray lines and $V_{VD}(p_2)$ is indicated by the black line segments), (b) an inward N-VD ($V_{VD}(p_1)$ is indicated by the gray line segments and $V_{VD}(p_2)$ is indicated by the black line segments). The links $(v_2 v_5)$, $(v_6 v_4)$ and $(v_3 v_6)$ are shared by $V_{VD}(p_2)$ of the outward and inward N-VDs (see Property 4.5).

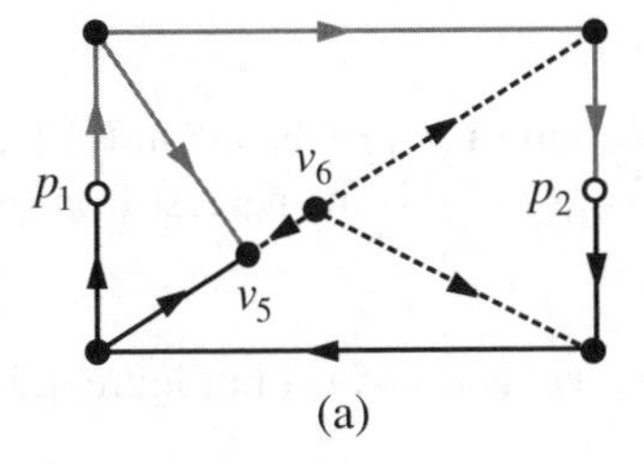

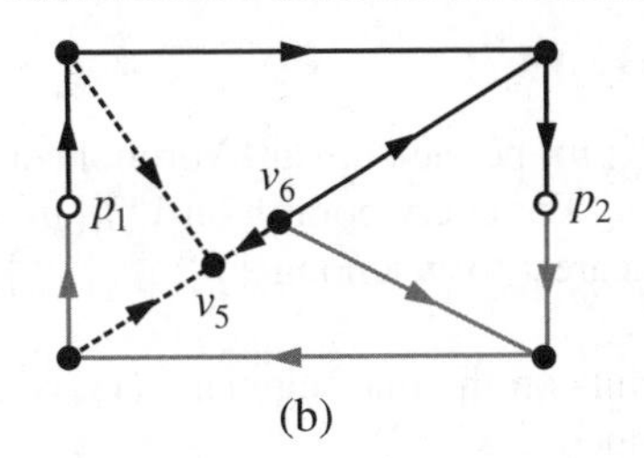

Figure 4.4 Directed network Voronoi diagrams (N-VDs) whose Voronoi cells are not collectively exhaustive (the points on the broken links belong to neither $V_{\text{VD}}(p_1)$ nor $V_{\text{VD}}(p_2)$): (a) an outward N-VD, (b) an inward N-VD.

We now compare the properties of directed N-VDs with those of an ordinary N-VD. In inspecting Figure 4.3, one may consider that Property 4.1 also holds for directed N-VDs, that is, the Voronoi cells of the directed N-VDs are mutually exclusive and collectively exhaustive. However, Property 4.1 does not always hold, even if a network has no critical nodes. Figure 4.4 provides some counterexamples, where the points on the broken links belong to neither $V_{\text{VD}}(p_1)$ nor $V_{\text{VD}}(p_2)$.

To deal with this nonexhaustiveness, we consider two types of node. If all links incident to a node are out from the node (e.g., v_6 in Figure 4.4a), we refer to the node as a *peak node* (note that if we regard direction as the downward direction of a slope, the term *peak* is understandable). Alternatively, if all links incident to a node are into the node (e.g., v_5 in Figure 4.4b), the node is a *bottom node*. In these terms, we describe a property of directed N-VDs that corresponds to Property 4.1 of the ordinary N-VD as follows:

Property 4.4

If an outward N-VD on a directed network $\vec{N} = (V, \vec{L})$ includes neither critical nor peak nodes, the Voronoi cells are mutually exclusive and collectively exhaustive for $\vec{N}$. If an inward N-VD on a directed network $\vec{N} = (V, \vec{L})$ includes neither critical nor bottom nodes, the Voronoi cells are mutually exclusive and collectively exhaustive for $\vec{N}$.

The directed N-VDs satisfy Properties 4.2 and 4.3 of the ordinary N-VD if we replace 'between ... and ...' with 'from ... to ...' for the outward N-VD and 'to ... from...' for the inward N-VD. For instance, Property 4.3 implies that the shortest-path distance from a generator point p_i in $P = \{p_1, \ldots, p_n\}$ to a point p on outward Voronoi cell $V_{\text{VD}}(p_i)$ is shorter than the shortest-path distances from the other generator points to the point p. However, because of the asymmetric property of the shortest-path distance on a directed network, it is not always true that the shortest-path distance from a point p to generator point p_i is shorter than the shortest-path distances from the point p to the other generator points. For example, in Figure 4.3a, the nearest generator point from the point p (the gray circle) in $V_{\text{VD}}(p_1)$ is not p_1, rather p_2. In this connection, it is worth noting the following property:

Property 4.5

Let $V_{\text{VD}}^{\text{out}}(p_i)$ be the outward Voronoi cell of p_i and $V_{\text{VD}}^{\text{in}}(p_i)$ be the inward Voronoi cell of p_i. For every point p on $V_{\text{VD}}^{\text{out}}(p_i) \cap V_{\text{VD}}^{\text{in}}(p_i)$, p is the nearest from p_i and p is the nearest to p_i among $\{p_1, \ldots, p_n\}$.

The points on the line segments (v_2, v_5), (v_3, v_6) and (v_6, v_4) in Figure 4.3 satisfy this property.

In the subsequent subsections, we formulate four types of N-VD on a nondirected network. While we could formulate these on a directed network, we do not go to that extent in this chapter.

4.2.2 Weighted network Voronoi diagram

A Voronoi diagram is the tessellation of a space that we may apply to market area tessellations. This market area tessellation results from consumer choice behavior. If consumers choose their nearest stores, the ordinary VD gives the market areas. However, consumers may choose stores in other ways, not only because they are the nearest. Consider, for instance, a pizza delivery service. Here, consumers may choose a pizza store after considering the transportation cost from the store to their home and the price of the pizza sold at the store. Importantly, the price and transportation cost may differ from store to store because of differences in their production and delivery technologies. Therefore, nearness, measured by physical distance (such as the Euclidean or shortest-path distance), is not always appropriate when estimating market area tessellation. An alternative is the 'weighted distance.'

Studies of market area tessellations have a long history. According to Shieh (1985), the earliest market area study dates to Rau (1841). Launhardt (1882) and Fetter (1924) are other forerunners, followed by Tuominen (1949), Hyson and Hyson (1950), and Gambini, Huff, and Jenks (1967). Boots (1980) theoretically as well as empirically examined market area tessellations in depth and proposed the weighted VD on a plane. Okabe *et al.* (2000) reviewed the related literature and provided a general framework for understanding the weighted VD on a plane (see Okabe *et al.* (2000, Section 3.1)). We extend the weighted VD on a plane to a network in this subsection.

We define a distance by $d_{\text{W}}(p, p_i) = \alpha_i d_{\text{S}}(p, p_i) + \beta_i$, where α_i and β_i are positive constants. In the above store choice example, the values of α_i and β_i reflect the unit transportation cost and the price of pizza sold at the store at p_i, respectively. We term the distance $d_{\text{W}}(p, p_i) = \alpha_i d_{\text{S}}(p, p_i) + \beta_i$ the *compoundly weighted shortest-path distance* or simply *weighted distance*. Using this distance, we define an N-VD by $\mathcal{V}(P) = \{V_{\text{VD}}(p_1), \ldots, V_{\text{VD}}(p_n)\}$, where $V_{\text{VD}}(p_i)$ is given by:

$$\begin{aligned} V_{\text{VD}}(p_i) &= \{p \mid d_{\text{W}}(p, p_i) \le d_{\text{W}}(p, p_j),\ j = 1, \ldots, n\} \\ &= \{p \mid \alpha_i d_{\text{S}}(p, p_i) + \beta_i \le \alpha_j\, d_{\text{S}}(p, p_j) + \beta_j,\ j = 1, \ldots, n\}. \end{aligned} \tag{4.5}$$

We term the set $\mathcal{V}(P)$ the *compoundly weighted N-VD* or the *weighted N-VD*. More specifically, we refer to $d_{\text{W}}(p, p_i) = d_{\text{S}}(p, p_i) + \beta_i$ as the *additively weighted*

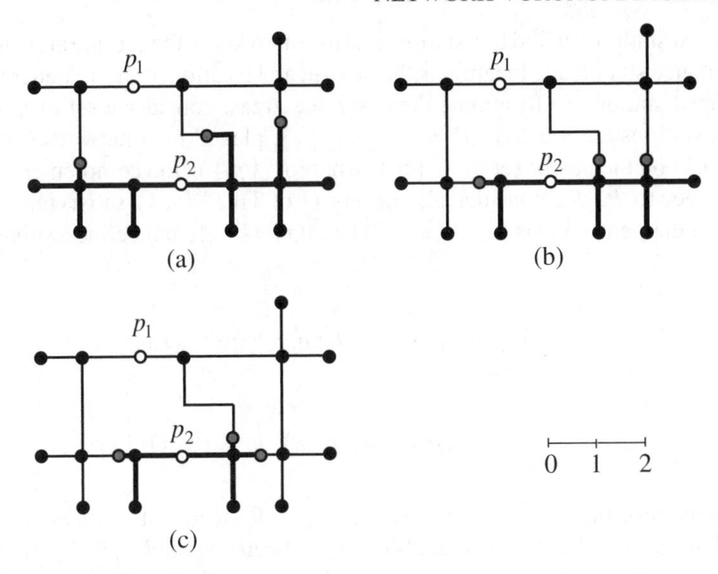

Figure 4.5 Weighted N-VDs: (a) an ordinary N-VD, (b) an additively weighted N-VD (where $d_W(p,p_1) = d_S(p,p_1) + 1$ and $d_W(p,p_2) = d_S(p,p_2) + 3$), (c) a multiplicatively weighted N-VD (where $d_W(p,p_1) = 2\,d_S(p,p_1)$ and $d_W(p,p_1) = 6\,d_S(p,p_1)$).

shortest-path distance and the $\mathcal{V}(P)$ defined with this distance as the *additively weighted N-VD*. We then term $d_W(p,p_i) = \alpha_i d_S(p,p_j)$ the *multiplicatively weighted shortest-path distance* and the $\mathcal{V}(P)$ defined with this distance as the *multiplicatively weighted N-VD*. Figure 4.5 depicts some simple examples, where panel (a) is an ordinary N-VD, panel (b) is an additively weighted N-VD (where $d_W(p,p_1) = d_S(p,p_1) + 1$ and $d_W(p,p_2) = d_S(p,p_2) + 3$), and panel (c) is a multiplicatively weighted N-VD (where $d_W(p,p_1) = 2\,d_S(p,p_1)$ and $d_W(p,p_2) = 6\,d_S(p,p_2)$). Actual applications of the weighed N-VD are in Figure 12.6 in Chapter 12 and Okabe *et al.* (2008, Figures 11 and 12).

We make a short remark on constants α_i and β_i. By the above definition, larger weights for these constants result in smaller Voronoi cells. In some cases, however, it may be intuitively more understandable to define the constants in such a way that larger weights of the constants result in larger Voronoi cells. For this reason, we sometimes write the weighted distance as $d_W(p,p_i) = (1/\alpha_i)\,d_S(p,p_i) - \beta_i$ ($\alpha_i > 0, \beta_i > 0$).

4.2.3 *k*-th nearest point network Voronoi diagram

Imagine that we telephone the nearest ambulance station for assistance, but all of the ambulance vehicles are on service and out of the station. We must then phone the next nearest station. To find this station, it is useful to know the service area in which an ambulance station at a location is the next-nearest station, or more generally, the k-th nearest station. To formulate these service areas, consider a set of points (at which n stations are located), $P = \{p_1, \ldots, p_n\}$, placed on a network N, and let $\Pi(P\backslash\{p_i\})$ be the power set of $P\backslash\{p_i\}$, where $P\backslash\{p_i\}$ is the complement of $\{p_i\}$ with respect to P. We consider all subsets Q of $\Pi(P\backslash\{p_i\})$ satisfying that the number of elements in Q is $k-1$ $(k \geq 2)$ i.e., $|Q| = k-1$, and define a subnetwork of N by:

$$V_{\mathrm{VD}}(p_i) = \bigcup_{Q\in\Pi(P\backslash\{p_i\}),|Q|=k-1} \{p | d_{\mathrm{S}}(p,p_h) \leq d_{\mathrm{S}}(p,p_i),\ p_h \in Q \text{ and}$$

$$d_{\mathrm{S}}(p,p_i) \leq d_{\mathrm{S}}(p,p_j),\ p_j \in P\backslash(Q \cup \{p_i\})\}. \quad (4.6)$$

We term the resulting set $\mathcal{V}(P) = \{V_{\mathrm{VD}}(p_1), \ldots, V_{\mathrm{VD}}(p_n)\}$ the *k-th nearest point N-VD* (Furuta, Uchida, and Suzuki, 2005). Specifically, when $k = 1$, $\mathcal{V}(P)$ refers to the *first-nearest point N-VD*, which is equivalent to the N-VD; when $k = n, \mathcal{V}(P)$ is the *farthest-point N-VD*. Figure 4.6 includes some simple examples.

Alternatively, we can define the k-th nearest point N-VD in terms of ordinary Voronoi diagrams. We illustrate this definition using a simple example in Figure 4.6. Panel (a) shows the ordinary N-VD generated by three generator points: $\mathcal{V}(\{p_1,p_2,p_3\})$, while panels (b), (c) and (d) show the ordinary N-VDs generated by two generator points: $\mathcal{V}(\{p_1,p_2\})$, $\mathcal{V}(\{p_1,p_3\})$ and $\mathcal{V}(\{p_2,p_3\})$, respectively. By definition, the complement of the Voronoi cell of p_1 in panel (a), denoted by $V_{\mathrm{VD}}(p_1|\{p_1,p_2,p_3\})$, (the thick black line segments in panel (a)) with respect to the Voronoi cell of p_1 in panel (b), denoted by $V_{\mathrm{VD}}(p_1|\{p_1,p_2\})$ (the thick black line segments in panel (b)), i.e., $V_{\mathrm{VD}}(p_1|\{p_1,p_2\})\backslash V_{\mathrm{VD}}(p_1|\{p_1,p_2,p_3\})$ indicates the subnetwork in which the first-nearest point is p_3 and the second nearest point is p_1 (the thick black line segments in panel (e)). Similarly, $V_{\mathrm{VD}}(p_2|\{p_1,p_2\})\backslash V_{\mathrm{VD}}(p_2|\{p_1,p_2,p_3\})$ in panel (e) indicates the subnetwork in which the first-nearest point is p_3 and the second-nearest point is p_2 (the thin black line segments panel (e)). We undertake a similar procedure for $\mathcal{V}(\{p_1,p_2,p_3\})$ and $\mathcal{V}(\{p_1,p_3\})$, along with $\mathcal{V}(\{p_1,p_2,p_3\})$ and $\mathcal{V}(\{p_2,p_3\})$. As a result, we obtain panels (f) and (g), where (p_i,p_j) indicates that the first-nearest point is p_i and the second-nearest point is p_j. It follows from these results that the second-nearest N-VD is given by panel (h). Using these results, we can obtain the farthest-point N-VD, as in Figure 4.6i. For example, the subnetwork in which p_1 is the farthest point is given by $V_{\mathrm{VD}}(p_1|\{p_1,p_2\})\backslash V_{\mathrm{VD}}(p_1|\{p_1,p_2,p_3\}) \cup V_{\mathrm{VD}}(p_3|\{p_1,p_3\})\backslash V_{\mathrm{VD}}(p_3|\{p_1,p_2,p_3\})$ (the thick black line segments in panel (h)), and so forth.

This definition can conceptually be extended to the k-th nearest point N-VDs for n generator points (for details, see Okabe *et al.* (2000, Section 3.2)). Note that in practice, this definition works for small k because of its combinatorial nature, and that the computational procedure varies according to k. In Section 4.3.5, we propose an algorithmically simpler definition, in which the computational procedure does not change according to k.

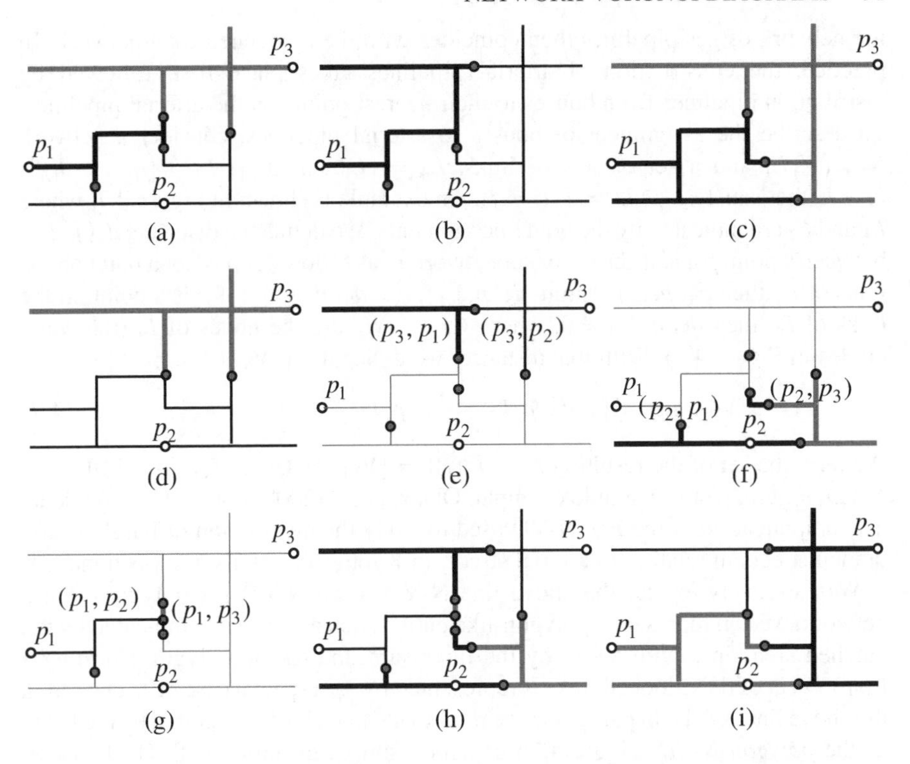

Figure 4.6 Construction of k-th nearest point network Voronoi diagrams (N-VDs): (a) the ordinary N-VD generated by p_1, p_2, and p_3, (b) that generated by p_1 and p_2, (c) that generated by p_1 and p_3, (d) that generated by p_2 and p_3, (e) the first and second nearest generator points from a point on the thick black line segments are p_3 and p_1, respectively (denoted by (p_3, p_1)), and for the thin black line segments, (p_3, p_2), (f) for the thick black line segments, (p_2, p_1), and for the thick gray line segments, (p_2, p_3), (g) for the thin black line segments, (p_1, p_2), and for the thick gray line segment, (p_1, p_3), (h) the second-nearest point N-VD generated by p_1, p_2 and p_3, (i) the farthest-point N-VD generated by p_1, p_2 and p_3 (the thick black line segments for p_1, the thin black line segments for p_2 and the thick gray line segments for p_3). The gray circles are boundary points.

Furuta, Uchida, and Suzuki (2005) include actual applications of the second-nearest point N-VD and the farthest-point N-VD. The k-th nearest point N-VD is an extension of the k-th nearest point Voronoi diagram on a plane. Studies of this began in the 1970s with Miles and Neyman (1970), Shamos and Hoey (1975), Shamos (1978), and Bentley and Maurer (1979), with many others following. For details, see Okabe *et al.* (2000, Section 3.1).

4.2.4 Line and polygon network Voronoi diagrams

Consider, for example, a network of gas pipelines. Pipelines typically comprise arterial and distribution pipelines. Further, because all gas pipelines are under roads, the network of gas pipelines then coincides with the associated road network. In practice, the construction of arterial pipelines takes place first, followed by distribution pipelines from houses to their nearest points on the arterial pipelines. To describe the assignment of houses to arterial pipelines, consider a network $N = (V, L)$ and a set of sets of links, $L_A = \{L_1, \ldots, L_{n_A}\}$, $L_i = \{l_{i1}, \ldots, l_{in_i}\}$, $l_{ij} \in L$, that satisfy $L_i \cap L_j = \varnothing$ $(i \neq j)$. An example is shown in Figure 4.7, where L_1 and L_2 are indicated by the bold line segments. We define the distance, $d_S(p, L_i)$, between a point p and a generator subnetwork L_i as follows: if p is not a point on the links of L_i, then $d_S(p, L_i) = \min\{d_S(p, v_{i1}), \ldots, d_S(p, v_{in_i})\}$; if p is a point on the links of L_i, then $d_S(p, L_i) = 0$, where $v_{i1}, \ldots, v_{in_i}$ are the nodes of L_i (the white circles in Figure 4.7). With this distance, we define a set, $V_{VD}(L_i)$, by:

$$V_{VD}(L_i) = \{p | d_S(p, L_i) \leq d_S(p, L_j), j = 1, \ldots, n_A\}. \tag{4.7}$$

We term the set of the resulting sets, $\mathcal{V}(P) = \{V_{VD}(p_1), \ldots, V_{VD}(p_{n_A})\}$, the *line N-VD*. Figure 4.7a is a simple example. Okabe *et al.* (2008, Figure 15) provide an actual application of the line N-VD used to study the distribution of burglaries on branch streets in relation to arterial streets in Kyoto (for details, see Section 5.3).

We can easily extend the above line N-VD to an N-VD for polygons. On a network, we can represent a polygon-like entity, such as a park (as indicated by the hatched areas in Figure 4.9b), by the links surrounding the polygon (forming a loop). A set of these looped links is represented by L_i, a special case of L_i defined in the above line N-VD. In particular, we refer to the line N-VD generated by the loops as the *polygon N-VD*. Figure 4.7b depicts a simple example, with Okabe *et al.* (2008) providing an actual application to elementary schools in Sagamihara, Japan.

Note that the line N-VD is an extension of the line Voronoi diagram on a plane, referred to as the *line P-VD*. Drysdake and Lee (1978), Drysdale (1979) and Kirkpatrick (1979), among others, have extensively studied this form of the P-VD since the late 1970s. Okabe *et al.* (2000, Section 3.5) summarizes the properties of the line P-VD.

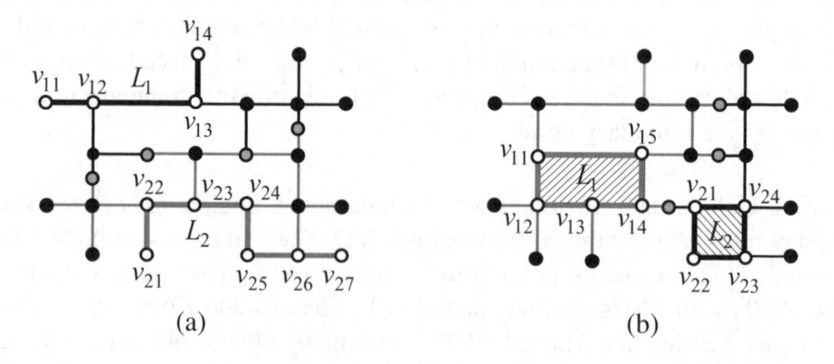

Figure 4.7 Network Voronoi diagrams (N-VDs) for lines: (a) a line N-VD, where the bold line segments are generator line segments, (b) a polygon N-VD, where the shaded polygons are generator polygons (the gray circles are the boundary points between the two Voronoi cells).

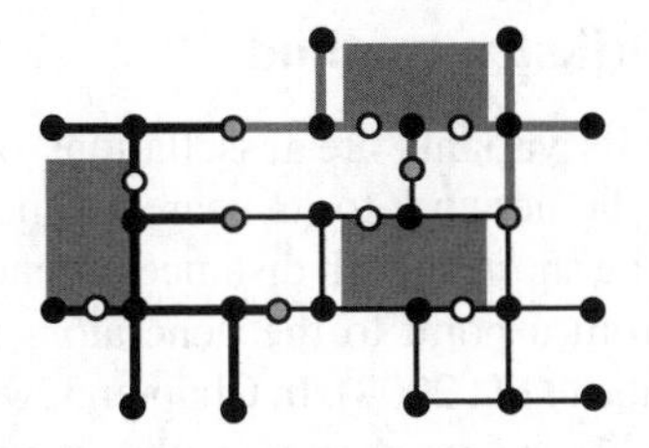

Figure 4.8 A point-set network Voronoi diagram. The white circles indicate, for example, entrances to the facilities (the shaded polygons); the gray circles indicate the boundary points between Voronoi cells.

4.2.5 Point-set network Voronoi diagram

In urbanized areas, a few chain stores, such as 7-Eleven and Lawson, compete in their service areas. If consumers go to their nearest store in an urbanized area, their service areas form a tessellation of the street network in the urbanized area. To be explicit, let $P_i = \{p_{i1}, \ldots, p_{im_i}\}$ be a set of points on $N = (V, L)$ belonging to the i-th class (stores of the i-th chain store), and $P = \{P_1, \ldots, P_n\}$. The distance between p and P_i is measured by $d_S(p, P_i) = \min\{d_S(p, p_{i1}), \ldots, d_S(p, p_{im_i})\}$. We define a set of points, $V_{VD}(P_i)$, as:

$$V_{VD}(P_i) = \{p | d_S(p, P_i) \le d_S(p, P_j), j = 1, \ldots, n\}. \tag{4.8}$$

We term the set of resulting sets $\mathcal{V}(P) = \{V_{VD}(P_1), \ldots, V_{VD}(P_n)\}$ the *point-set N-VD*. In addition to the market areas of chain stores, we can also use this diagram to estimate the service areas of large facilities with several entrances, such as department stores in a downtown area (the white circles and the shaded areas in Figure 4.8) or large parks accessible through several gates, where the set of entrances of the i-th facility is represented by $P_i = \{p_{i1}, \ldots, p_{im_i}\}$. Figure 4.8 illustrates a simple example of the point-set N-VD. An actual example is shown in Okabe *et al.* (2008, Figure 17) for chain stores, 7-Eleven, Lawson, Circle K and others, in Kyoto.

4.3 Computational methods for network Voronoi diagrams

In the preceding sections, we formulated N-VDs conceptually. In practice, if we wish to use these N-VDs for analyzing actual data, we require computational methods for their construction. In this section, we develop such methods. We first develop an algorithm commonly used in computational methods for N-VDs. We then develop individual computational methods for the ordinary, directed, weighted, k-th nearest point, line, and point-set N-VDs.

4.3.1 Multisource Dijkstra method

The ordinary N-VD and its variants are tessellations of a network obtained by assigning every point on the network to its nearest generator, where we measure nearness as a function of the shortest-path distances. Hence, the computation of the shortest-path distances from a point to the generators is the most basic task in constructing N-VDs (Okabe *et al.*, 2008). In Chapter 3, we have already provided a method for computing single-source shortest paths, known as the *Dijkstra method* (Dijkstra, 1959). In the current section, we provide a modification, known as the *multisource Dijkstra method*, which is simultaneously able to treat multiple sources.

To formulate the multisource Dijkstra method, let $N = (V, L)$ be a network, $P \subset V$ be a generator set, and $c(v, v')$ be the length of the link connecting nodes v and v', or more generally, a cost (weight) assigned to the link. In the following algorithm (Algorithm 4.1), we employ the same notation as the single-source Dijkstra method, namely, $D(v)$ denotes the length of the path (i.e., the shortest-path distance) from node v to the nearest source, pred(v) is the immediate predecessor node on the path from the nearest source to v, and A is a storage for storing nodes whose neighbors have not yet been checked. In addition, we introduce a new value, $T(v)$, that indicates the source node nearest to v.

Algorithm 4.1 is similar to the single-source shortest-path algorithm (Algorithm 3.5) save two respects: the number of sources is more than one, and we include $T(v)$. Note that the meaning of each step in the following is almost the same as that in Algorithm 3.5 in Section 3.4.1 of Chapter 3.

Algorithm 4.1 (Multisource Dijkstra method)

Input: network $N=(V, L)$ with node set V and link set L, the length $c(l)$ for links $l \in L$, and the set $P \subset V$ of the start nodes.
Output: shortest-path distance $D(v)$ from P to v, and the immediate predecessor node pred(v) of node v on the shortest path from P to v, for all $v \in V$.
Procedure:

1. $D(v) \leftarrow 0$ and $T(v) \leftarrow v$ for $v \in P$.
2. $D(v) \leftarrow \infty$ for all $v \in V \backslash P$.
3. $A \leftarrow P$.
4. Choose and delete the node v with the smallest $D(v)$ from A. (Note that if there are two or more nodes with the same smallest value, choose any one.)
5. For each node v' adjacent to v, if
   $$D(v') > D(v) + c(v', v),$$
   then
   $$D(v') \leftarrow D(v) + c(v', v)$$
   $$\text{pred}(v') \leftarrow v$$
   $$T(v') \leftarrow T(v)$$
   and insert v' to A
   endif
6. If A is not empty, go to Step 4.
   Otherwise stop.

Steps 1, 2, and 3 are the initialization. Steps 4 and 5 are the main computation, which is the same as the single-source algorithm explained in Section 3.4.1 of Chapter 3. Step 6 is the termination condition. When this algorithm terminates, for every node $v \in V$, the resulting $T(v)$ indicates the nearest generator point, and the resulting $D(v)$ shows the shortest-path distance from v to $T(v)$. In subsequent subsections, we commonly use these values for computing the N-VDs.

The derivation of this time complexity is almost identical to that of the single-source shortest-path algorithm in Section 3.4.1 of Chapter 3. Algorithm 4.1 runs in $O(n_L \log n_V)$ time, where n_L is the number of links and n_V is the number of nodes.

4.3.2 Computational method for the ordinary network Voronoi diagram

Using the multisource Dijkstra method, we first develop a computational method for constructing the ordinary N-VD formulated in Section 4.1. Once the output of Algorithm 4.1 is given (for instance, Figure 4.9), we can readily know that node $v \in V$ belongs to the Voronoi cell of the generator point indicated by $T(v)$. Therefore, the remaining computation for the ordinary N-VD is to divide links according to which Voronoi cells they belong to. To do so, let $l = (v, v') \in L$ be a link connecting two nodes v and v'. If $T(v) = T(v')$ (e.g., the link (v_1, v_2) in Figure 4.9), the whole link belongs to the Voronoi cell of $T(v)$ and we do not divide it; otherwise, i.e., $T(v) \neq T(v')$ (e.g., the link (v_2, v_3)), we divide the link at the point, $p \in l$, satisfying that:

$$D(v) + d_S(v, p) = D(v') + d_S(v', p), \tag{4.9}$$

where $d_S(q, p)$ represents the distance from p to q measured along link l. Note that there always exists the point p that satisfies equation (4.9), because v and v' belong

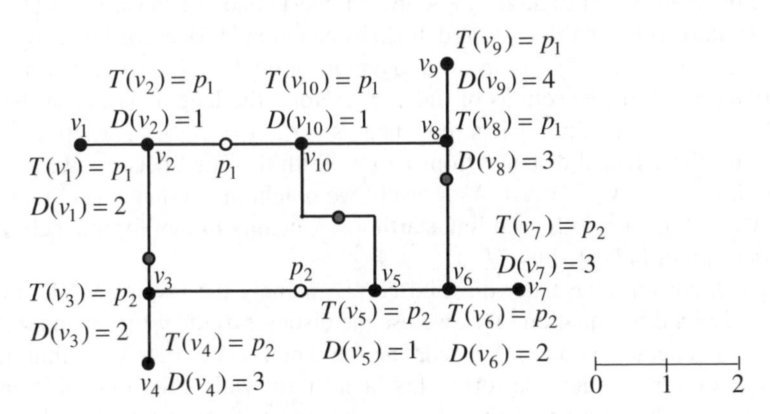

Figure 4.9 Data $T(v_i)$ and $D(v_i)$ obtained from Algorithm 4.1 for constructing an ordinary network Voronoi diagram (the white circles are generator points and the gray circles are boundary points between Voronoi cells).

to distinct Voronoi cells, and hence $|D(v) - D(v')| \leq d(v, v')$ holds. We divide the link into two shorter links by inserting a new node at p (e.g., the gray circles in Figure 4.9), and regard the link from v to p as part of the Voronoi cell of $T(v)$, and the link from v' to p as part of the Voronoi cell of $T(v')$. Consequently, the new node at p is the boundary point of the two Voronoi cells. We do the same procedure for every link $l = (v, v')$ satisfying $T(v) \neq T(v')$. As a result, we obtain the ordinary N-VD generated by P. Thus, the time complexity of the ordinary N-VD is $O(n_L \log n_V)$. This computational method is implemented in the SANET software package (Okabe, Okunuki, and Shiode, 2006a, 2006b; Okabe and Satoh, 2009) with an application in Chapter 12.

4.3.3 Computational method for the directed network Voronoi diagram

On a nondirected network, the length $c(v, v')$ of a link connecting v and v' indicates both distances from v to v' and from v' to v. On a directed network, on the other hand, we distinguish these distances by direction. In general, a directed network contains both directed and undirected links, such as a road network partly containing one-way streets. However, for simplicity we first assume that all the links are directed; a network containing both directed and undirected links will be considered a little later.

To develop a computational method for constructing the directed N-VD in Section 4.2.1, let $\vec{N} = (V, \vec{L})$ be a network in which all the links are directed, and assume that the network does not contain critical nodes, peak nodes or bottom nodes. In this case, all of each link belongs to one and the same Voronoi cell, and hence, unlike the nondirected N-VD, we need not divide links on their midways (see Figure 4.3). In other words, the multisource Dijkstra method itself can construct the directed N-VD; no postprocessing is necessary.

The directed N-VD includes the outward N-VD and the inward N-VD. First, consider the former. In this case, as defined by equation (4.3), the distance $\vec{d}_S(p_i, p)$ from a generator node p_i to an arbitrary point p on $\vec{N} = (V, \vec{L})$ is measured in accordance with the directions of links. Therefore, the length $c(v', v)$ of the link from v to v' in Algorithm 4.1 is redefined as $c(v', v) = \vec{d}_S(v, v')$ where $\vec{d}_S(v, v')$ represents the directed distance from v to v'. With this modification, we perform Algorithm 4.1 on $\vec{N} = (V, \vec{L})$. As a result, we obtain the output saying that each node $v \in V$ together with each link starting at v belong to the Voronoi cell of the generator point indicated by $T(v)$.

Second, consider the latter directed N-VD, namely the inward N-VD. In this case, as defined by equation (4.4), we use the distance $\vec{d}_S(p, p_i)$ from an arbitrary point to a generator node p_i. We redefine the length $c(v', v)$ in Algorithm 4.1 as $c(v', v) = \vec{d}_S(v', v)$ and perform this algorithm with this modification on $\vec{N} = (V, \vec{L})$. The resulting output states that each node $v \in V$, together with each link ending at v, belong to the Voronoi cell of the generator point indicated by $T(v)$.

Until now, we have assumed that all of the links are directed. We now relax this assumption and consider a network containing both directed and undirected links. In this case, the processing is twofold: primary processing and postprocessing. In primary processing, we replace each undirected link with two mutually opposite directed links, redefine length $c(v', v)$ according to which directed N-VD we want to construct, outward or inward, and apply Algorithm 4.1 to the directed network. The nodes and the directed links of the original directed network are assigned to the Voronoi cells in the same way as described above.

In postprocessing, we treat the undirected links of the original directed network. Let l be an undirected link connecting v and v' in the original directed network (recall that this link was replaced with two directed links when we applied Algorithm 4.1). There are two cases, Case 1: $T(v) = T(v')$ and Case 2: $T(v) \neq T(v')$. In the former case, we assign link l to the Voronoi cell of the generator point $T(v)$. In Case 2, we divide link l at the point p that satisfies equation (4.9) as in the same manner as the nondirected N-VD. We assign the link connecting v and p to the Voronoi cell of $T(v)$, and the link connecting p and v' to the Voronoi cell of $T(v')$. Through these processing, we can obtain the directed N-VD for a network containing both directed and undirected links. The time complexity for computing the directed N-VD is also $O(n_L \log n_V)$.

4.3.4 Computational method for the weighted network Voronoi diagram

We develop in this subsection a computational method for constructing the weighted N-VD formulated in Section 4.2.2 with the weighted distance $d_W(p, p_i) = \alpha_i d_S(p, p_i) + \beta_i$, where generator node $p_i \in P$ has a multiplicative weight $\alpha_i \geq 0$ and an additive weight $\beta_i \geq 0$. We can obtain the algorithm for the weighted N-VD by modifying Algorithm 4.1 in three respects.

First, we replace Step 1 of Algorithm 4.1 with:

$$1': D(p_i) \leftarrow \beta_i \quad \text{and} \quad T(p_i) \leftarrow i \quad \text{for} \quad p_i \in P.$$

With this replacement, the shortest-path distance from p_i in Algorithm 4.1 becomes longer than the original distance by β_i, and hence the additively weighted distance is realized in Step 1'.

Second, we replace length $c(v', v)$ in Step 5 of Algorithm 4.1 with $\alpha_i c(v', v_i)$, where $i = T(v)$. This implies that the original length of the link is multiplicatively weighted according to the weight of the start node of the currently considered path. With this modification, the multiplicative weight is realized in Step 5.

The third modification requires the postprocessing in Section 4.3.3. Let l be a link connecting v and v'. Suppose $T(v) \neq T(v')$ in the output of Algorithm 4.1. Then we divide l into two shorter links so that they belong to distinct Voronoi cells. To be explicit, let $i = T(v)$ and $j = T(v')$. We then replace equation (4.9) with:

$$D(v) + \alpha_i d(v, p) = D(v') + \alpha_j d(v', p), \tag{4.10}$$

and divide l at the point p that satisfies this equation (the gray circles in Figure 4.5).

Algorithm 4.1 with these three modifications computes the compoundly weighted N-VD. Note that if we want an additively weighted N-VD (in this case, $\alpha_i = 1$ for all i), the second and the third modifications are not necessary. If we want a multiplicatively weighted N-VD (in this case, $\beta_i = 0$ for all i), the first modification is not necessary. All this additional processing can be done in $O(n_V)$ time, and hence the time complexity for constructing the weighted N-VD is also $O(n_L \log n_V)$. The computational method for the additively weighted N-VD is implemented in the SANET software package (Okabe, Okunuki, and Shiode, 2006a, 2006b; Okabe, Satoh, and Shiode, 2009) and its application to the street network in Shibuya ward in Tokyo is shown in Chapter 12.

4.3.5 Computational method for the k-th nearest point network Voronoi diagram

In the above N-VDs, we used the nearest generator points, but in the k-th nearest point N-VD formulated in Section 4.2.3, we use not only the first, but also the second, third, ..., and k-th nearest generator points. To construct this N-VD in this section, we devote a little attention to a simple method which can quickly retrieve the k-th nearest point N-VD for any given k. The procedure consists of preprocessing and main processing.

In preprocessing, we undertake two tasks. First, we apply the single-source shortest-path method repeatedly to every generator point as a source node. Mathematically, let $p_i \in P$ be a generator point, $v \in V$ be an arbitrary node, and $(p_i, D(v, p_i))$ be the pair of generator node p_i and the shortest-path distance $D(v, p_i)$ from p_i to v. For example, $D(v, p_1)$, $D(v, p_2)$ and $D(v, p_3)$ are shown in Figure 4.10a–c respectively. We then assign $(p_i, D(v, p_i))$ to node v for $i = 1, 2, \ldots, n$, where n is the number of points in P. As a result, each node has the list of n pairs. Second, for each node v, we sort the list of the n pairs in increasing order with respect to $D(v, p_i)$. The resulting sorted list is denoted by $(p_{(1)}, D(v, p_{(1)}))$, ..., $(p_{(n)}, D(v, p_{(n)}))$, where $p_{(k)} \in P$ (note that the correspondence between the original suffix, say p_i and $p_{(k)}$, is recorded), $D(v, p_{(k)}) \in \{D(v, p_{(1)}), \ldots, D(v, p_{(n)})\}$, and $D(v, p_{(1)}) \leq \cdots \leq D(v, p_{(n)})$. Let $Q(v, k) = (p_{(k)}, D(v, p_{(k)}))$ be the k-th pair in the resulting sorted list. For example, $Q(v_i,1)$, $Q(v_i,2)$ and $Q(v_i,3)$ for $i = 1, 2, 3$ are shown in Figure 4.10d. Then, $Q(v, k) = (p_{(k)}, D(v, p_{(k)}))$ tells us that the k-th nearest generator node from v is $p_{(k)}$ and the shortest-path distance to $p_{(k)}$ is given by $D(v, p_{(k)})$. In preprocessing, we are required to repeat the single-source Dijkstra algorithm n times, and hence its time complexity is $O(n\, n_L \log n_V)$.

The main processing for a given k $(1 \leq k \leq n)$ runs in the following way. For each $v \in V$, if $Q(v, k) = (p_{(k)}, D(v, p_{(k)}))$, we assign v to the Voronoi cell of $p_{(k)}$. For each $l \in L$ connecting v and v', suppose that $Q(v, k) = (p_{(k)}, D(v, p_{(k)}))$ and $Q(v', k) = (p_{(h)}, D(v', p_{(h)}))$. If $p_{(k)} = p_{(h)}$ (e.g., for (v_1, v_2) in Figure 4.10, $Q(v_1,2) = (p_1,2.5)$ and $Q(v_2,2) = (p_1,1.5)$, where p_1 is common), then we assign the entire link l (e.g., (v_1, v_2)) to the Voronoi cell of $p_{(k)}$ (e.g., p_1). If $p_{(k)} \neq p_{(h)}$

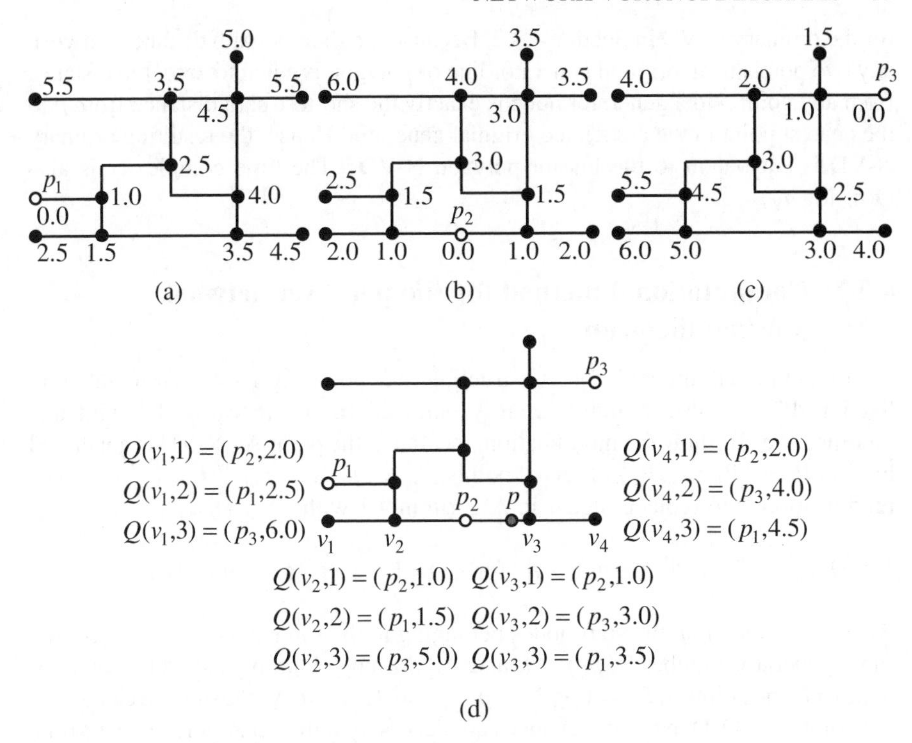

Figure 4.10 Data for constructing the first, second and third nearest point N-VDs: (a) $(p_1, D(v, p_1))$, (b) $(p_2, D(v, p_2))$, (c) $(p_3, D(v, p_3))$, (d) $Q(v, k) = (p_{(k)}, D(v, p_{(k)}))$.

(e.g., for (v_2, v_3), $Q(v_2, 2) = (p_1, 1.5)$ and $Q(v_3, 2) = (p_3, 3.0)$, where $p_1 \neq p_3$), we divide l at the point, p (the gray circle), that satisfies $D(v, p_{(k)}) + d(v, p) = D(v', p_{(h)}) + d(v', p)$, and assign the link from v to p to the Voronoi cell of $p_{(k)}$ and that from p to v' to the Voronoi cell of $p_{(h)}$. The preprocessing and main processing above provide the k-th nearest point N-VD. The time complexity of the main processing is $O(n_L)$ because $n_V \leq n_L$.

4.3.6 Computational methods for the line and polygon network Voronoi diagrams

We can straightforwardly construct the line and polygon N-VDs formulated in Section 4.2.4 if we apply the computational method for the ordinary N-VD in Section 4.3.2 with a single modification. The modification is that for each generator, we change the lengths of the links belonging to generator $L_i = \{l_{i1}, \ldots, l_{in_i}\}$ to 0s, $i = 1, \ldots, n_A$ (the bold line segments in Figure 4.7), and choose an arbitrary node in $\{v_{i1}, \ldots, v_{in_i}\}$ (the nodes of the links in L_i; the white circles in Figure 4.7) as a generating node. We then apply the computational method for the ordinary N-VD in Section 4.3.2. Because the shortest-path distance between any two points in an original generator line or polygon is equal to zero, the distance from any point p to a generator node is exactly the shortest-path distance from p to the nearest point in the associated original generator. Hence, the resulting ordinary N-VD is equivalent to the line or polygon N-VDs. The time complexity is also $O(n_L \log n_V)$.

4.3.7 Computational method for the point-set network Voronoi diagram

A computational method for constructing the point-set N-VD formulated in Section 4.2.5 is also straightforwardly obtained from Algorithm 4.1 with one modification. To show the modification, recall that the point-set N-VD is generated by a set $P_i = \{P_1, \ldots, P_n\}$ of sets of points, $P_i = \{p_{i1}, \ldots, p_{im_i}\}, i = 1, \ldots, n$. The modification is to replace Step 1 in Algorithm 4.1 with:

$$1'': \; D(p_{ij}) \leftarrow 0 \quad \text{and} \quad T(p_{ij}) \leftarrow i \quad \text{for} \quad i = 1, \ldots, n \quad \text{and} \quad j = 1, \ldots, m_i.$$

By this modification, the start nodes belonging to the same generator set have the same generator number $T(p_{ij}) = i$, and Algorithm 4.1 propagates this generator number to the points of $P_i = \{p_{i1}, \ldots, p_{im_i}\}, i = 1, \ldots, n$. As a result, we can obtain the point-set N-VD. No other changes are necessary either in Algorithm 4.1 or the postprocessing. Therefore, the time complexity is also $O(n_L \log n_V)$.

Note that SANET does not directly provide tools for the line N-VD and the point-set N-VD, but we can construct them with the ordinary N-VD tool with a few modifications.

5

Network nearest-neighbor distance methods

One underlying concept common to nearly all spatial methods is *neighborhood*. In fact, the Voronoi subnetwork presented in the preceding chapter is a mathematical representation of this concept. Also all of the methods in the following chapters are formulated by specifying the concept of neighborhood in different ways. More fundamentally, neighborhood is a basic concept of abstract topological space (see, e.g., Pervin (1964, Chapter 3)), from which many spaces used in spatial analysis, including Euclidean and network spaces, stem. In this chapter, we present the *network-constrained nearest-neighbor distance method* or *network nearest-neighbor distance method*, hereafter *network NND method*, where we define neighborhood in terms of the shortest-path distance from an event to the next nearest event. In practice, we employ the network NND method for examining the questions raised in Chapter 1, including Q2 (Do boutiques tend to stand side-by-side alongside streets in a downtown area?) and Q3 (Do street burglaries tend to take place near railway stations?). We generalize such questions as:

Q2′: Given a set of points on a network, how can we test whether the shortest-path distance from each point to the next nearest point is significantly short (or long)?

Q3′: Given two sets of points (of types A and B) on a network, how can we test whether the shortest-path distance from each point of type A to the nearest point of type B is significantly short (or long)?

Spatial Analysis along Networks: Statistical and Computational Methods, First Edition.
Atsuyuki Okabe and Kokichi Sugihara.

The network NND method responds statistically to these questions.

The network NND method is an extension of the NND method as defined on a plane (referred to as the *planar NND method*) to that defined on a network. Formulated in the mid-twentieth century by, among others, Cottam and Curtis (1949), Hopkins and Skellam (1954), Clark and Evans (1954), Moore (1954), Morisita (1954), and Pielou (1959) (see Greig-Smith (1957) for an excellent review), the planar NND method is one of the most traditional methods in spatial analysis. Consequently, many texts already describe the method in depth; for example, King (1969, Section 5.2), Pielou (1977, Chapter 10), Getis and Boots (1978, Section 2.2.2), Ripley (1981, Section 8.2), Diggle (1983, Section 2.3), Upton and Fingleton (1985, Section 2.4), Cressie (1991, Section 8.2.6), Bailey and Gatrell (1995, Section 3.4.3), and Illian *et al.* (2008, Section 2.7.3).

The network NND method itself includes two types. The first type deals with points of the same kind, such as fast-food shops, and examines the spatial pattern of shops in terms of the shortest-path distance from every fast-food shop to the next nearest fast-food shop. The second type deals with two sets of points of different kinds, for instance, fast-food shops and railway stations, and examines the distribution of shops in terms of the shortest-path distance from every fast-food shop to the nearest railway station. To distinguish between these two types, we refer to the former as the network *auto* NND method and the latter as the network *cross* NND method (by convention, the network *auto* NND method is simply referred to as the *network* NND method in the literature). We present these two network NND methods in Sections 5.1 and 5.2, a variant in Section 5.3, and the necessary computational methods in Section 5.4. Note that in what follows, we refer to 'shortest-path distance' as simply 'distance,' and omit 'network,' as in 'network NND method,' except for definitions and headings.

5.1 Network auto nearest-neighbor distance methods

We consider a network $N = N(V, L)$ composed of a set of nodes $V = \{v_1, \ldots, v_{n_V}\}$ and a set of links $L = \{l_1, \ldots, l_{n_L}\}$. Let $\tilde{L} = \bigcup_{i=1}^{n_L} l_i$ be a set of points forming the links (including the nodes) of N, and make the following assumption.

Assumption 5.1

Points $p_1, \ldots, p_n$ are generated according to a stochastic point process on $\tilde{L}$ with a fixed number n of points.

In theory, we could assume that the number of points is probabilistic. Instead, for a few reasons, we assume it is fixed. First, we suppose that all the entities under study, as represented by the points $p_1, \ldots, p_n$, are known. Consequently, a collection of entities forms a statistical population, and they are not samples from a population. While this assumption may sound restrictive, it is not always so because detailed location data on individual facilities are now increasingly available. For instance, we can obtain digital location data for almost every store in a district from the telephone book through address matching (Chapter 12 includes many examples). Second, we assume cross-sectional data; time series data, where the number of points varies stochastically over time, are beyond the scope of this volume.

We now specify Assumption 5.1 by assuming that the points in $P = \{p_1, \ldots, p_n\}$ follow the homogeneous binomial point process (Section 2.4.2 in Chapter 2). This process implies the *complete spatial randomness* (*CSR*) *hypothesis*: the points of P are independently and identically distributed according to the same uniform distribution over $\tilde{L}$ (equation (2.4) in Section 2.4.2). To test this hypothesis, we use statistics that are a function of the distance from every point in P to the next nearest point in P. We first focus on a specific point p_i in P, say, the i-th fast-food shop, and formulate a method for testing whether the i-th fast-food shop and the next nearest fast-food shop are significantly close (or distant). We next consider all points in P and formulate a method for testing whether the average distance from every point in P to the next nearest point in P is significantly short (or long). The NND method for testing the former is termed the *network local auto NND method* and the latter the *network global auto NND method*. We formulate these methods in the following two subsections. Note that while the literature rarely refers to the local NND method, we describe it here because the global NND method is derived from it.

5.1.1 Network local auto nearest-neighbor distance method

For a specific point p_i in P, we consider a subnetwork of $\tilde{L}$, denoted by $\tilde{L}(t|p_i)$, consisting of points on $\tilde{L}$ that are within distance t from p_i, or mathematically, $\tilde{L}(t|p_i) = \{p|d_S(p_i, p) \leq t, p \in \tilde{L}\}$ (for example, $\tilde{L}(2|p_5)$ is indicated by the bold line segments in Figure 5.1). Note that conceptually, $\tilde{L}(t|p_i)$ may correspond to the disk centered at p_i with radius t on a plane; on a network, we sometimes refer to it as the *buffer network* centered at p_i with width t. We refer to the distance from p_i to the next nearest point in P as the *nearest-neighbor distance* (*NN distance* in short) from p_i and denote it by $d_S(p_i, p_i^*)$, where p_i^* is the nearest point of p_i in P (e.g., p_5^* is p_4 in

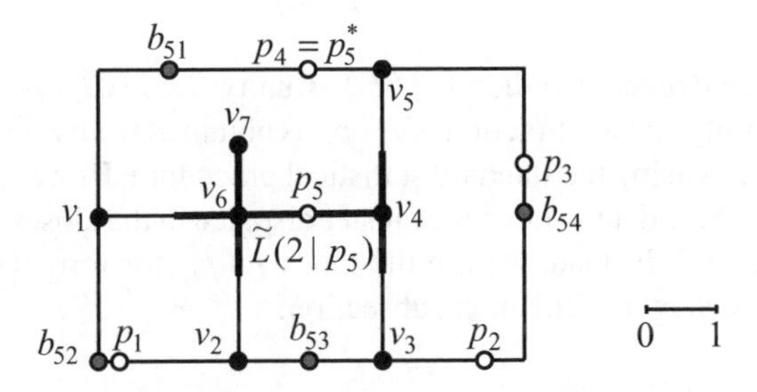

Figure 5.1 The set of points $\tilde{L}(2|p_5)$ that are within distance 2 from p_5 or the buffer network centered at p_5 with width 2 (the bold line segments). The black circles are the nodes of the network formed by the line segments, the white circles are the points of P and the gray circles are the breakpoints of the extended shortest-path tree rooted at p_5.

Figure 5.1). It should be noted that in the following, $d_S(p_i, p_i^*)$ may indicate an observed value or may be a random value of the NN distance, depending on the context where $d_S(p_i, p_i^*)$ is used.

For statistical tests, we require the probability distribution function $F_i(t|p_i)$ of random variable $d_S(p_i, p_i^*)$ under the CSR hypothesis. This function is given by the probability, $\Pr[d_S(p_i, p_i^*) \leq t]$, that $d_S(p_i, p_i^*)$ is less than or equal to t, which is obtained from:

$$\begin{aligned}
&\Pr[d_S(p_i, p_i^*) \leq t] \\
&= 1 - \Pr[\text{all the points of } P \text{ except } p_i \text{ are farther than } t \text{ from } p_i] \\
&= 1 - \Pr[\text{the } n-1 \text{ points, } p_1, \ldots, p_{i-1}, p_i, \ldots, p_n \text{ are placed on } \tilde{L}\backslash\tilde{L}(t|p_i)],
\end{aligned}$$

where $\tilde{L}\backslash\tilde{L}(t|p_i)$ means the complement of subnetwork $\tilde{L}(t|p_i)$ with respect to the entire network $\tilde{L}$; for example, the thin lines in Figure 5.1. Therefore, function $F_i(t|p_i)$ is written as:

$$F_i(t|p_i) = 1 - \left(\frac{|\tilde{L}| - |\tilde{L}(t|p_i)|}{|\tilde{L}|}\right)^{n-1}, \tag{5.1}$$

where $|\tilde{L}|$ denotes the length of $\tilde{L}$. We can obtain the exact value of $F_i(t|p_i)$ with respect to t using the computational method shown in Section 5.4. Consequently, we can compute the exact expected value, $\mu(p_i)$, and variance, $\sigma^2(p_i)$, of random variable $d_S(p_i, p_i^*)$ using equation (5.1). Note that equation (5.1) exactly takes the bounded nature of a network into account. Therefore, as discussed in Section 2.4.3 of Chapter 2, if we apply this function to a naturally bounded network, we do not have edge effects.

For a statistical test, we define an index, $I_L(p_i)$, as the ratio of the value of $d_S(p_i, p_i^*)$ to its expected value, or mathematically:

$$I_L(p_i) = \frac{d_S(p_i, p_i^*)}{\mu(p_i)}. \tag{5.2}$$

By definition, the expected value of $I_L(p_i)$ is unity; i.e., $E(I_L(p_i)) = 1$. Because the exact probability density function of $I_L(p_i)$ is obtained from $F_i(t|p_i)$, we can test the CSR hypothesis using the standard statistical procedure. However, we rarely use this test, except where determining the cluster distance in the closest-pair clustering method in Chapter 7. Instead, we use the index $I_L(p_i)$ for formulating the global NND method, as shown in the next subsection.

5.1.2 Network global auto nearest-neighbor distance method

In this subsection, we extend the above local index to a global index. We consider the NN distance not only from p_i but also from those of $p_1, \ldots, p_{i-1}, p_{i+1}, \ldots, p_n$, and average the resulting NN distances across all points in P; i.e., $\sum_{i=1}^{n} d_S(p_i, p_i^*)/n$. Let μ be the expected value of random variable $\sum_{i=1}^{n} d_S(p_i, p_i^*)/n$ under the CSR hypothesis (i.e., $p_1, \ldots, p_n$ follows the homogeneous binomial point process). Noting that $E(\sum_{i=1}^{n} d_S(p_i, p_i^*)/n)$ is the expected value of $d_S(p_i, p_i^*) = \mu(p_i)$ across $p_i \in \tilde{L}$, the value of μ is obtained from the integration:

$$\mu = \frac{1}{|\tilde{L}|} \int_{p_i \in \tilde{L}} \mu(p_i) \mathrm{d}p_i, \tag{5.3}$$

where the integral is the integration of $\mu(p_i)$ along $\tilde{L}$. Because this integration is difficult to compute analytically, in practice, we employ Monte Carlo simulation described in Section 3.4.4 of Chapter 3. Given μ, we define an index as the ratio of the average of $d_S(p_1, p_i^*), \ldots, d_S(p_n, p_i^*)$ to its expected value, or mathematically:

$$I_G = \frac{1}{\mu} \frac{\sum_{i=1}^{n} d_S(p_i, p_i^*)}{n}. \tag{5.4}$$

The index I_G is an extension of the Clark–Evans index (Clark and Evans, 1954) as defined on a plane to that defined on a network (Okabe, Yomono, and Kitamura, 1995). By definition, the expected value of random variable I_G is unity under the CSR hypothesis.

For statistical tests, we require the probability density function of random variable $\sum_{i=1}^{n} d_S(p_i, p_i^*)/n$, but this is difficult to obtain analytically. Alternatively, we can employ the property that $\sum_{i=1}^{n} d_S(p_i, p_i^*)/n$ approaches the normal distribution as n becomes large because the central limit theorem applies to the average of the random variables $d_S(p_1, p_1^*), \ldots, d_S(p_n, p_n^*)$ (for the central limit theorem, see, for example, Wilks (1962, Section 9.2)). In practice, if an observed number n is large enough, we can test the CSR hypothesis using a normal distribution with the expected value and variance obtained from Monte Carlo simulation. To be explicit, let I_G^* and I_G^{**} be the lower and upper critical values of a significance level α. We can then conclude at confidence level $1 - \alpha$ that if an observed value of I_G is shorter (or longer) than I_G^* (or I_G^{**}), the observed average NN distance is significantly shorter (or longer) than the expected value obtained under the CSR hypothesis, implying that the points tend to cluster (or be dispersed). An actual example is shown in Section 12.2.2 of Chapter 12.

In the above test statistic I_G, as shown in equation (5.4), the observed NN distances $d_S(p_i, p_i^*)$, $i = 1, \ldots, n$ are aggregated in a single value $\sum_{i=1}^{n} d_S(p_i, p_i^*)/n$. However, we may wish instead to examine the observed values in a more disaggregated manner. In such a case, we can use the function defined by the number, $n(t)$, of points satisfying $d_S(p_i, p_i^*) \leq t$ with respect to t. The function defined by $G(t) = n(t)/n$ is termed the *G function* (O'Sullivan and Unwin, 2003). Once again, it is difficult to obtain analytically the expected value and variance of $n(t)$. In practice, we compute the upper and lower envelope curves of $n(t)$ for a given significance level α with respect to t using Monte Carlo simulation, as implemented in the SANET software package (Okabe, Okunuki, and Shiode, 2006a,b; Okabe and Satoh, 2009). We provide an application of this software to actual data in Chapter 12.

5.2 Network cross nearest-neighbor distance methods

In the preceding section, we considered only a single type of point. In this section, we consider two types of point: a set of type A points, $P_{\mathrm{A}} = \{p_{\mathrm{A}1}, \ldots, p_{\mathrm{A}n_{\mathrm{A}}}\}$, and a set of type B points, $P_{\mathrm{B}} = \{p_{\mathrm{B}1}, \ldots, p_{\mathrm{B}n_{\mathrm{B}}}\}$, and make the following assumptions.

Assumption 5.2

Points $p_{\mathrm{A}1}, \ldots, p_{\mathrm{A}n_{\mathrm{A}}}$ are generated according to a stochastic point process on $\tilde{L}$ with a fixed number n_{A} of type A points.

Assumption 5.3

Points of $p_{\mathrm{B}1}, \ldots, p_{\mathrm{B}n_{\mathrm{B}}}$ are deterministically placed on N with a fixed number n_{B} of type B points.

Note that Assumption 5.2 and Assumption 5.1 are identical (Section 5.1 discussed the implications). Note also that Assumption 5.3 does not restrict the distribution pattern of type B points. That is, any configuration is allowed, including regularly spaced points of type B on $\tilde{L}$. Assumptions 5.2 and 5.3 also imply that type A points are temporal, while type B points are stable over time. In reality, type A points are, for instance, fast-food shops, which appear and disappear in a fairly short time period, and their locations are probabilistic. In contrast, type B points are, for example, railway stations, which last for a long time, and their locations are fixed. In theory, we can expect that type B points also follow a stochastic point process, such as a the Cox process (Cox, 1955). However, we do not address this possibility in this book.

We now specify the stochastic point process in Assumption 5.2 with the homogeneous binomial point process (Section 2.4.2 in Chapter 2). This process implies the CSR hypothesis that type A points are distributed independently of the configuration of type B points. To test this hypothesis, we develop statistics as a function of the distance from every point of type A to the nearest point of type B. Recalling Property 4.3 in Section 4.1.2 of Chapter 4, we note that the nearest points can be efficiently identified with the aid of the ordinary network Voronoi diagram. Let $V_{\mathrm{VD}}(p_{\mathrm{B}1}), \ldots, V_{\mathrm{VD}}(p_{\mathrm{B}n_{\mathrm{B}}})$ be Voronoi cells (subnetwork) of the ordinary network Voronoi diagram generated by the points of type B on $\tilde{L}$. Because Voronoi cell $V_{\mathrm{VD}}(p_{\mathrm{B}i})$ is a subnetwork of N, we alternatively denote it with $N_i = N_i(V_i, L_i)$ or $\tilde{L}_i = \bigcup_{j=1}^{n_{Li}} l_{ij}$, where $V_i = \{v_{i1}, \ldots, v_{in_{Vi}}\}$ and $L_i = \{l_{i1}, \ldots, l_{in_{Li}}\}$. Property 4.3 states that $p_{\mathrm{B}i}$ is the nearest generator point from a point $p_{\mathrm{A}j}$ if and only if $p_{\mathrm{A}j}$ is on $\tilde{L}_i$. In the following two subsections, we first focus on a local network (i.e., a subnetwork) N_i and then the entire network N.

5.2.1 Network local cross nearest-neighbor distance method

We consider the local network $N_i = N_i(V_i, L_i)$ or $\tilde{L}_i$ (a set of points forming N_i). For illustrative purposes, Figure 5.2 depicts a simple example where $V_{\mathrm{VD}}(p_{\mathrm{B}1})$ and

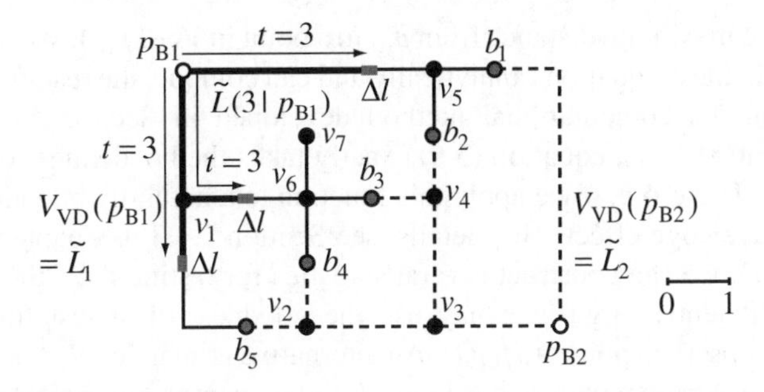

Figure 5.2 Network $N = N(V, L)$ (the continuous and broken line segments), line segments Δl on N (the gray short line segments), the set of nodes $V = \{v_1, \ldots, v_7\}$ (the black circles), Voronoi cells $V_{VD}(p_{B1})$ (the continuous line segments) and $V_{VD}(p_{B2})$ (the broken line segments) generated by p_{B1} and p_{B2} (the white circles), respectively, $\tilde{L}(3|p_{B1})$ (the bold line segments) and boundary points $b_1, \ldots, b_5$ between the Voronoi cells (the gray circles).

$V_{VD}(p_{B2})$ are the Voronoi cells generated by p_{B1} and p_{B2}, points $b_1, \ldots, b_5$ are boundary points between the Voronoi cells, and $V = \{v_1, \ldots, v_7\}$.

Let $m_{Bi}(t)$ be the number of different points p on $\tilde{L}_i = \bigcup_{j=1}^{n_{Li}} l_{ij}$, each of which satisfies $d_S(p, p_{Bi}) = t$, and denote a small line segment at p by Δl (e.g., in Figure 5.2, $t = 3$, $m_{Bi}(t) = 3$ and Δl indicated by the gray short line segments). Here $m_{Bi}(t)$ is the number of different links on which there exists a point p satisfying $d_S(p, p_{Bi}) = t$. The probability of a point p_{Aj} of type A being placed on the small line segments Δl at t is given by $m_{Bi}(t)|\Delta l|/|\tilde{L}_i|$ under the CSR hypothesis. Therefore, the probability density function, $f_{Bi}(t)$, that point p_{Aj} of type A is placed at distance t from p_{Bi}, i.e., $d_S(p_{Aj}, p_{Bi}) = t$, is given by:

$$f_{Bi}(t) = \frac{m_{Bi}(t)}{|\tilde{L}_i|}. \tag{5.5}$$

Alternatively, let $\tilde{L}(t|p_{Bi}) = \{p | d_S(p_{Bi}, p) \leq t, p \in \tilde{L}_i\}$ be the subnetwork in which the distance from p_{Bi} is less than or equal to t (the bold line segments in Figure 5.2), and then the probability distribution function, $F_{Bi}(t)$, of $f_{Bi}(t)$, or equivalently, the probability that a point of type A is placed on $\tilde{L}(t|p_{Bi})$, is written as:

$$F_{Bi}(t) = \frac{1}{|\tilde{L}_i|} \int_0^t m_{Bi}(s) \mathrm{d}s = \frac{|\tilde{L}(t|p_{Bi})|}{|\tilde{L}_i|}. \tag{5.6}$$

Therefore, we obtain the expected value and variance from:

$$\mathrm{E}(d_S(p_{Aj}, p_{Bi})) = \mu_{Bi} = \frac{1}{|\tilde{L}_i|} \int_0^{t_i^*} t\, m_{Bi}(t) \mathrm{d}t, \tag{5.7}$$

$$\mathrm{Var}(d_S(p_{Aj}, p_{Bi})) = \sigma_{Bi}^2 = \frac{1}{|\tilde{L}_i|} \int_0^{t_i^*} t^2\, m_{Bi}(t) \mathrm{d}t - \mu_{Bi}^2, \tag{5.8}$$

where t_i^* is the maximum distance from p_{Aj} to a point in $V_{VD}(p_{Bi})$. We can solve the integrations in these equations analytically and can compute the resulting functions exactly using the computational method developed in Section 5.4. Note that likewise equation (5.1), equation (5.6) exactly takes the boudedness of a network into account. Therefore, if we apply this function to a naturally bounded network, we do not have edge effects (for details, see Section 2.4.3 in Chapter 2).

Using $f_{Bi}(t)$, we can construct several statistics for testing the CSR hypothesis. If n_{Ai} is sufficiently large, we may use the goodness-of-fit test for fitting the histogram of observed points to $f_{Bi}(t)$. An alternative is an index, I_{Bi}, defined by the average distance from type A points in $V_{VD}(p_{Bi})$ to their nearest points in $V_{VD}(p_{Bi})$ to its expected value, or in mathematical terms:

$$I_{Bi} = \frac{1}{\mu_{Bi}} \frac{\sum_{j=1}^{n_{Ai}} d_S(p_{Aj}, p_{Bi})}{n_{Ai}}. \tag{5.9}$$

Before we proceed, we make several remarks on the use of this local index. First, index I_{Bi} is meaningless when $n_{Ai} = 0$, which may take place under the CSR hypothesis. Second, I_{Bi} is a conditional index in the sense that the index is defined for a given $n_{Ai} \geq 1$. Noting that n_{Ai} is probabilistic according to a binomial distribution with two parameters $n_A = \sum_{i=1}^{n_B} n_{Ai}$ (n_A is fixed, as assumed in Assumption 5.1) and $|\tilde{L}_i|/|\tilde{L}|$, we can define an unconditional index of I_{Bi} for $n_{Ai} \geq 1$. Third, by definition, E(I_{Bi}) $= 1$ for a given $n_{Ai} \geq 1$. In theory, we can obtain the exact variance of conditional index I_{Bi} from equations (5.7) and (5.8). In practice, an approximation is acceptable if n_{Ai} is large enough. In fact, for a sufficiently large n_{Ai}, the value of I_{Bi} approximately follows a normal distribution with E(I_{Bi}) $= 1$ and Var(I_{Bi}) $= \sigma_{Bi}^2/(\mu_{Bi}^2 n_{Ai})$, because I_{Bi} is the average of n_{Ai} independent random variables generated by the same bounded probability density function (the central limit theorem is applicable). Therefore, using index I_{Bi}, we can test whether the distribution of the points of P_A on $\tilde{L}_i$ is independent of the location of p_{Bi}, or whether the location of point p_{Bi} affects the distribution of the points of P_A on $\tilde{L}_i$. More explicitly, we conclude at the 0.95 confidence level that the points of type A on $\tilde{L}_i$ tend to cluster around point p_{Bi} if $z_{Bi} = \mu_{Bi}\sqrt{n_{Ai}}(I_{Bi} - 1)/\sigma_{Bi} \leq -1.96$, where I_{Bi} is an observed value.

In conjunction with the G function in Section 5.1.2, one may also speculate an alternative statistic by defining the function, $n_{Ai}(t)$, by the number of points satisfying $d_S(p_{Aj}, p_{Bi}) \leq t$ for type A points in $V_{VD}(p_{Bi})$. However, as this function is equivalent to the K function presented in Chapter 6, we do not discuss it here.

5.2.2 Network global cross nearest-neighbor distance method

The extension of the local cross NND method to the global cross NND method is straightforward. In fact, the formulation of the global index corresponding to the local index I_{Bi} is almost parallel. Let $d_S(p, p_B^*)$ be the distance from a point p on $\tilde{L}$ to

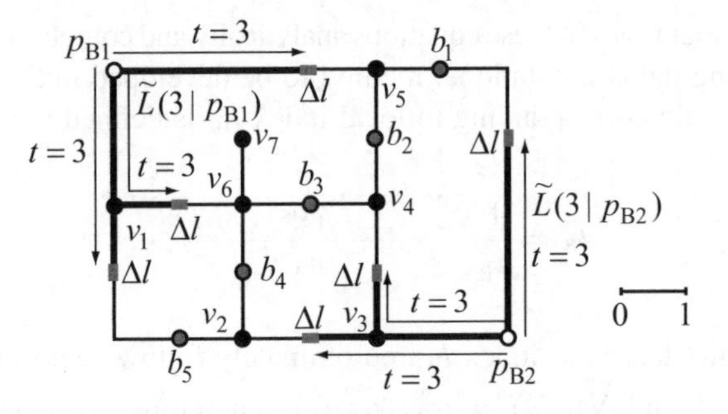

Figure 5.3 Derivation of the probability density function of $f_B(t)$ (for notation, see Figure 5.2).

the nearest point p_B^* of type B (note that $d_S(p, p_B^*) = d_S(p, p_{Bi})$ for p on $\tilde{L}_i = V_{VD}(p_{Bi})$ because Property 4.3 in Chapter 4 holds), and let $m_B(t)$ be the number of different points p on $\tilde{L}$, each of which satisfies $d_S(p, p_B^*) = t$ (the gray small line segments in Figure 5.3). Noting that $m_{Bi}(t)(i = 1, \ldots, n_B)$ is the number of different points p on $\tilde{L}_i$ satisfying that $d_S(p, p_{Bi}^*) = t$, $\tilde{L} = \bigcup_{i=1}^{n_B} \tilde{L}_i$, $\tilde{L}_i \cap \tilde{L}_j (i \neq j)$ contains at most finite number of points, and $|\tilde{L}| = \sum_{i=1}^{n_B} |\tilde{L}_i|$, $m_B(t)$ is given by $\sum_{i=1}^{n_B} m_{Bi}(t)$. Therefore, the probability density function, $f_B(t)$, of the distance from a point p_{Aj} of type A to the nearest point of type B being equal to t, i.e., $d_S(p_{Aj}, p_B^*) = t$, is given by:

$$f_B(t) = \frac{m_B(t)}{|\tilde{L}|} = \frac{\sum_{i=1}^{n_B} m_{Bi}(t)}{|\tilde{L}|}. \tag{5.10}$$

Likewise the derivation from equations (5.5) to (5.6), the probability distribution function, $F_B(t)$, of $f_B(t)$ is obtained from equation (5.10) as:

$$F_B(t) = \frac{1}{|\tilde{L}|} \sum_{i=1}^{n_B} \int_0^s m_{Bi}(s) \mathrm{d}s = \frac{1}{|\tilde{L}|} \sum_{i=1}^{n_B} |\tilde{L}(t|p_{Bi})|. \tag{5.11}$$

Therefore, the expected value and variance are:

$$\mathrm{E}(d_S(p_{Aj}, p_B^*)) = \mu_B = \frac{1}{|\tilde{L}|} \sum_{i=1}^{n_B} \int_0^{t_i^*} t\, m_{Bi}(t) \mathrm{d}t, \tag{5.12}$$

$$\mathrm{Var}(d_S(p_{Aj}, p_B^*)) = \sigma_B^2 = \frac{1}{|\tilde{L}|} \sum_{i=1}^{n_B} \int_0^{t_i^*} t^2\, m_{Bi}(t) \mathrm{d}t - \mu_B^2. \tag{5.13}$$

We solve the integrations in these equations analytically and compute equations (5.12) and (5.13) using the computational method to be developed in Section 5.4.

The test statistic corresponding to local index I_{Bi} is defined by:

$$I_B = \frac{1}{\mu_B} \frac{\sum_{i=1}^{n_B} \sum_{j=1}^{n_{Ai}} d_S(p_{Aj}, p_{Bi})}{n_A}. \tag{5.14}$$

For a sufficiently large n_A, index I_B approximately follows a normal distribution with $E(I_B) = 1$ and $Var(I_B) = \sigma_B^2/(\mu_B^2 n_A)$. Therefore, we can test the CSR hypothesis with the standard normal distribution with $z = \mu_B^2 \sqrt{n_A}(I_B - 1)/\sigma_B$. If the observed value of I_B is smaller than $1 - 1.96\sigma_B/(\mu_B^2 \sqrt{n_A})$, the observed points of type A are significantly close on average to the nearest points of type B with confidence level 0.95.

We can formulate an alternative statistic in terms of the number, $n_A(t)$, of type A points p_{Aj} on $\tilde{L}$ satisfying $d_S(p_{Aj}, p_B^*) \leq t$. Mathematically, the function $F^*(t) = n_A(t)/n_A$ corresponds to the F *function* referred to in O'Sullivan and Unwin (2003). However, it should be noted that the F^* function in this chapter assumes the type B points are fixed, whereas the F function in O'Sullivan and Unwin (2003) assumes that the type B points are uniformly random. For a given t, the probability that the distance from a point of type A to the nearest point of type B is less than or equal to t is given by $F_B(t)$ in equation (5.11). Therefore, for a given t, the random number $n_A(t)$ under the CSR hypothesis follows a binomial distribution with parameters n_A and $F_B(t)$, and the expected value and variance are given by:

$$E(n(t)) = \mu_{n(t)} = n_A \frac{1}{|\tilde{L}|} \sum_{i=1}^{n_B} |\tilde{L}(t|p_{Bi})| \tag{5.15}$$

$$Var(n(t)) = \sigma_{n(t)}^2 = n_A \left(\frac{1}{|\tilde{L}|} \sum_{i=1}^{n_B} |\tilde{L}(t|p_{Bi})| \right) \left(1 - \frac{1}{|\tilde{L}|} \sum_{i=1}^{n_B} |\tilde{L}(t|p_{Bi})| \right). \tag{5.16}$$

We can compute these values exactly using the techniques shown in Section 5.3. Therefore, we can obtain the upper and lower envelope curves of function $n_A(t)$ (or the F^* function) for a given significance level with respect to t from a binomial distribution with parameters $\mu_{n_A(t)}$ and $\sigma_{n_A(t)}^2$. If n_A is sufficiently large, we can approximate the binomial distribution using a normal distribution with parameters $\mu_{n_A(t)}$ and $\sigma_{n_A(t)}^2$. The SANET software package implements this test statistic, and we illustrate its application to actual data in Chapter 12. Note that if we apply the global cross NND methods (including the index I_B and the function $n_A(t)$ or F^*) to naturally bounded networks, we are not bothered with the edge effect, because those methods exactly take boundedness into account, as discussed in Section 2.4.3 in Chapter 2.

5.3 Network nearest-neighbor distance method for lines

In the preceding sections, we formulated the network NND methods with distances from certain points to the next nearest points. In this section, we formulate an NND method with distances from points to their nearest links that form a subset of the links of network $N = (V, L)$. This method is potentially useful for testing, for example, why street crimes tend to take place on branch streets located near arterial streets.

To develop a statistic for examining this phenomenon, we consider a set of points $P = \{p_1, \ldots, p_n\}$ (e.g., the white circles in Figure 5.4) and a subset $L_B = \{l_{B1}, \ldots, l_{Bm}\}$ (e.g., arterial streets in the above example; the bold line segments in Figure 5.4) of $L = \{l_1, \ldots, l_{n_L}\}$. Let $\tilde{L}_B = \bigcup_{i=1}^{m} l_{Bi}$, and $\tilde{L} \backslash \tilde{L}_B$ be the complement of $\tilde{L}_B$ with respect to $\tilde{L}$. We make the following assumption.

Assumption 5.4

Points $p_1, \ldots, p_n$ are generated according to a stochastic point process on $\tilde{L} \backslash \tilde{L}_B$ with a fixed number of points.

We can assume that the points are generated on the entire network $\tilde{L}$, but for ease of explanation, assume $\tilde{L} \backslash \tilde{L}_B$ for the present. We consider the distance from p_i to the nearest point on $\tilde{L}_B$, denoted by $d_S(p_i, l_i^*)$, where $l_i^* \in L_B$. It is obvious that the nearest point on $\tilde{L}_B$ from p_i is one of the end nodes of links $l_{B1}, \ldots, l_{Bm}$. Let $V_B = \{v_{B1}, \ldots, v_{Bk}\}$ be the set of those nodes (the gray circles for the bold line segments in Figure 5.4). The distance from p_i to the nearest point on $\tilde{L}_B$ is then equal to the distance from p_i to the nearest node in V_B, denoted by $d_S(p_i, v_i^*)$, where $v_i^* \in V_B$. Therefore, testing the CSR hypothesis that the points of P are distributed independently of the configuration of links L_B is equivalent to testing the CSR hypothesis that the points of P are distributed independently of the configuration of the nodes of V_B. We note from the discussion in the preceding section that this hypothesis can be tested by using the global cross NND method because the NND

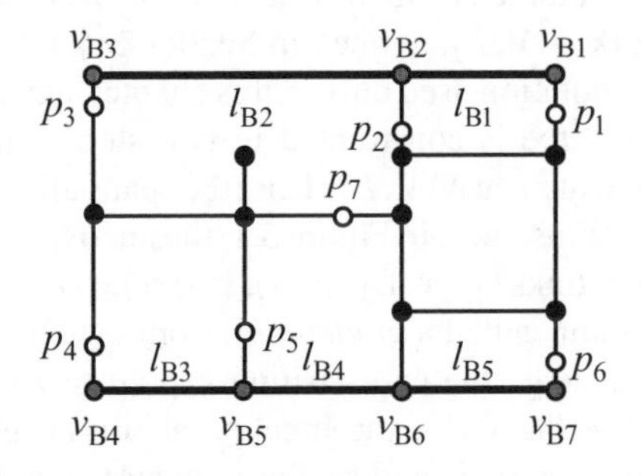

Figure 5.4 The network nearest-neighbor distance method for lines (the bold line segments indicate generator line segments).

method for lines is exactly the same as the global cross NND method, except that P_A and P_B are replaced with P and V_B, respectively.

Satoh and Okabe (2006) applied the NND method for lines to the spatial analysis of street crime. They first examined whether street burglaries tended to take place on arterial $\tilde{L}_B$ or branch $\tilde{L} \backslash \tilde{L}_B$ streets in Kyoto using a binomial distribution with parameters n_A and $|\tilde{L}_B|/|\tilde{L}|$, where n_A is the number of burglaries. They then examined the distribution of burglaries on the branch streets $\tilde{L} \backslash \tilde{L}_B$. Because Kyoto has many one-way streets, Satoh and Okabe (2006) considered the outward NN distance $\vec{d}_S(l^*_{i\rightarrow}, p_i)$ from arterial streets to each incidence spot and the inward NN distance $\overleftarrow{d}_S(p_i, l^*_i)$ from each incidence spot to the arterial streets (for the directed distances, see Section 4.2.1 in Chapter 4) and applied the NND method for lines with these directed distances to the distribution of burglaries on Kyoto branch streets. The results indicated that the average inward NN distance was significantly short, but the average outward NN distance was not. This could indicate that thieves consider it easier to flee from crime spots on branch streets to get to arterial streets than to escape from arterial streets to get to branch streets.

5.4 Computational methods for the network nearest-neighbor distance methods

Having formulated the network NND methods in the preceding sections, in this section, we discuss how to compute these methods in practice. The section consists of two subsections. The first subsection develops the computational methods for the auto NND methods, and the second subsection develops the computational methods for the cross NND method.

5.4.1 Computational methods for the network auto nearest-neighbor distance methods

The statistics formulated in Section 5.1 are derived from equation (5.1), where the key term is $|\tilde{L}(t|p_i)|$, as obtained from the *extended shortest-path tree*, T_i, rooted at a point $p_i \in P$ on network $N(V, L)$, defined in Section 3.4.1.3 of Chapter 3. For the purpose of introducing notation used only in this chapter, again, we describe how to construct this tree. The tree is constructed in two steps. First, we construct the shortest-path tree rooted at p_i on $N(V, L)$. This tree spans all nodes of V but does not cover some links in L; for instance, in Figure 5.1, the uncovered links of the shortest-path tree rooted at p_5 are links $(v_1, v_5), (v_1, v_2), (v_2, v_3), (v_3, v_5)$. On each uncovered link, there exists one point, called a *breakpoint*, from which there are two different equal-length paths to p_i (e.g., the path visiting b_{51}, v_1, v_6, p_5 and the path visiting b_{51}, v_5, v_4, p_5). Let B_i be the set of the breakpoints on all uncovered links (e.g., $B_3 = \{b_{51}, b_{52}, b_{53}, b_{54}\}$, as indicated by the gray circles in Figure 5.1). We then break the j-th uncovered link at its breakpoint b_{ij} into two pieces (e.g., the link (v_2, v_3) is divided into the links (v_2, b_{53}) and (v_3, b_{53}) in Figure 5.1), and add new nodes b'_{ij} and b''_{ij} at the end nodes of the resulting pieces (note that the location of b'_{ij} is the same as that of b''_{ij}, but they are treated as separated nodes). Let $B'_i = \{b'_{ij}, b''_{ij} | b_{ij} \in B_i\}$ and L' be the set of refined links resulting from inserting B'_i. The refined network is denoted by $N' = (V', L') = (V \cup B'_i, L')$. On this network N', we again construct the shortest-path tree rooted at p_i. This tree, known as the *extended shortest-path tree*, covers all the links in L' or L. A method for computing the extended shortest-path tree is shown in Section 3.4.1.3 of Chapter 3 in details. Using this method, we illustrate computational methods for the local auto nearest-neighbor distance method in Section 5.1.1 and the global auto nearest-neighbor distance method in Section 5.1.2.

5.4.1.1 Computational methods for the network local auto nearest-neighbor distance method

The local auto NND method is, as shown in equation (5.1), obtained from the function $|\tilde{L}(t|p_i)|$. To construct this function for nonnegative real t, we define a function, $m_i(t)$, by the number of different points p on $T(p_i)$, each of which satisfies $d_S(p, p_i) = t$. To illustrate how we compute $m_i(t)$, we depict Figure 5.5, where

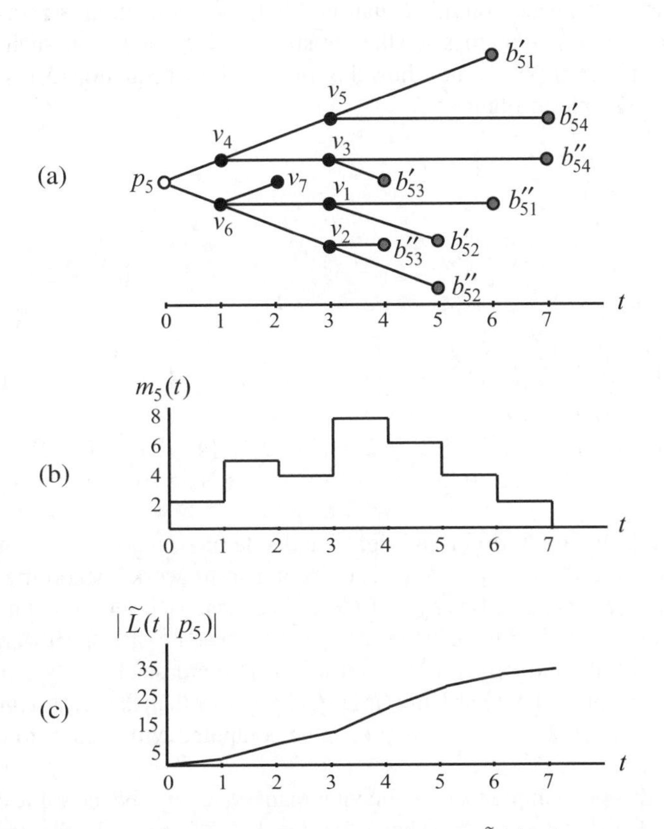

Figure 5.5 The method for computing $m_5(t)$ and $\tilde{L}(t|p_5)$: (a) the extended shortest-path tree rooted at p_5 in Figure 5.1, (b) the functions $m_5(t)$ and (c) $|\tilde{L}(t|p_5)|$ derived from the extended shortest-path tree of (a).

panel (a) shows the extended shortest-path tree rooted at p_5 on the network in Figure 5.1 (the horizontal axis in Figure 5.5 indicates the shortest-path distance from p_5 to every node).

For all nodes v of T_i, we list the distances $d_S(p_i, v)$ from node p_i and sort the nodes in nondecreasing order of $d_S(p_i, v)$. Let $(v^{(0)}, v^{(1)}, \cdots, v^{(n_{V'})})$ be the resulting sorted list of nodes where $v^{(0)} = p_i$, and $n_{V'}$ is the number of all nodes in T_i (for example, $(p_5, v_4, v_6, v_7, \cdots, b'_{54})$ in panel (a) of Figure 5.5). Furthermore, the corresponding list of distances from p_i is $(t^{(0)}, t^{(1)}, \cdots, t^{(n_{V'})})$, where $t^{(0)} = 0$ (in panel (a), $t^{(0)} = 0$, $t^{(1)} = d_S(p_5, v_4), t^{(2)} = d_S(p_5, v_6), t^{(3)} = d_S(p_5, v_7), \cdots$) and $\delta^{(j)}$ is the degree of node v_j for $j = 0, 1, \ldots, n_{V'}$ (in panel (a), $\delta^{(0)} = 2, \delta^{(1)} = 3, \delta^{(2)} = 4, \delta^{(3)} = 1, \cdots$). Then, for each interval $t^{(j)} < t < t^{(j+1)}, j = 0, 1, \ldots, n_{V''}$ (if it exists), $m_i(t)$ has a constant value, which is computed by:

$$m_i(t) = \delta^{(0)} + \sum_{k=1}^{j}(\delta^{(k)} - 2) \quad \text{for } t^{(j)} < t < t^{(j+1)}. \tag{5.17}$$

Two remarks are needed on this equation. First, the summation is zero for $j = 0$. Second, if $t^{(j)} = t^{(j+1)}$, there is no t that satisfies $t^{(j)} < t < t^{(j+1)}$. In such a case, we need not compute $m_i(t)$. To show how the computation of equation (5.17) proceeds, we use an example in Figure 5.5.

For $t^{(0)} = 0 < t < t^{(1)} = 1$,
$$m_5(t) = \delta^{(0)} = 2;$$
for $t^{(1)} = t^{(2)} = 1 < t < t^{(3)} = 2$,
$$\begin{aligned} m_5(t) &= \delta^{(0)} + \textstyle\sum_{k=1}^{2}(\delta^{(k)} - 2) = \delta^{(0)} + (\delta^{(1)} - 2) + (\delta^{(2)} - 2) \\ &= 2 + (3 - 2) + (4 - 2) = 5; \end{aligned}$$
for $t^{(3)} = 2 < t < t^{(4)} = 3$,
$$m_5(t) = \delta^{(0)} + \textstyle\sum_{k=1}^{3}(\delta^{(k)} - 2) = 2 + (3 - 2) + (4 - 2) + (1 - 2) = 4;$$
for $t^{(4)} = t^{(5)} = t^{(6)} = t^{(7)} = 3 < t < t^{(8)} = 4$,
$$\begin{aligned} m_5(t) = \delta_0 + \textstyle\sum_{k=1}^{7}(\delta_k - 2) &= 2 + (3 - 2) + (4 - 2) + (1 - 2) + (3 - 2) \\ &+(3 - 2) + (3 - 2) + (3 - 2) = 8; \text{ and so on.} \end{aligned}$$

In general, the first term on the right-hand side of equation (5.17) implies that $m_i(t) = \delta^{(0)}$ for $t^{(0)} = 0 < t < t_1$. The second term works according to: (i) if $\delta^{(k)} = 1$, then $m_i(t)$ decreases by 1 at $t = t^{(k)}$ because v_k is a terminal node; (ii) if $\delta^{(k)} = 2$, then $m_i(t)$ does not change at $t = t^{(k)}$ because $v^{(k)}$ has one inward link and one outward link; and (iii) if $\delta^{(k)} \geq 3$, then $m_i(t)$ increases by $\delta_k - 2$ at $t = t^{(k)}$, because the number of outward links is larger by $\delta_k - 2$ than the inward link. Hence, $m_i(t)$ in each interval $t^{(j)} < t < t^{(j+1)}$ can be computed with respect to increasing $j = 0, 1, 2, \ldots$.

To state this procedure in an incremental manner, let $m_i^{(j)}$ be the value of $m_i(t)$ in the interval $t^{(j)} < t < t^{(j+1)}$. Then, for $j = 1, 2, \ldots, n_{V'} - 1$, the incremental process is written as:

Table 5.1 Computation of $m_i^{(j)}$ using equation (5.18).

j	$\delta^{(j)}$	$\delta^{(j)}-2$	$m_i^{(j)}$	range of t
0	2	0	2	$0<t<1$
1	3	1	3	—
2	4	2	5	$1<t<2$
3	1	−1	4	$2<t<3$
4	3	1	5	—
5	3	1	6	—
6	3	1	7	—
7	3	1	8	$3<t<4$

$$m_i^{(j)} = m_i^{(j-1)} + (\delta_j - 2) \quad \text{for } t^{(j)} < t < t^{(j+1)}. \tag{5.18}$$

In reference to the example in Figure 5.3, we carry out the computation of $m_i{}^{(j)}$ in equation (5.18) through the process shown in Table 5.1. The leftmost column provides the index $j = 0, 1, 2, \ldots$, indicating the process, and the second and third columns show the degrees $\delta^{(j)}$ and $\delta^{(j)} - 2$, respectively, of the j-th node, the fourth column shows the value of $m_i^{(j)}$, and the rightmost column shows the range of t for which $m_i{}^{(j)}$ is valid. Note that for some values of j (for example, $j = 1, 4, 5, 6$), the ranges are empty. However, this is not a problem because what we need are the values of $m_i^{(j)}$ such that $t^{(j)} < t < t^{(j+1)}$ is not empty.

Next, let $|\tilde{L}^{(j)}(t|p_i)|$ be the value of $|\tilde{L}(t|p_i)|$ in the interval $t^{(j)} < t < t^{(j+1)}$. Then, $|\tilde{L}^{(j)}(t|p_i)|$ is computed by:

$$\left|\tilde{L}^{(j)}(t|p_i)\right| = \sum_{k=1}^{j-1}(t^{(k+1)} - t^{(k)})m_i^{(k)} + (t - t^{(j)})m_i^{(j)} \text{ for } t^{(j)} < t < t^{(j+1)}. \tag{5.19}$$

Using equations (5.18) and (5.19), we compute $F_i(t|p_i)$, from which we obtain the expected value $\mu(p_i)$, the variance $\sigma^2(p_i)$ of $d_S(p_i, p_i^*)$, and index $I_L(p_i)$.

We now evaluate the time complexity of the above computational methods. First, note that $|L| = O(n_V)$ because the network is planar, and that $|V'| = O(n_V)$ and $|V''| = O(n_V)$ because the number of breakpoints is not greater than $|L|$, where $|L|$ indicates the number of elements in the set L. The extended shortest-path tree can be constructed in $O(n_V \log n_V)$ time (Section 3.4.1.3 in Chapter 3). Recall that when we construct an extended shortest-path tree, we also obtain the distances from root p_i to all of the nodes and breakpoints. Sorting the n_V distances requires $O(n_V \log n_V)$ time. The value of $m_i^{(j)}$ for all j can be computed by equations (5.17) and (5.18) in $O(n_V)$ time. Hence, we can also obtain by equation (5.19) the value $|\tilde{L}^{(j)}(t|p_i)|$ as a function of t for $t^{(j)} < t < t^{(j+1)}$ for all j in $O(n_V)$ time. In summary, we can compute $|\tilde{L}^{(j)}(t|p_i)|$ for all j in $O(n_V \log n_V)$ time.

5.4.1.2 Computational methods for the network global auto nearest-neighbor distance method

We now consider a computational method for the global auto NND method. As mentioned in Section 5.1, we are obliged to employ Monte Carlo simulation, and hence what we need to compute is, for a given set $P = \{p_1, \ldots, p_n\}$ of data points, to compute the distances from p_i to the nearest point for all $i = 1, 2, \ldots, n$. For this purpose, we first insert the points of P in N and denote the resulting refined network by $N' = (V', L')$, where $V' = V \cup P$ (note that $N' = (V', L')$ should be distinguished from that in Section 5.4.1). Next, for each point p_i, we compute the shortest paths from p_i to the other points until we reach the nearest point in $P \backslash \{p_i\}$. The shortest-path search from one point can be done in $O(n_V \log n_V)$ time (Section 3.4.2 in Chapter 3), and hence the total time complexity for computing the shortest-path distances from p_i to the nearest point for all $p_i \in P$ is $O(nn_V \log n_V)$. This computation time is required for one set P. In the Monte Carlo simulation, we generate set P many times (say m), to obtain a reliable approximation of the probability distribution function $F_i(t|p_i)$. Therefore the time complexity to obtain $F_i(t|p_i)$ is of $O(m\,n\,n_V \log n_V)$.

5.4.2 Computational methods for the network cross nearest-neighbor distance methods

We now turn to a computational method for the other NND method; namely, the cross NND method. We first consider a local method and then a global method.

5.4.2.1 Computational methods for the network local cross nearest-neighbor distance method

As noted in Section 5.2.1, the local cross NND method is a conditional method in the sense that the distribution of $n_{\mathrm{A}i}$ points of type A on the Voronoi cell $V_{\mathrm{VD}}(p_{\mathrm{B}i})$ of the i-th point $p_{\mathrm{B}i}$ of type B is examined assuming that $n_{\mathrm{A}i}$ is fixed. Thus, we can regard $V_{\mathrm{VD}}(p_{\mathrm{B}i})$ as the entire network such that the total number of points is given by $n_{\mathrm{A}i}$. This implies that we can almost straightforwardly apply the computational method for the local auto NND method found in the preceding section to this conditional method. The only difference is that a root node of the shortest-path tree is not chosen from a point in $n_{\mathrm{A}i}$ points but is already fixed as $p_{\mathrm{B}i}$. We detail the procedure for this method as follows.

We denote the Voronoi subnetwork $V_{\mathrm{VD}}(p_{\mathrm{B}i})$ by N_i (Chapter 4). First, construct the extended shortest-path tree $T_{\mathrm{B}i}$ of N_i with respect to the root node at $p_{\mathrm{B}i}$. Let $n_{\mathrm{B}i}$ be the number of nodes of $V_{\mathrm{VD}}(p_{\mathrm{B}i})$. For descriptive simplicity, in this subsection, we use n for $n_{\mathrm{B}i}$. The number of nodes of $T_{\mathrm{B}i}$ is then $n + 1$ (i.e., $p_{\mathrm{B}i}$ and the n nodes of $V_{\mathrm{VD}}(p_{\mathrm{B}i})$). Second, sort the nodes of $T_{\mathrm{B}i}$ in nondecreasing order of the distance from $p_{\mathrm{B}i}$. Let $(v_{\mathrm{B}i}^{(0)}, v_{\mathrm{B}i}^{(1)}, \cdots, v_{\mathrm{B}i}^{(n)})$ be the resulting list of nodes where the associated list of the distance from $p_{\mathrm{B}i}$ is $(t_{\mathrm{B}i}^{(0)}, t_{\mathrm{B}i}^{(1)}, \cdots, t_{\mathrm{B}i}^{(n)})$, where $v_i^{(0)} = p_{\mathrm{B}i}$ and $t_{\mathrm{B}i}^{(0)} = 0$, and the degree of $v_{\mathrm{B}i}^{(j)}$ is denoted by $\delta_{\mathrm{B}i}^{(j)}$ for $j = 1, 2, \ldots, n_{\mathrm{A}i}$. For each interval,

$t_{\mathrm{B}i}^{(j)} < t < t_{\mathrm{B}i}^{(j+1)}$, we denote the number of points p such that $d_{\mathrm{S}}(p_{\mathrm{B}i}, p) = t$ by $m_{\mathrm{B}i}^{(j)}$. Third, compute:

$$m_{\mathrm{B}i}^{(j)} = \delta_{\mathrm{B}i}^{(0)} + \sum_{k=1}^{j}(\delta_{\mathrm{B}i}^{(k)} - 2) \text{ for } t_{\mathrm{B}i}^{(j)} < t < t_{\mathrm{B}i}^{(j+1)}. \tag{5.20}$$

Let $|\tilde{L}_i(t|p_{\mathrm{B}i})|$ be the total length of the links of N_i that are within distance t from $p_{\mathrm{B}i}$. We then write $|\tilde{L}_i(t|p_{\mathrm{B}i})|$ for $t_{\mathrm{B}i}^{(j)} < t < t_{\mathrm{B}i}^{(j+1)}$ as:

$$|\tilde{L}_i(t|p_{\mathrm{B}i})| = \sum_{k=0}^{j-1}(t_{\mathrm{B}i}^{(k+1)} - t_{\mathrm{B}i}^{(k)})m_{\mathrm{B}i}^{(k)} + (t - t_{\mathrm{B}i}^{(j)})m_{\mathrm{B}i}^{(j)}. \tag{5.21}$$

Fourth, compute $F_{\mathrm{B}i}(t)$ using equation (5.6). Fifth, compute the expected value $\mu_{\mathrm{B}i}$ and variance $\sigma_{\mathrm{B}i}^2$ using equations (5.7) and (5.8) with the resulting $F_{\mathrm{B}i}(t)$. Sixth, compute index $I_{\mathrm{B}i}$ by equation (5.9). Then, as described in Section 5.2.1, we can test the CSR hypothesis with μ_{Bi}, $\sigma_{\mathrm{B}i}^2$ and an observed value of $I_{\mathrm{B}i}$.

The time complexity of this method can be evaluated in the following manner. Recall that the Voronoi diagram for $P_{\mathrm{B}i}$ as generators can be computed in $\mathrm{O}(n_V \log n_V)$ time (Section 4.3.2 in Chapter 4). Once we obtain the Voronoi subnetwork N_i for $p_{\mathrm{B}i} \in P_{\mathrm{B}}$, $|\tilde{L}_i(t|p_{\mathrm{B}i})|$ can be computed in the same manner as in Section 5.4.1.1 in $\mathrm{O}(n_{V_i} \log n_{V_i})$ time, where n_{V_i} is the number of nodes in the Voronoi subnetwork N_i.

5.4.2.2 Computational methods for the network global cross nearest-neighbor distance method

When we are interested in a global trend and not just the local trend, we use the global cross NND method. As noted in equations (5.10)–(5.16), the basic functions to compute are $m_{\mathrm{B}}(t)$ and $L_{\mathrm{B}}(t|p_{\mathrm{B}i})$. Because the global cross NND method is, as described in Section 5.2.2, formulated through aggregating the Voronoi subnetworks $N_i = V_{\mathrm{VD}}(p_{\mathrm{B}i})$, $i = 1, 2, \ldots, n_B$, a computational method for the global NND method uses the extended shortest-path trees $T_{\mathrm{B}i}$ rooted at $p_{\mathrm{B}i}$ in $V_{\mathrm{VD}}(p_{\mathrm{B}i})$, $i = 1, 2, \ldots, n_{\mathrm{B}}$ (for instance, $T_{\mathrm{B}1}$ and $T_{\mathrm{B}2}$ of Figure 5.2 shown in Figure 5.6). In addition, we use a 'trick,' which makes the computation of the global cross NND method almost identical to that of the local cross NND method.

The 'trick' is made by the introduction of a dummy node, v_0^* (e.g., the broken line circle in Figure 5.6), and n_{B} dummy links connecting v_0^* and the nodes in P_B (the broken line segments). We regard the lengths of all these dummy links as zero. Then, we obtain the tree rooted at v_0^* such that the roots of the extended shortest-path trees $T_{\mathrm{B}i}$ are the other end node of the link connected to v_0^*. Now denote this tree by T^* and the set of all the nodes of T^* by V^*, and let $n^* + 1$ be the number of nodes in V^*; i.e., the dummy node v_0^* and n^* real nodes. Then, we can compute $m_{\mathrm{B}}(t)$ and $L_{\mathrm{B}}(t|p_{\mathrm{B}i})$ in almost the same manner as the local cross NND method. Put simply, we sort those vertices in nondecreasing order of the

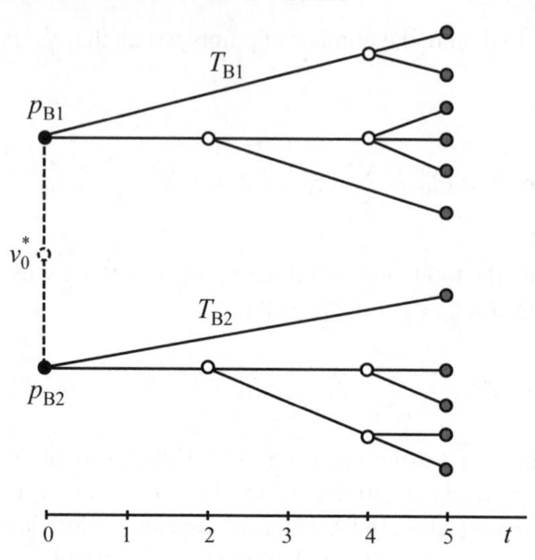

Figure 5.6 The extended shortest-path trees rooted at p_{B1} in $V_{VD}(p_{B1})$ and p_{B2} in $V_{VD}(p_{B2})$ of the network shown in Figure 5.2 (the broken line circle is dummy node v_0^*, and the broken line segments are dummy links).

distances from v_0^*. Let $(v_B^{(0)}, v_B^{(1)}, \cdots, v_B^{(n^*)})$ be the resulting list of vertices, $(t_B^{(0)}, t_B^{(1)}, \cdots, t_B^{(n^*)})$ be the associated list of distances and $(\delta_B^{(0)}, \delta_B^{(1)}, \cdots, \delta_B^{(n^*)})$ be the list of degrees of the vertices, where $v_B^{(0)} = v_0^*$, $t_B^{(0)} = 0$ and $\delta_B^{(0)} = n_B$. The subsequent procedure is then the same as that of the local cross NND method by replacing $(v_{Bi}^{(0)}, v_{Bi}^{(1)}, \cdots, v_{Bi}^{(n)})$, $(t_{Bi}^{(0)}, t_{Bi}^{(1)}, \cdots, t_{Bi}^{(n)})$, $(\delta_{Bi}^{(0)}, \delta_{Bi}^{(1)}, \cdots, \delta_{Bi}^{(n^*)})$ with $(v_B^{(0)}, v_B^{(1)}, \cdots, v_B^{(n^*)})$, $(t_B^{(0)}, t_B^{(1)}, \cdots, t_B^{(n^*)})$, $(\delta_B^{(0)}, \delta_B^{(1)}, \cdots, \delta_B^{(n^*)})$, respectively. Using this computational method, we can obtain the probability distribution function of the distance from random points of type A to their nearest points of type B, (i.e., $F_B(t)$ of equation (5.11)), the expected value μ_B and variance σ_B^2 and index I_B of equation (5.14). Therefore, we can test the CSR hypothesis with an observed value of I_B, μ_B and σ_B^2 as described in Section 5.2.2. We can also test the CSR hypothesis by the F^* function using $F_B(t)$ of equation (5.11). Because of the above trick, the time complexity for computing $m_B(t)$ and $F_B(t)$ is $O(n_{V_i} \log n_{V_i})$.

The SANET software package (Okabe, Okunuki, and Shiode, 2006a, 2006b; Okabe, Satoh, and Shiode, 2009) implements the computational methods in this section. We provide a practical application of these methods with the SANET software in Chapter 12.

6

Network *K* function methods

The two main distance-based methods for point pattern analysis in spatial analysis are the (network) *nearest-neighbor distance method*, as presented in the preceding chapter, and the (*network*) K *function method*, as illustrated in this chapter. We potentially use both methods for responding to similar questions, such as Q2 and Q3 in Chapter 1, that is, Q2: Do boutiques tend to stand side-by-side alongside streets in a downtown area, and Q3: Do street burglaries tend to take place near railway stations? However, their ways of answering these questions are different. More particularly, the network nearest-neighbor distance method specifies these questions as in Q2′ and Q3′ in Chapter 5, whereas the network *K* function method specifies these questions as follows:

Q2″: Given a set of points on a network, how can we test whether the number of points within a distance from each point is significantly many (or few)?

Q3″: Given two sets of points, type A and type B, on a network, how can we test whether the number of type A points within a distance from each point of type B is significantly many (or few)?

The network *K* function method statistically responds to these general questions.

The network *K* function method is an extension of the *K* function method defined on a plane, referred to as the *planar* K *function method*, to a network. Bartle (1964) provided the initial idea, while Ripley (1976) noted its potential for widespread use and developed it in depth (Ripley, 1977, 1979, 1981). Note that while the term '*K* function' is also in use in mathematics, it is a very different function (a generalization of the hyperfactorial to complex numbers; see Whittaker and Watson

Spatial Analysis along Networks: Statistical and Computational Methods, First Edition.
Atsuyuki Okabe and Kokichi Sugihara.

(1990)). Accordingly, in order to distinguish between these alternative K functions, the K function in spatial statistics is also sometimes referred to as *Ripley's* K *function* (Dixon, 2002).

In all likelihood, the planar K function method is perhaps one of the most frequently used methods in spatial analysis. In fact, many textbooks describe it in detail, including, for example, Ripley (1981, Section 8.3), Diggle (1983, Section 5.2), Upton and Fingleton (1985, Section 1.5) under combined count and distance analysis, Cressie (1993, Section 8.4.3), Stoyan and Stoyan (1994, Section 14.4), Bailey and Gatrell (1995, Section 3.4.4), Stoyan, Kendall, and Mecke (1995, Section 4.5), and Illian *et al.* (2008, Section 4.3.3). In terms of the network K function method, Okabe and Yamada (2001) provide an extension of the planar K function method, and this has been applied to the analysis of traffic accidents (Yamada and Thill, 2004) and landscapes (Spooner *et al.*, 2004), among other topics. Okabe, Okunuki, and Shiode (2006a, b) and Okabe and Satoh (2009) review the geographic information system (GIS)-based tools for performing the network K function method, and we illustrate these applications in Chapter 12 of this volume.

The network K function method includes the network *auto* K function method (which Dixon, 2002, calls the *self* K function) and the network *cross* K function method. The main distinction is that the *auto* K function method (customarily referred to as the K function method) deals with a set of points of a single kind (e.g., fast-food shops) and considers the distances between these points. In contrast, the *cross* K function method deals with two sets of points of different kinds (e.g., fast-food shops and railway stations) and instead considers the distances between the two different kinds of points. Note that in the following, we refer to 'shortest-path distance' as 'distance' and omit 'network' from 'network K function method' (except for headings and definitions) unless specified otherwise.

6.1 Network auto *K* function methods

We consider a network $N = N(V, L)$ composed of a set of nodes $V = \{v_1, \ldots, v_{n_V}\}$ and a set of links $L = \{l_1, \ldots, l_{n_L}\}$ and let $\tilde{L} = \bigcup_{i=1}^{n_L} l_i$ be a set of points forming the links of N (including the nodes). In this section, we deal with a set of points $P = \{p_1, \ldots, p_n\}$ of a single kind, and accept Assumption 1 in Chapter 5, that is, points $p_1, \ldots, p_n$ are generated according to a stochastic point process on $\tilde{L}$ with a fixed number n of points (the validity of this assumption is discussed in Section 5.1). In particular, we are concerned with the homogeneous binomial point process or the complete spatial randomness (CSR) hypothesis (Section 2.4.2 in Chapter 2). To test this hypothesis, we develop two statistics: the first is formulated with the number of points within a distance t from a specific point in P to the other points in P, and the second is formulated with the average number of points within distance t from every point in P to the other points in P. Because the former views the configuration of points from a local point while the latter views it as a whole, we refer to them as the *network local* auto K *function method* and the *network global* auto K *function method*, respectively. We discuss these methods in the following two subsections.

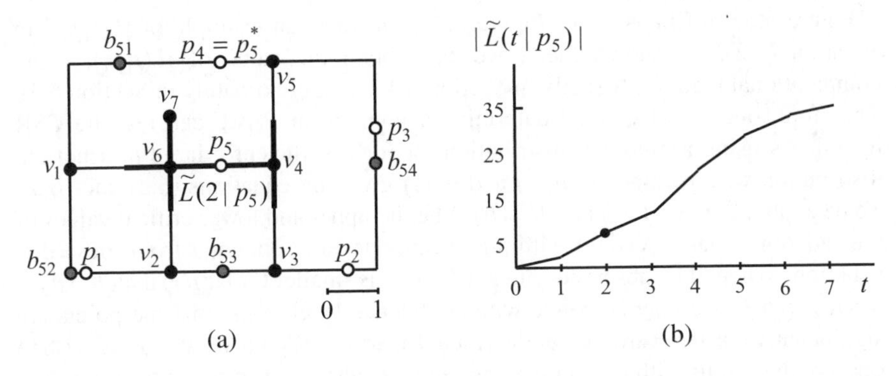

Figure 6.1 The derivation of $|\tilde{L}(t|p_5)|$: (a) the buffer network $\tilde{L}(2|p_5)$ centered at p_5 with width 2 (the bold line segments), (b) the function $|\tilde{L}(t|p_5)|$ with respect to t.

Note that Getis and Franklin (1987) first proposed, but did not uniquely name, the concept of the *local* K function on a plane.

6.1.1 Network local auto *K* function method

For a specific point p_i in P, we consider a subnetwork of $\tilde{L}$, denoted by $\tilde{L}(t|p_i)$, in which the distance from every point in $\tilde{L}(t|p_i)$ to p_i is less than or equal to t, i.e., $\tilde{L}(t|p_i) = \{p|d_S(p_i,p) \leq t, p \in \tilde{L}\}$. As noted in Section 5.1.1 of Chapter 5, the subnetwork $\tilde{L}(t|p_i)$ conceptually corresponds to the disk centered at p_i with radius t on a plane; on a network, we sometimes refer to it as the *buffer network* centered at p_i with width t. The bold line segments in Figure 5.1 in the preceding chapter, which is reproduced in Figure 6.1a, indicate an example: $\tilde{L}(2|p_5)$.

Let $n(t|p_i)$ be the number of points of P included in $\tilde{L}(t|p_i)$ except for p_i. With $n(t|p_i)$, we define a function, $K(t|p_i)$, by:

$$K(t|p_i) = \frac{1}{\rho} n(t|p_i), \tag{6.1}$$

where $\rho = (n-1)/|\tilde{L}|$ is the density of the points (note that the numerator is $n-1$ because p_i is excluded from P; $|\tilde{L}|$ is the length of $\tilde{L}$). To test the CSR hypothesis, we generate $n-1$ points uniformly and independently over the whole network $\tilde{L}$. Then the variable $n(t|p_i)$ is a random variable that counts the points included in $\tilde{L}(t|p_i)$, and hence the probability distribution function of $n(t|p_i)$ is a binomial distribution with parameters $n-1$ and $|\tilde{L}(t|p_i)|/|\tilde{L}|$. Consequently, under the CSR hypothesis, variable $K(t|p_i)$ is a random variable following a binomial distribution with the expected value and variance given by:

$$\mathrm{E}(K(t|p_i)) = \mu_i(t) = |\tilde{L}(t|p_i)|, \tag{6.2}$$

$$\mathrm{Var}(K(t|p_i)) = \sigma_i^2(t) = \frac{1}{n-1}\left(|\tilde{L}|\right)\left(|\tilde{L}(t|p_i)|\right)\left(1 - \frac{|\tilde{L}(t|p_i)|}{|\tilde{L}|}\right). \tag{6.3}$$

Figure 5.5c in Chapter 5 or Figure 6.1b illustrates an example of $|\tilde{L}(t|p_i)|$ in equation (6.2). As shown there, we can obtain the value of $|\tilde{L}(t|p_i)|$ using computational methods (exactly in Section 5.4 and approximately in Section 6.4). Therefore, once we fix t and determine a point p_i in P, we can test the CSR hypothesis using a binomial distribution, or for a sufficiently large n, a normal distribution with parameters $\mu_i(t)$ and $\sigma_i^2(t)$ given by equations (6.2) and (6.3). To be explicit, let $K^*(t|p_i)$ and $K^{**}(t|p_i)$ be the upper and lower critical values of the random variable $K(t|p_i)$ with significance level α obtained from these distributions. Then, if an observed value of $K(t|p_i)$ is smaller (or larger) than $K^*(t|p_i)$ (or $K^{**}(t|p_i)$), we may conclude with confidence level $1-\alpha$ that the points are significantly clustered around (or distributed apart from) point p_i. When we wish to observe the results with respect to consecutive t values, we compute the upper and lower envelope curves $K^*(t|p_i)$ and $K^{**}(t|p_i)$ with respect to t and examine whether an observed curve of $K(t|p_i)$ is outside or inside these envelope curves. Chapter 12 provides actual examples of these envelope curves.

It should be noted that the above local auto K function method and the K function methods to be formulated in the following sections exactly take the boundedness of a network into account. Therefore, as discussed in Section 2.4.3 of Chapter 2, if we apply these methods to a naturally bounded network, we are not bothered with edge effects.

6.1.2 Network global auto *K* function method

The statistic given by equation (6.1) is a local statistic in the sense that it is formulated with the variable $n(t|p_i)$ for a specific point p_i. In turn, we formulate a global statistic $K(t)$ by averaging $n(t|p_i)$ over $i = 1, \ldots, n$. Mathematically, we define $K(t)$ as:

$$K(t) = \frac{1}{\rho} \frac{\sum_{i=1}^{n} n(t|p_i)}{n}. \tag{6.4}$$

To test the CSR hypothesis, we are supposed to compute $\mathrm{E}(K(t))$ and $\mathrm{Var}(K(t))$. However, their analytical derivations are difficult. In practice, we employ Monte Carlo simulation, as implemented in the SANET software package (Okabe, Okunuki, and Shiode, 2006a,b; Okabe and Satoh, 2009). In Chapter 12, we provide an application to the distribution of preparatory schools across Shibuya ward in Tokyo (Figures 12.11 and 12.12).

6.2 Network cross *K* function methods

The auto K function method discussed in the preceding section deals with only one type of point, whereas the cross K function method in this section considers two types of point: type A points $P_{\mathrm{A}} = \{p_{\mathrm{A}1}, \ldots, p_{\mathrm{A}n_{\mathrm{A}}}\}$ and type B points $P_{\mathrm{B}} = \{p_{\mathrm{B}1}, \ldots, p_{\mathrm{B}n_{\mathrm{B}}}\}$

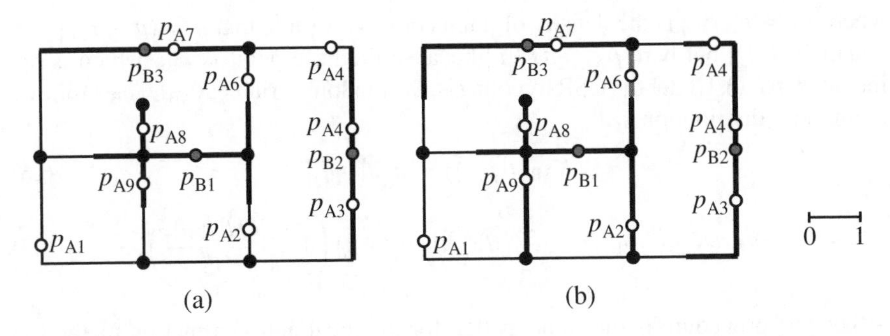

Figure 6.2 The buffer networks $\tilde{L}(t|p_{B1})$, $\tilde{L}(t|p_{B2})$, and $\tilde{L}(t|p_{B3})$ centered at p_{B1}, p_{B2}, and p_{B3}, respectively with width (a) $t=2$ and (b) $t=3$, indicated by the bold line segments (the gray bold line segment is shared with $\tilde{L}(3|p_{B1})$ and $\tilde{L}(3|p_{B2})$).

(e.g., the white and gray circles, respectively in Figure 6.2). To formulate this method, we make Assumptions 2 and 3 as in Section 5.2 of the preceding chapter, that is, points $p_{A1}, \ldots, p_{An_A}$ are generated according to a stochastic point process on $\tilde{L}$ with a fixed number n_A of type A points, while points $p_{B1}, \ldots, p_{Bn_B}$ are deterministically placed on $\tilde{L}$ with a fixed number n_B of type B points. We discuss the implications of these assumptions in Section 5.2 of Chapter 5. In brief, type A points represent location events of relatively temporary facilities, such as fast-food shops, whereas type B points represent those of longer-lived infrastructural facilities, such as railway stations.

To formulate a statistical test for the two types of point, we specify the general stochastic point process in Assumption 2 as the homogeneous binomial point process defined in Section 2.4.2 of Chapter 2. This process implies the CSR hypothesis in that type A points are distributed independently of the configuration of type B points. We test this hypothesis in two ways. The first method examines the number of type A points within a distance t from a specific point p_{Bi} in P_B (e.g., each set of the connected bold line segments in Figure 6.2). The second method examines the average number of type A points within a distance t from every point in P_B. Corresponding to the local and global auto K function methods, we refer to these methods as the *network local cross* K *function method* and the *network global cross* K *function method*, respectively. We discuss these methods in the subsequent two subsections.

6.2.1 Network local cross *K* function method

Mathematically, the local cross K function method is the same as the local auto K function method except for the notation. Let $n(t|p_{Bi})$ be the number of points of P_A included in $\tilde{L}(t|p_{Bi})$ and:

$$K_{AB}(t|p_{Bi}) = \frac{1}{\rho_A} n(t|p_{Bi}), \tag{6.5}$$

where $\rho_A = n_A/|\tilde{L}|$ is the density of points of type A (note that $\rho = (n-1)/|\tilde{L}|$ in equation (6.1), but here $\rho_A = n_A/|\tilde{L}|$ because the base point is p_{Bi}, which is not included in P_A). Under the CSR hypothesis, the random variable $K_{AB}(t|p_{Bi})$ follows a binomial distribution with:

$$\mathrm{E}(K_{AB}(t|p_{Bi})) = |\tilde{L}(t|p_{Bi})|, \tag{6.6}$$

$$\mathrm{Var}(K_{AB}(t|p_{Bi})) = \frac{1}{n_A}\left(|\tilde{L}|\right)\left(|\tilde{L}(t|p_{Bi})|\right)\left(1 - \frac{|\tilde{L}(t|p_{Bi})|}{|\tilde{L}|}\right). \tag{6.7}$$

The test procedure is the same as that for the local auto K function method in Section 6.1.1 and so we do not repeat it here. In Chapter 12, we detail an application of the local cross K function method to the distribution of churches in relation to railway stations in Shibuya ward in Tokyo (Figures 12.16 and 12.17).

6.2.2 Network global cross *K* function method

The global cross K function is derived from averaging the local cross K functions $K_{AB}(t|p_{Bi})$ with respect to $p_{B1}, \ldots, p_{Bn_B}$. Mathematically, the function is:

$$K_{AB}(t) = \frac{1}{\rho_A}\frac{\sum_{i=1}^{n_B} n(t|p_{Bi})}{n_B}. \tag{6.8}$$

The derivation of the expected value and variance of $K_{AB}(t)$ varies according to the range of t. Let t_{max} be the maximum value of t such that the subnetworks $\tilde{L}(t|p_{Bi}), i = 1, \ldots, n_B$ do not overlap each other except at boundary points (for example, in Figure 6.2, $t_{max} = 2.5$). When $t \leq t_{max}$ holds (as in Figure 6.2a, where $t = 2$), the subnetworks $\tilde{L}(t|p_{Bi}), i = 1, \ldots, n_B$ do not intersect. Therefore, the summation of the numbers of type A points in the subnetworks is equal to the number of type A points in the union of the subnetworks, that is, $\bigcup_{i=1}^{n_B}\tilde{L}(t|p_{Bi})$. For example, in Figure 6.2a, $n(t|p_{B1}) + n(t|p_{B2}) + n(t|p_{B3}) = 3 + 2 + 1 = 6$, and the number of type A points in $\tilde{L}(t|p_{B1}) \cup \tilde{L}(t|p_{B2}) \cup \tilde{L}(t|p_{B3})$ is also 6. Therefore, under the CSR hypothesis, the probability that the random variable $\sum_{i=1}^{n_B} n(t|p_{Bi})$ takes a specific value, say k, is equal to the probability of k points of type A being located in $\bigcup_{i=1}^{n_B}\tilde{L}(t|p_{Bi})$. This means that $\sum_{i=1}^{n_B} n(t|p_{Bi})$ follows a binomial distribution with parameters n_A and $\left|\bigcup_{i=1}^{n_B} L(t|p_{Bi})\right|/|\tilde{L}| = \sum_{i=1}^{n_B}|L(t|p_{Bi})|/|\tilde{L}|$. Therefore, we obtain the distribution of $K_{AB}(t)$ as a binomial distribution with the expected value and variance given by:

$$\mathrm{E}(K_{AB}(t)) = \sum_{i=1}^{n_B} |\tilde{L}(t|p_{B_i})|, \tag{6.9}$$

$$\mathrm{Var}(K_{AB}(t)) = \frac{1}{n_A}|\tilde{L}|\sum_{i=1}^{n_B}|\tilde{L}(t|p_{B_i})|\left(1 - \frac{\sum_{i=1}^{n_B}|\tilde{L}(t|p_{B_i})|}{|\tilde{L}|}\right). \tag{6.10}$$

When $t > t_{\max}$ holds (as in Figure 6.2b, where $t = 3$), the derivation of $\mathrm{E}(K_{\mathrm{AB}}(t))$ and $\mathrm{Var}(K_{\mathrm{AB}}(t))$ becomes complicated, because $\sum_{i=1}^{n_{\mathrm{B}}} n(t|\,p_{\mathrm{B}i}) = k$ does not always mean that k points of type A are located in $\bigcup_{i=1}^{n_{\mathrm{B}}} \tilde{L}(t|\,p_{\mathrm{B}i})$, i.e., some points may be counted more than once in $\sum_{i=1}^{n_{\mathrm{B}}} n(t|\,p_{\mathrm{B}i})$. For instance, point $p_{\mathrm{A}6}$ in Figure 6.2b is counted not only in $n(t|\,p_{\mathrm{B}1})$ but also in $n(t|\,p_{\mathrm{B}3})$. As a result, $n(t|\,p_{\mathrm{B}1}) + n(t|\,p_{\mathrm{B}2}) + n(t|\,p_{\mathrm{B}3}) = 5 + 3 + 2 = 10$, but the number of type A points in $\tilde{L}(t|\,p_{\mathrm{B}1}) \cup \tilde{L}(t|\,p_{\mathrm{B}2}) \cup \tilde{L}(t|\,p_{\mathrm{B}3})$ is 9. To take multiple counts into account, let $\tilde{L}_{(k)}(t|P_{\mathrm{B}})$ be the line segments on $\tilde{L}$ where exactly k subnetworks of n_{B} subnetworks $\tilde{L}(t|\,p_{\mathrm{B}j})$, $j = 1, \ldots, n_{\mathrm{B}}$ overlap. For instance, in Figure 6.2b, $\tilde{L}_{(1)}(t|\,p_{\mathrm{B}_1})$ and $\tilde{L}_{(2)}(t|\,p_{\mathrm{B}_2})$ overlap on the gray line segment (note that $\tilde{L}_{(0)}(t|P_{\mathrm{B}})$ denotes the line segments not covered by any subnetwork $\tilde{L}(t|\,p_{\mathrm{B}j})$, for example, the thin line segments in Figure 6.2b). Let $n_{(k)}(t|P_{\mathrm{B}})$ be the number of type A points included in $\tilde{L}_{(k)}(t|P_{\mathrm{B}})$. Then $\sum_{i=1}^{n_{\mathrm{B}}} n(t|\,p_{\mathrm{B}i}) = \sum_{k=0}^{n_{\mathrm{B}}} k\, n_{(k)}(t|P_{\mathrm{B}})$ holds. Under the CSR hypothesis, the probability that the random variable $\sum_{k=0}^{n_{\mathrm{B}}} k\, n_{(k)}(t|P_{\mathrm{B}})$ equals m is given by:

$$\Pr\left[\sum_{k=0}^{n_{\mathrm{B}}} k\, n_{(k)}(t|P_{\mathrm{B}}) = m\right] = \sum \frac{n_{\mathrm{B}}!}{r_0!r_1! \cdots r_{n_{\mathrm{B}}}!} \left(\frac{|\tilde{L}_{(0)}(t|P_{\mathrm{B}})|}{|\tilde{L}|}\right)^{r_0} \times \left(\frac{|\tilde{L}_{(1)}(t|P_{\mathrm{B}})|}{|\tilde{L}|}\right)^{r_1} \cdots \left(\frac{|\tilde{L}_{(n_{\mathrm{B}})}(t|P_{\mathrm{B}})|}{|\tilde{L}|}\right)^{r_{n_{\mathrm{B}}}}, \quad (6.11)$$

where the summation is over all possible nonnegative integers $r_0, r_1, \ldots, r_{n_{\mathrm{B}}}$ such that $\sum_{i=0}^{n_{\mathrm{B}}} r_i = n_{\mathrm{B}}$ and $\sum_{i=0}^{n_{\mathrm{B}}} i\, r_i = m$. This is the *univariate multinomial distribution* (Johnson, Kotz, and Kemp, 1992, Section 11.18). The expected value and variance of this distribution are respectively given by:

$$\mathrm{E}(K_{\mathrm{AB}}(t)) = \sum_{k=1}^{n_{\mathrm{B}}} k \frac{|\tilde{L}_{(k)}(t|P_{\mathrm{B}})|}{|\tilde{L}|}, \quad (6.12)$$

$$\mathrm{Var}(K_{\mathrm{AB}}(t)) = \frac{1}{n_{\mathrm{A}}} |\tilde{L}| \sum_{k=1}^{n_{\mathrm{B}}} k^2 |\tilde{L}_{(k)}(t|P_{\mathrm{B}})| - \frac{1}{n_{\mathrm{A}}} \left(\sum_{k=1}^{n_{\mathrm{B}}} k |\tilde{L}_{(k)}(t|P_{\mathrm{B}})|\right)^2. \quad (6.13)$$

We can test the CSR hypothesis using a univariate multinomial distribution with parameters n_{A} and $|\tilde{L}_{(k)}(t|P_{\mathrm{B}})|/|\tilde{L}|, k = 1, \ldots, n_{\mathrm{B}}$. When n_{A} is sufficiently large, the univariate multinomial distribution may be approximated by a normal distribution with expected value and variance given by equations (6.12) and (6.13). Thus, for a specific t, we can test the CSR hypothesis using this normal distribution. The procedure for this test is the same as that for the local auto K function method in Section 6.1.1, and so we do not repeat it here.

If we wish to examine the CSR hypothesis across consecutive t values from zero to the possible maximum t^* of t, as shown in Section 6.1.1, we compute the upper and lower envelope curves $K^*_{\mathrm{AB}}(t)$ and $K^{**}_{\mathrm{AB}}(t)$ of the random variable $K_{\mathrm{AB}}(t)$ with significance level α with respect to t, and compare an observed curve of $K_{\mathrm{AB}}(t)$ with the envelope curves $K^*_{\mathrm{AB}}(t)$ and $K^{**}_{\mathrm{AB}}(t)$ across $0 < t < t^*$. In Chapter 12, we detail an actual example of these envelope curves as applied to the distribution of aromatherapy houses in relation to railway stations in Shibuya ward in Tokyo (Figures 12.13 and 12.14).

6.2.3 Network global Voronoi cross *K* function method

The global cross K function is formulated with the number of type A points included in subnetworks $\tilde{L}(t|p_{\mathrm{B}i}), i = 1, \ldots, n_{\mathrm{B}}$. As shown in the derivation of equations (6.12) and (6.13) above, the method for counting these points is complicated, because the points of type A included in only one subnetwork $\tilde{L}(t|p_{\mathrm{B}_i})$ are counted once, while those included in two subnetworks are counted twice, and those included in three subnetworks are counted thrice, and so on. This implies that a point of type A that is within distance t from exactly k points of type B is weighted k times. An alternative way of counting the number of type A points included in the subnetworks $\tilde{L}(t|p_{\mathrm{B}i}), i = 1, \ldots, n_{\mathrm{B}}$ is to count them without weighting. That is to say, we count the number $n(t|P_{\mathrm{B}})$ of type A points p_{A_j} included in at least one of the subnetworks $\tilde{L}(t|p_{\mathrm{B}i}), i = 1, \ldots, n_{\mathrm{B}}$, or mathematically $p_{\mathrm{A}_j} \in \bigcup_{i=1}^{n_{\mathrm{B}}} \tilde{L}(t|p_{\mathrm{B}i})$. If we adopt this way of counting, the test for the CRS hypothesis becomes simple. A global statistic alternative to equation (6.8) is then:

$$K'_{\mathrm{AB}}(t) = \frac{1}{\rho_{\mathrm{A}}} n(t|P_{\mathrm{B}}). \tag{6.14}$$

Note that $K'_{\mathrm{AB}}(t) = K_{\mathrm{AB}}(t)$ for $t \leq t_{\max}$.

Alternatively, we can write this function in terms of the network Voronoi diagram presented in Chapter 4. Let $\mathcal{V} = \{V_{\mathrm{VD}}(p_{\mathrm{B}1}), \ldots, V_{\mathrm{VD}}(p_{\mathrm{B}n_{\mathrm{B}}})\}$ be the network Voronoi diagram consisting of Voronoi subnetworks (cells) generated by the points of P_{B}, and let $\tilde{L}_{\mathrm{VD}}(t|p_{\mathrm{B}i}) = \tilde{L}(t|p_{\mathrm{B}i}) \cap V_{\mathrm{VD}}(p_{\mathrm{B}i})$ be the subnetwork of $\tilde{L}(t|p_{\mathrm{B}i})$ included in $V_{\mathrm{VD}}(p_{\mathrm{B}i})$. Then, $\bigcup_{i=1}^{n_{\mathrm{B}}} \tilde{L}(t|p_{\mathrm{B}i}) = \bigcup_{i=1}^{n_{\mathrm{B}}} \tilde{L}_{\mathrm{VD}}(t|p_{\mathrm{B}i})$ and $\tilde{L}_{\mathrm{VD}}(t|p_{\mathrm{B}i}) \cap \tilde{L}_{\mathrm{VD}}(t|p_{\mathrm{B}j})$ includes at most a finite number of points for $i \neq j$, because Voronoi cells $V_{\mathrm{VD}}(p_{\mathrm{B}i}), i = 1, \ldots, n_{\mathrm{B}}$ are mutually exclusive (except for the boundary points) and collectively exhaustive. Therefore, $|\bigcup_{i=1}^{n_{\mathrm{B}}} \tilde{L}(t|p_{\mathrm{B}i})|$ is given by $\sum_{i=1}^{n_{\mathrm{B}}} |\tilde{L}_{\mathrm{VD}}(t|p_{\mathrm{B}i})|$, and the expected value and variance of $K'_{\mathrm{AB}}(t)$ are written as:

$$\mathrm{E}(K'_{\mathrm{AB}}(t)) = \sum_{i=1}^{n_{\mathrm{B}}} |\tilde{L}_{\mathrm{VD}}(t|p_{\mathrm{B}i})|, \tag{6.15}$$

$$\mathrm{Var}(K'_{\mathrm{AB}}(t)) = \frac{1}{n_{\mathrm{A}}} (|\tilde{L}|) \left(\sum_{i=1}^{n_{\mathrm{B}}} |\tilde{L}_{\mathrm{VD}}(t|p_{\mathrm{B}i})| \right) \left(1 - \frac{\sum_{i=1}^{n_{\mathrm{B}}} |\tilde{L}_{\mathrm{VD}}(t|p_{\mathrm{B}i})|}{|\tilde{L}|} \right). \tag{6.16}$$

We call function $K'_{\mathrm{AB}}(t)$ the *network global Voronoi cross* K *function* after the *planar global Voronoi cross* K *function method* proposed by Okabe, Boots, and Satoh (2010).

This alternative *K* function not only is simple in computation but also reveals another facet of a point pattern. That is, the global cross *K* function is concerned with the distances from every point of type A to every point of type B. In contrast, the global Voronoi cross *K* function is concerned with the distance from every point of type A to the whole collection of type B points, or equivalently, the distances from each point of type A to its nearest neighbor in the collection of type B points. This implies that the global Voronoi cross *K* function examines a point distribution in a shorter range of t than the global cross *K* function. In fact, the maximum range of t for the former is mathematically given by $\max\{\max_i\{d_{\mathrm{S}}(p_{\mathrm{B}i}, p), p \in V_{\mathrm{VD}}(p_{\mathrm{B}i})\}\}$, whereas that for the latter is given by $\max\{\max_i\{d_{\mathrm{S}}(p_{\mathrm{B}i}, p), p \in \tilde{L}\}\}$. Thus, because $V_{\mathrm{VD}}(p_{\mathrm{B}i}) \subset L$, the range of t is considerably smaller for the Voronoi cross *K* function. This observation reminds us of its relation to the global nearest-neighbor distance method in Chapter 5. In fact, comparison of equations (5.15) and (5.16) and equations (6.15) and (6.16) shows that $K'_{\mathrm{AB}}(t)$ is almost the same as the F^* function defined in Section 5.2.2, with the exception of the constant terms. In Figure 12.15 in Chapter 12, we provide an application of the global Voronoi *K* function method to the distribution of aromatherapy houses on the street network in relation to railway stations across Shibuya ward in Tokyo. This contrasts sharply with the global *K* function method as applied to the same data set.

6.3 Network *K* function methods in relation to geometric characteristics of a network

The basis of all kinds of the *K* function methods presented in this chapter is the shortest-path distances between points on a network. The same holds for the nearest-neighbor distance methods in the preceding chapter. Moreover, the formulation of almost all the methods in this volume is in terms of shortest-path distances on networks. Therefore, to apply network spatial methods appropriately, it is important to understand the nature of the shortest-path distance in relation to the geometric characteristics of a network. In the literature, there are few studies on this subject, with the exception of Morita and Okunuki (2006) and Morita (2008). In this section, we first discuss the relationship between the shortest-path distance and the Euclidean distance, and then outline Morita's (2008) study, which examined the effect of the level of detail of a network on the results obtained by the global *K* function method.

6.3.1 Relationship between the shortest-path distance and the Euclidean distance

A simple geometric property of the shortest-path distance is that the corresponding Euclidean distance bounds the shortest-path distance between two points on a

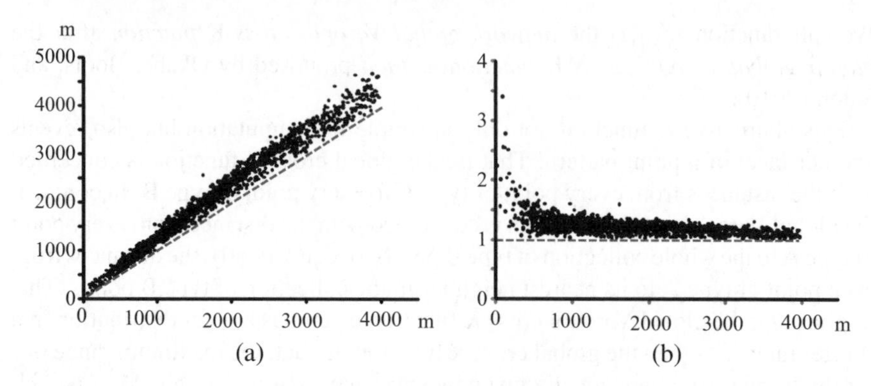

Figure 6.3 The relationships between shortest-path distances (the vertical axis) and their corresponding Euclidean distances (the horizontal axis) in Kokuryo, a suburb of Tokyo: (a) the scatter diagram of the shortest-path distances and their corresponding Euclidean distances, (b) the ratio of the shortest-path distance to its corresponding Euclidean distance (the gray broken lines indicate the lower bounds).

network. Therefore, there is potentially a relationship between these two distances. Figure 1.5 in Chapter 1 provides an empirical example in which we observe a weak relationship between the two distances if the Euclidean distance is less than 400 meters in Kokuryo, a suburb of Tokyo. Figure 6.3a depicts the scatter diagram of shortest-path distances and the corresponding Euclidean distances in the same region but for 0–4000 meters. In this range, we find a strong correlation between them, or numerically, 0.992. We also note a fan-shaped cloud of dots in panel (a), which implies that the absolute difference between the two distances becomes larger as the Euclidean distance increases.

Alternatively, Figure 6.3b views the same data from a different perspective, that is, the relative ratio of the Euclidean distance to its corresponding shortest-path distance (part of which is shown in Figure 1.5). This figure demonstrates that the ratio decreases as the Euclidean distance increases (for instance, the ratio is about 1.2 at 2 km and 1.1 at 3 km), and it eventually appears to approach unity. According to Marcon and Puech (2003), in practice the *K* function serves for the distance for less than about half of the distance between the two farthest points in a study area. In the study area in Kokuryo, the maximum distance was about 4 km. Therefore, the relative difference between the two distances is fairly large (i.e., the ratio is larger than 1.2), implying that the network *K* function method would provide a more appropriate method for analyzing events along networks in that area than the planar *K* function method.

Having observed the trends in Figure 6.3, one might consider that the shortest-path distance approaches the corresponding Euclidean distance as the density of line segments (i.e., the length of the line segments forming a network per unit area) increases. However, in general, this does not always hold. Figure 6.4 provides some counterexamples. In panel (a), in which the network has a radial pattern,

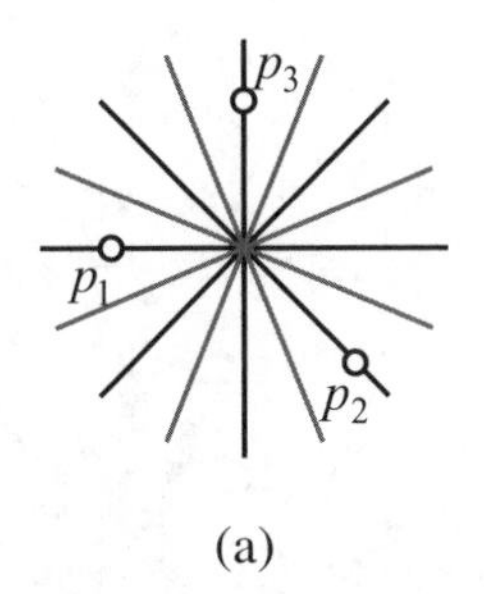

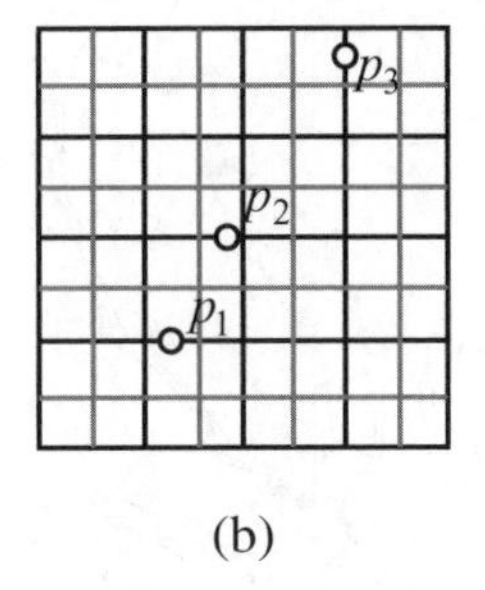

Figure 6.4 The shortest-path distances $d_S(p_1,p_2)$, $d_S(p_1,p_3)$ and $d_S(p_2,p_3)$ in panel (a) (a radial pattern) and $d_S(p_1,p_3)$, $d_S(p_2,p_3)$ in panel (b) (a grid pattern) do not change when the networks are refined.

the shortest-path distances $d_S(p_1,p_2)$, $d_S(p_1,p_3)$, and $d_S(p_2,p_3)$ between the white circles on the black line segments remain the same even if the gray line segments are added. It might then appear that the shortest-path distance approaches the corresponding Euclidean distance as the density of nodes, (i.e., the number of nodes of a network per unit area) increases. Once again, this is not always the case, as in panel (b), in which the network has a grid pattern. The shortest-path distance between p_1 and p_2 on the black line segments changes when the gray line segments are added (the node density increases) but the shortest-path distance between p_1 and p_3 and that between p_2 and p_3 are invariant. These examples imply that network spatial methods would be more appropriate than the planar alternative in the case of radial and grid networks.

6.3.2 Network global auto *K* function in relation to the level-of-detail of a network

Suppose that we wish to examine the distribution of facilities alongside a road network using the global *K* function method, and that we have two possible choices of networks: a network consisting of only arterial roads and a network including not only arterial roads but also branch roads. We are concerned with the extent to which the results differ according to each network's level of detail.

Morita (2008) examined this question by applying the global *K* function method to the distribution of elementary schools in Hachimancho, a rural area in Gifu Prefecture, Japan. Panel (a) in Figure 6.5 illustrates the network of arterial roads and panel (b) the network of all roads in the region including the arterial and branch roads. The black circles indicate elementary schools located alongside the arterial roads.

Morita (2008) applied the network global auto *K* function method to those two sets of network data. Figure 6.6 shows the expected *K* function obtained under the CSR hypothesis and the observed *K* function. The difference between the resulting *K* functions in panels (a) and (b) is distinct. When the method is applied to the network of arterial roads, as shown in panel (a), the observed *K* function $K(t)$ is

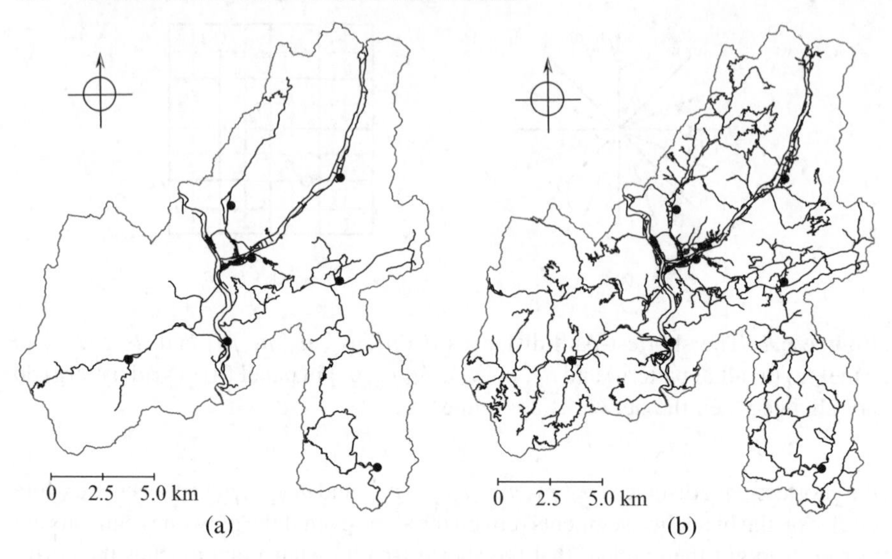

Figure 6.5 The road networks in Hachimancho (the area bounded by the gray lines), a rural area in Gifu Prefecture, Japan: (a) the network of arterial roads, (b) the network of all roads (including arterial and branch roads). The black circles indicate elementary schools (the GIS data are provided by M. Morita).

below the expected K function $\mathrm{E}(K(t))$ across almost all of the domain of t. This trend means that the distribution of elementary schools tends to be rather dispersed over the arterial road network (Figure 6.5a). In contrast, when the method is applied to the network including all roads, the opposite relation is observed in panel (b) in Figure 6.6, that is to say, the distribution of elementary schools tends to be fairly clustered on the network in Figure 6.5b. This finding means that the results of the K function method vary according to the level of detail of a network. In addition, we

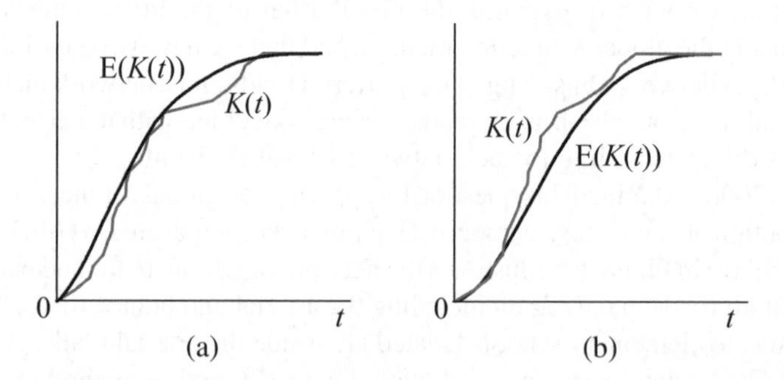

Figure 6.6 The observed network global auto K function $K(t)$ (the gray curves) and the expected network global K function $\mathrm{E}(K(t))$ (the black curves): (a) the results for the road network in Figure 6.5a, (b) the results for the road network in Figure 6.5b (the data are provided by M. Morita).

may obtain another implication from this finding if we suppose that the regions in panels (a) and (b) are different regions with different densities of line segments or nodes. This finding may then indicate that the behavior of the *K* function method varies according to the line or node density of a network. While theoretical examination of this empirical finding appears difficult, an initial investigation was attempted by Morita (2008), who examined the lower bound of $\mathrm{E}(K(t))$ for a radial network (as in Figure 6.4a) and proved that there exists a lower bound that is independent of the density of the line segments forming a radial network.

So far, we have discussed the nature of *K* function methods in relation to the density of line segments, the density of nodes, and the level of detail of a network. However, these are not the only geometric factors of a network influencing the results of *K* function methods. Further theoretical and empirical work is required.

6.4 Computational methods for the network *K* function methods

Having established the auto and cross *K* function methods in Sections 6.1 and 6.2, we now develop their computational methods in this section. Recalling the computational methods developed in Chapter 5, we recognize that those methods are straightforwardly applicable to the local auto *K* function method (Section 6.1.1), the global auto *K* function method (Section 6.1.2), and the local cross *K* function method (Section 6.2.1). However, they are not applicable to the global cross *K* function method (Section 6.2.2). Accordingly, in this section we develop a new method for the global cross *K* function. However, because computation of this method includes combinatorial computation, exact computation is difficult in practice. We thus develop an approximation method. In this regard, we also develop approximation methods for the local auto *K* function, the global auto *K* function, and the local cross *K* function as alternatives to the exact methods in Chapter 5, because the implementation of these approximation methods is simple in practice.

The major tasks commonly required for computing the *K* functions are first to count the number of points in a subnetwork satisfying a certain condition (i.e., $n(t_j|p_i)$, $n(t|p_{\mathrm{B}i})$, and $n_{(k)}(t|P_{\mathrm{B}}))$ and then to compute the length of the subnetwork (i.e., $|\tilde{L}(t_j|p_i)|$, $|\tilde{L}(t|p_{\mathrm{B}i})|$, and $|\tilde{L}_{(k)}(t|P_{\mathrm{B}})|$). We refer to these as Task 1 and Task 2, respectively. Tasks 1 and 2 vary slightly from function to function, as shown in the following subsections.

6.4.1 Computational methods for the network auto *K* function methods

The auto *K* function method includes two types: the *local* auto *K* function method and the *global* auto *K* function method. We present the former in Section 6.4.1.1 and the latter in Section 6.4.1.2.

6.4.1.1 Computational methods for the network local auto *K* function method

We suppose that the points of P have been already inserted in a given network $N(V, L)$ as nodes, and hence $P \subset V$ holds (note that the number of nodes is denoted by n_V). In order to compute the local auto K function $K(t|p_i)$ for a fixed point p_i, we approximate this continuous function by the discrete function $K(t_j|p_i), j = 1, 2, \ldots, m$, where continuous t is discretized as $0 < t_1 < t_2 < \cdots < t_m$ (intervals are determined according to the required precision level), as shown in Figure 6.7.

Task 1 for the local auto K function is to count $n(t_j|p_i)$ as defined in Section 6.1.1. To do this, we first construct the extended shortest-path tree rooted at p_i. An example of the extended shortest-path tree rooted at p_5 on the network in Figure 5.1 in the preceding chapter is in Figure 6.7 in this chapter (the first paragraph of Section 5.4.1 in Chapter 5 describes the means of constructing this tree, and its computational method is in Section 3.4.1.3 in Chapter 3). Second, for a fixed $t = t_j (1 \leq j \leq m)$, we enumerate the points $p \in P \backslash \{p_i\}$ satisfying the condition that distances $d(p, p_i)$ are shorter than or equal to t_j by use of the extended shortest-path tree (Section 3.4.1.3). Finally, we undertake the same task for $j = 1, \ldots, m$. As a result, we obtain $n(t_j|p_i), j = 1, 2, \ldots, m$. In the example in Figure 6.7, $n(t_1|p_5) = 0$, $n(t_2|p_5) = 0$, $n(t_3|p_5) = 3$, and $n(t_4|p_5) = 4$. The resulting function divided by $\rho = (n-1)/|\tilde{L}|$ gives the local auto K function defined by equation (6.1).

Task 2 is to compute the length $|\tilde{L}(t_j|p_i)|$ as defined in Section 6.1.1. To do this, we order the distances from p_i to every node $v \in V$ using the extended shortest-path tree rooted at p_i and denote the resulting distances by $d(v_{(1)}, p_i)$, $d(v_{(2)}, p_i)$, ..., $d(v_{(n_V-1)}, p_i)$. For example, in Figure 6.7, the root is p_5, and $v_{(1)} = v_4$, $v_{(2)} = v_6$, $v_{(3)} = v_7, \ldots$. For a given t_j, we then introduce a new variable, D, with the initial value $D = 0$, which is used for summing the lengths of the links consisting of points that are within the distance t_j from $v_{(k)} \in V$, $k = 1, \ldots, n_V - 1$. Suppose that we have just reached $v_{(k)}$, and let l be the link incident to $v_{(k)}$ on the path from p_i to $v_{(k)}$ of the extended shortest-path tree (e.g., for $v_{(2)} = v_6$, l is the link joining p_5 and v_6 in Figure 6.7). At $v_{(k)}$, we meet two cases: $d(v_{(k)}, p_i) \leq t_j$ and $d(v_{(k)}, p_i) > t_j$. In the former (e.g., for $v_{(2)} = v_6$, $d(v_{(2)}, p_5) = d(v_6, p_5) \leq t_1$ in Figure 6.7), the whole part

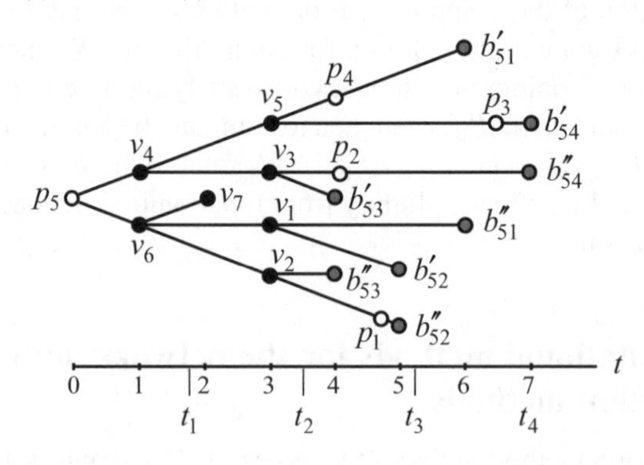

Figure 6.7 The extended shortest-path tree rooted at p_5 for the network shown in Figure 6.1a.

of the link l is within distance t_j from p_i. We then update D by $D \leftarrow D + |l|$. (where $|l|$ denotes the length of l) and move on to $v_{(k+1)}$. In the latter (e.g., for $v_{(3)} = v_6$, $d(v_{(3)}, p_5) = d(v_7, p_5) > t_1$ in Figure 6.7), part of l is within the distance t_j from p_i, and the other end node w of l satisfies $d(w, p_i) > t_j$. Hence, there is the point p on l satisfying $d(p, p_i) = t_j$. We then update D by $D \leftarrow D + (t_j - d(w, p_i))$ and move on to $v_{(k+1)}$. We undertake this procedure for $k = 1, \ldots, n_V - 1$ and $j = 1, \ldots, m$. As a result, we obtain the function $|\tilde{L}(t_j | p_i)|$ with respect to t_j, $k = 1, \ldots, n_V - 1$ for p_i, with which we can compute the expected value and variance of $K(t | p_i)$ given by equations (6.2) and (6.3).

The time complexity of the local auto K function depends on that used for constructing the extended shortest-path tree and for achieving Tasks 1 and 2. The time complexity of the extended shortest-path tree is $O(n_L \log n_L)$ for each root node p_i in P (for details, see Section 3.4.1.3 in Chapter 3). Task 1 (i.e., counting the number of points in $P \backslash \{p_i\}$) requires $O(n)$ for each value of t_j. Task 2 (i.e., computing the total length $|\tilde{L}(t | p_i)|$ of links within the distance t_j from p_i) requires $O(n_L)$ for each value of t_j. Noting $n = O(n_L)$, the total time complexity for computing the local auto K function for p_i and testing the CSR hypothesis is $O(n_L \log n_L + mn_L)$, where m is the number of the discrete values $t_1, t_2, \ldots t_m$ of t.

6.4.1.2 Computational methods for the network global auto *K* function method

We now turn to the global auto K function method. Task 1 for the global auto K function is to count $\sum_{i=1}^{n} n(t_j | p_i)$ as defined in Section 6.1.2. The computational method for achieving this task is the same as that for the local auto K function. However, we do not face Task 2 but instead compute the expected value and variance of $K(t_j)$. As noted in Section 6.1.2, the analytical formulae for these values are unknown. Therefore, we employ Monte Carlo simulation. We can do this by repeating the method of computing $K(t_j)$ for a set P of random points a large number of times, say, 1000. The SANET software implements this simulation method, and Chapter 12 illustrates its application.

6.4.2 Computational methods for the network cross *K* function methods

In the preceding section, we dealt with a single type of point; in this section, we deal with two types of point: $P_A = \{p_{A1}, \ldots, p_{An_A}\}$ and $P_B = \{p_{B1}, \ldots, p_{Bn_B}\}$. Specifically, we show computational methods for the local cross K function in Section 6.4.2.1, global K function in Section 6.4.2.2 and global Voronoi cross K function in Section 6.4.2.3.

6.4.2.1 Computational methods for the network local cross *K* function method

Computational methods for the local cross K function method, specifically those for Task 1 (counting $n(t | p_{Bi})$) and Task 2 (computing $|\tilde{L}(t | p_{Bi})|$), are almost the same as those developed in Section 6.4.1. The only difference is that a specific point is chosen from P_{B} and counting is for the points in the other set P_{A}. Hence, we can straightforwardly apply the computational methods in Section 6.4.1 to the local cross K function. Using these methods, once observed values are given we can compute the local cross K function of equation (6.5). We can also obtain the expected value of equation (6.6) and the variance of equation (6.7), with which we can test the CSR hypothesis.

6.4.2.2 Computational methods for the network global cross *K* function method

Task 1 for the global cross K function method is to count the number $\sum_{i=1}^{n_{\mathrm{B}}} n(t_j|p_{\mathrm{B}\,i}) = \sum_{k=0}^{n_{\mathrm{B}}} k\, n_{(k)}(t_j|P_{\mathrm{B}})$ of points in P_{A} that are within distance t_j from points in P_{B}, in such a way that if a point $p_{\mathrm{A}i} \in P_{\mathrm{A}}$ is within that distance from k points in P_{B}, then $p_{\mathrm{A}i}$ is counted k times. We can easily perform this counting task by repeating the counting method in Section 6.4.2.1 n_{B} times, that is, $n(t_j|\,p_{\mathrm{B}i})$ for each specific point $p_{\mathrm{B}j} \in P_{\mathrm{B}}$, and taking their sum. Once $\sum_{i=1}^{n_{\mathrm{B}}} n(t_j|\,p_{\mathrm{B}i})$ is obtained, we can compute the value of $K_{\mathrm{AB}}(t_j)$ by equation (6.8).

Task 2 is to compute the value of $|\tilde{L}_{(k)}(t_j|P_{\mathrm{B}})|$ for $k = 1, 2, \ldots, n_{\mathrm{B}}$, which is obtained by the sum of the lengths of the links consisting of points that are within the distance t_j from exactly k out of n_{B} points in P_{B}. The time complexity of this computation varies according to $t_j \leq t_{\max}$ or $t_j > t_{\max}$. When $t_j \leq t_{\max}$, the value of k takes only $k = 1$ and so $|\tilde{L}_{(k)}(t_j|P_{\mathrm{B}})| = \sum_{i=1}^{n_{\mathrm{B}}} |\tilde{L}(t_j|\,p_{\mathrm{B}i})|$. Almost the same computational method used in the preceding subsection applies to this computation, and we can obtain the expected value of equation (6.9) and the variance of equation (6.10), with which we can test the CSR hypothesis. When $t_j > t_{\max}$, the exact computation of $|\tilde{L}_{(k)}(t_j|P_{\mathrm{B}})|$ is rather complicated. Alternatively, we formulate an approximation method in the following, which while not always exact, is much simpler.

We assume that all the points in P_{A} and P_{B} are inserted as nodes in the network $N(V, L)$, and hence $P_{\mathrm{A}} \cup P_{\mathrm{B}} \subset V$. In addition, we insert sufficiently many points as new nodes in $N(V, L)$ so that the lengths of all the links are sufficiently small. We first compute the extended shortest-path trees of N rooted at $p_{\mathrm{B}i}$ for all $p_{\mathrm{B}i} \in P_{\mathrm{B}}$. As a result, each node $v \in V$ is associated with n_{B} different values of the shortest-path distances from all the points in P_{B} to v. Second, for each node v, we sort the distance values in increasing order and denote the resulting list of distance values by $(d_v^{(1)}, d_v^{(2)}, \ldots, d_v^{(n_{\mathrm{B}})})$ with $d_v^{(0)} = 0$, implying $0 = d_v^{(0)} \leq d_v^{(1)} \leq d_v^{(2)} \leq \cdots \leq d_v^{(n_{\mathrm{B}})}$. For node $v \in V$, we define $Q(v, t)$ as the number of points in P_{B} that are within distance t from v, or mathematically, $Q(v, t) = \max\{j | d_v^{(j)} \leq t\}$. Because we have already sorted the distances from node v to all the points in P_{B}, we can easily obtain $Q(v, t)$ for any given t.

Now suppose that we are given a specific value of t, say t_j. In order to compute $|\tilde{L}_{(k)}(t_j|P_{\mathrm{B}})|$ for $k = 0, 1, 2, \ldots, n_{\mathrm{B}}$, we introduce n_{B} variables, $D_0, D_1, D_2, \ldots, D_{n_{\mathrm{B}}}$, and initialize them to zero. We then undertake the following procedure for every link l (its end nodes are u and v) in L.

(a)

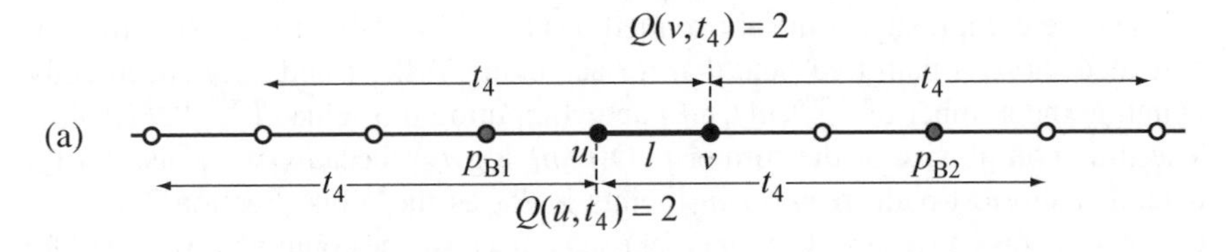

(b)

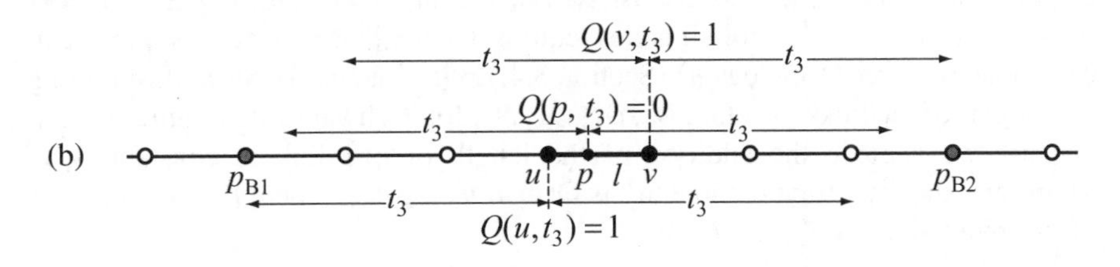

Figure 6.8 An example of computation of $|\tilde{L}_{(k)}(t_4|P_{\mathrm{B}})|$: (a) a well-worked case for $Q(u, t_4) = Q(v, t_4)$, (b) a poorly worked case for $Q(u, t_3) = Q(v, t_3)$.

We check whether $Q(u, t) = Q(v, t)$ or $Q(u, t) \neq Q(v, t)$ one by one for every $l \in L$. In the former, we may expect in most cases that there are exactly $Q(u, t)$ points of P_{B} that are within distance t from every point on the link l. An example is depicted in Figure 6.8a, where there are exactly two points of P_{B} (i.e., p_{B1} and p_{B2}) that are within distance t_4 from every point on l. Hence, whenever $Q(u, t) = Q(v, t)$ holds for $l \in L$, we sum the length of l in $D_{Q(u,t)}$, or in algorithmic terms, we update $D_{Q(u,t)} \leftarrow D_{Q(u,t)} + |l|$, where $|l|$ denotes the length of link l.

However, in some cases, this may be unexpected. A counter example is in Figure 6.8b. In this case, because $Q(u, t_3) = Q(v, t_3) = 1$, the above procedure executes $D_{Q(u,t_3)} \leftarrow D_{Q(u,t_3)} + |l|$, but point p on l shown in Figure 6.8b is not within distance t_3 from either p_{B1} or p_{B2}, and consequently $Q(p, t_3) = 0$. Therefore, we should not do $D_{Q(u,t_3)} \leftarrow D_{Q(u,t_3)} + |l|$. However, we consider this to occur only rarely in practice, and so the above procedure gives a good approximation.

Now consider the second case, i.e., $Q(u, t) \neq Q(v, t)$. Without loss of generality, we assume that $Q(v, t) > Q(u, t)$. We then adopt the following conventional method according to $Q(v, t) - Q(u, t) = 1$ or $Q(u, t) - Q(v, t) = q \geq 2$. In the former case, we divide the length of l into halves and add one half to $D_{Q(u,t)}$ and one half to $D_{Q(v,t)}$. In the latter case, we divide the length $|l|$ into $|l|/(2(q-1)), |l|/(q-1), \ldots, |l|/(q-1), |l|/(2(q-1))$, and add them to $D_{Q(u,t)}, D_{Q(u,t)+1}, \ldots, D_{Q(v,t)}$, respectively. In other words, we split the length $|l|$ of link l into equal portions and add $|l|/(q-1)$ to the values D_j for j satisfying $Q(u, t) < j < Q(v, t)$ and $|l|/(2(q-1))$ to $D_{Q(u,t)}$ and $D_{Q(v,t)}$. This method is also a rough approximation, but it can be computed easily because we do not consider the detailed structure of the actual values of $Q(p, t)$ on point p on l. We proceed to apply this method for every link of l in L and eventually obtain the values of $|\tilde{L}_{(k)}(t|P_{\mathrm{B}})| = D_k$ for $k = 0, 1, \ldots, n_{\mathrm{B}}$. Substituting these values into equations (6.12) and (6.13), we obtain the expected value and variance of $K_{\mathrm{AB}}(t)$, with which we can test the CSR hypothesis.

The time complexity of the above methods consists of that for constructing the extended shortest-path trees and that for achieving Tasks 1 and 2, or specifically, counting the number of $\sum_{i=1}^{n_B} n(t|p_{Bi})$ and computing the value of $\sum_{i=1}^{n_B} |\tilde{L}(t|p_{Bi})|$. The time complexity of the former is $O(n_B n_L \log n_L)$, because we construct the extended shortest-path trees for n_B points in P_B as the roots (Section 3.4.1.3 in Chapter 3). Task 1 requires $O(n_A n_B)$ for each t, because the counting is repeated for all points (as roots) in P_B. Task 2 consists of two subtasks. First, sorting the distances into $(d_v^{(1)}, d_v^{(2)}, \ldots, d_v^{(n_B)})$ for all nodes requires $O(n_V n_B \log n_B)$, because sorting at each node requires $O(n_B \log n_B)$ (Section 3.4.1.6 in Chapter 3). Second, summing the lengths of the links to obtain $D_0, D_1, \ldots, D_{n_B}$ for each value of t requires $O(n_L)$, because we achieve the addition of the length of one link in constant time. Therefore, the total time complexity is $O(m\, n_V n_B \log n_B)$, where m is the number of discrete values $t_1, t_2, \ldots, t_m$ for t.

6.4.2.3 Computational methods for the network global Voronoi cross *K* function method

The computational method for the global Voronoi cross K function method presented in Section 6.2.3 is almost the same as that for the global cross K function method developed in Section 6.4.2.2 save for a few respects.

We are meant to achieve Tasks 1 and 2. Task 1 is to count the number $n(t|P_B)$ of points in P_A within distance t from at least one point in P_B. We can obtain the number by enumerating nodes $v \in V$ that satisfy $d_v^{(1)} \leq t$ with respect to every $v \in V$, using $0 = d_v^{(0)} \leq d_v^{(1)} \leq d_v^{(2)} \leq \cdots \leq d_v^{(n_B)}$ derived from the extended shortest-path trees rooted at points in P_B. Task 2 is to compute the length $|\bigcup_{i=1}^{n_B} \tilde{L}(t|p_B)|$. Let u and v be the end nodes of a link $l \in L$ satisfying $d_u^{(1)} \leq d_v^{(1)}$. First, we initialize $D = 0$. Next, for each link $l \in L$: if $d_u^{(1)} \leq t$ and $d_v^{(1)} \leq t$, we update D by $D \leftarrow D + |l|$; if $d_u^{(1)} \leq t$ and $d_v^{(1)} > t$, we update D by $D \leftarrow D + |l|/2$; and if $d_u^{(1)} > t$ and $d_v^{(1)} > t$, we do nothing. When we have done this for all $l \in L$, we obtain D as an approximation of $|\bigcup_{i=1}^{n_B} \tilde{L}(t|p_{Bi})|$. Because $|\bigcup_{i=1}^{n_B} \tilde{L}(t|p_{Bi})| = \sum_{i=1}^{n_B} |\tilde{L}_{VD}(t|p_{Bi})|$ (shown in Section 6.2.3), the substitution of $|\bigcup_{i=1}^{n_B} \tilde{L}(t|p_{Bi})|$ into equations (6.15) and (6.16) gives the expected value and variance of K'_{AB} of equation (6.14), with which we can test the CSR hypothesis.

The SANET software package (Okabe, Okunuki, and Shiode, 2006a, 2006b; Okabe, Satoh, and Shiode, 2009) implements the computational methods in Section 6.4 using approximation with Monte Carlo simulation. Chapter 12 illustrates the usage of this computational tool with actual examples.

7

Network spatial autocorrelation

In this chapter, we present statistical and computational methods for examining *spatial autocorrelation* on a network, that is, the correlation between attribute values of the same kind at different locations on a network or on different subnetworks (including line segments and links) forming the network. The statistics can answer, for example, question Q4 in Chapter 1: Is the roadside land price of a street segment similar to those of the adjacent street segments? In general terms,

Question Q4′: Given a set of attribute values of spatial units on a network (which may be represented by points, line segments, subnetworks) with the degrees of closeness between those spatial units (which may be categorical or numerical), are attribute values similar if their spatial units are close each other?

In the real world, besides roadside land prices, there are many examples on networks as well as in regions in which an attribute value at a location is strongly affected by those at its neighborhood locations. Tobler (1970) generally stated these geographical phenomena as the *first law of geography*: everything is related to everything else, but near things are more related than distant things (for discussion, see Miller (2004), Sui (2004), Tobler (2004)). The spatial autocorrelation indexes introduced in this chapter statistically examine this kind of spatial phenomena.

One of the earliest contributors to the study on spatial autocorrelation was Moran (1948), whose statistics are referred to as *Moran's I* statistics. Since then, many

Spatial Analysis along Networks: Statistical and Computational Methods, First Edition.
Atsuyuki Okabe and Kokichi Sugihara.

similar statistics (including spatial association) have been proposed in the literature; for example, Geary's *c* (Geary, 1954), Getis–Ord's *G* (Getis and Ord, 1992), Tango's C_F (Tango, 1995, 2000), Oden's *Ipop* (Oden, 1995), and a spatial version of the local score statistics by Lawson (2001) and Waller and Gotway (2004). Spatial autocorrelation is also incorporated in econometrics, establishing a distinct field in spatial analysis, *spatial econometrics* (Anselin, 1988; Tiefelsdorf, 2000; Fotheringham, Brunsdon, and Charlton, 2002; Arbia, 2006; LeSage and Pace, 2009). Originally, statistics of spatial autocorrelation were developed for *areal data*, that is, data consisting of attribute values of subareas forming a tessellation of a region, such as the numbers of inhabitants in administrative districts. We refer to the autocorrelation of areal data as *planar spatial autocorrelation*.

An extension from planar spatial autocorrelation to spatial autocorrelation on networks, referred to as *network spatial autocorrelation*, has been developed by a number of researchers since the early 1980s. One of the earliest uses of the term *network autocorrelation* is found in White, Burton, and Dow (1981). They were concerned with *social networks*, such as trade or market networks, networks of political relationships and networks of historical linguistic relationships. They extended the traditional spatial autocorrelation analysis of Cliff and Ord (1973) to a more abstract formulation of autocorrelation, in which societies were seen as occupying positions in networks of relations among societies, and autocorrelation exists among societies that are close in the networks, where the closeness is measured by not only spatial distance but also historical relatedness of languages (White, Burton, and Dow, 1981, p. 832). Their study was followed by Dow, Burton, and White (1982), Dow *et al.* (1984), Doreian, Teuter, and Wang (1984) and others. The studies in the 1980s were reviewed by Doreian (1990) and recent progress is seen in Dow (2007). Black (1992) notes that the network autocorrelation referred to in those studies might be better named *social autocorrelation*.

In contrast to autocorrelation on social networks, Goodchild (1987) noted another type of network autocorrelation: autocorrelation between the attribute values of the links of a network. One of the earliest applications was developed by Black (1992). He distinguished *spatial* autocorrelation and *network* autocorrelation, and explained that *spatial* autocorrelation usually concerns variables at given locations being influenced by variables at nearby or contiguous locations in a spatial context; *network* autocorrelation concerns the dependence of variable values on given links to similar values on other links to which the link is connected in a network context (Black, 1992, p. 207). He applied *network* autocorrelation analysis to the migration flows between 1965 and 1970 for the nine major census regions of the United States. Following Black (1992), Berglund and Karlstrom (1999) examined the *network* autocorrelation of migration flows in east central Sweden, Stockholm county and surrounding regions. Chun (2008) applied *network* autocorrelation analysis to the migration flows between 1995 and 2000 for the 48 conterminous US states and Washington DC. Note that Getis (1991), Bolduc, Laferriere, and Santarossa (1992), Getis (1991), Bolduc, Laferriere, and Santarossa (1995), Fisher and Griffith (2008) and others also studied the autocorrelation of travel flows or interaction flows; they did not use the term *network* autocorrelation but *spatial* autocorrelation.

The network autocorrelation defined by Black (1992) is not restricted to migration flows. Black (1991) and Black and Thomas (1998) studied network autocorrelations of accidents on Belgium's motorways. In their studies, *network* autocorrelation means the correlation between the value of a variable on a given segment of a network (not necessarily a link between intersections of a road network) and the values of that variable on contiguous segments of the network. Flahaut, Mouchart, and Martin (2003) detected *black zones* (concentrations of road accidents) on one road segment, the N29 in Belgium, using a *spatial* autocorrelation index, where accidents were aggregated across the basic spatial units (100 m road segments), and the performance of this index was compared with the kernel density estimation method (Chapter 9). Steenberghen *et al.* (2004) also examined the existence of black zones in a road network in Mechelen near Brussels by applying a *spatial* autocorrelation index to accidents on road segments of the network. The resulting black zones are referred to as *linear clusters*, which were compared with the black zones in the region, referred to as *two-dimensional clusters*, obtained from the kernel density estimation method (Chapter 9) when it was applied to accident points in the region (points on a plane, not on a network). Yamada and Thill (2010) examined the local *spatial* autocorrelation of highway vehicle crashes on road segments in Buffalo, NY in 1997. Shiode (2008) studied the *network* spatial autocorrelation of commercial facilities alongside a street network in Shibuya, Tokyo with 17 sizes of equal-length network cells. She also studied the *planar* spatial autocorrelation within the same dataset, and showed a notable finding that the results obtained from the network spatial autocorrelation and those from the planar spatial autocorrelation are distinctively different when the network cell size is large (Shiode and Shiode, 2008, Figure 9). It should be noted that all the above studies use Moran's I statistics, which are illustrated in this chapter.

The chapter consists of four sections. The first section formulates a general framework for studying autocorrelation, and classifies the various kinds of network autocorrelations reviewed above under this framework. The second section introduces two concepts of spatial randomness of attribute values on a network: *permutation spatial randomness* and *normal variate spatial randomness*. The third section describes two types of Moran's I statistic. The first, termed *local Moran's I* statistic, analyzes *local spatial autocorrelation*, i.e., the correlation between the attribute value of a specific subnetwork and those of its surrounding subnetworks. The second, termed *global Moran's I* statistic, analyzes *global spatial autocorrelation*, i.e., the average level of spatial autocorrelation over the whole network. The statistical properties of these statistics under the permutation spatial randomness and the normal variate spatial randomness are presented. The last section describes methods for computing those statistics.

7.1 Classification of autocorrelations

As noted above, the concept of network autocorrelation varies from researcher to researcher, and the term *network autocorrelation* can have several meanings. In this

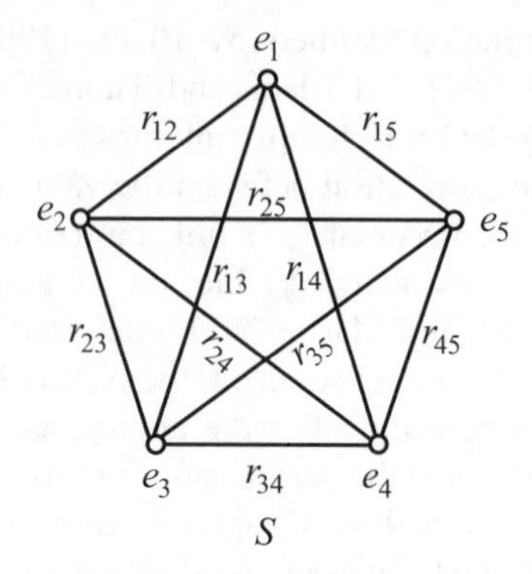

Figure 7.1 A relational network for an abstract autocorrelation in a space S, where entities and their relations are represented by nodes $e_1, \ldots, e_n$ and links $r_{ij}, i \neq j, i, j = 1, \ldots, n$, respectively.

section, we clarify these differences by introducing a framework for autocorrelation (a similar classification is presented in Peeters and Thomas (2009)). We first formulate an abstract form of autocorrelation, and next we describe several types of autocorrelation by embodying the abstract form.

We consider a set of entities, $e_1, \ldots, e_n$ (where an entity e_i may consist of one entity or more, say $e_i = \{e_{i1}, e_{i2}\}$; see the general discussion on the concept of entities in Chapter 2), and let x_i be an attribute value of e_i. We next consider the *relations* $r_{ij}, i \neq j, i, j = 1, \ldots, n$ between entities $e_1, \ldots, e_n$, and let w_{ij} be the *relational value* of r_{ij}, i.e., a numerical value indicating the strength of relatedness. To show the relations diagrammatically, we regard entities $e_1, \ldots, e_n$ as nodes and relations $r_{ij}, i \neq j, i, j = 1, \ldots, n$ as links. Then, we can represent relations $r_{ij}, i \neq j, i, j = 1, \ldots, n$ between entities $e_1, \ldots, e_n$ by a network, $N_R = (\{e_1, \ldots, e_n\}, \{r_{ij} | i \neq j, i, j = 1, \ldots, n\})$, termed a *relational network* (Figure 7.1). Entities $e_1, \ldots, e_n$ are embedded in a space, S, or more broadly in a context, where their relations are described by N_R. In these terms, we define an *abstract form of autocorrelation* or *abstract autocorrelation* by the autocorrelation between attribute values $x_1, \ldots, x_n$ of entities $e_1, \ldots, e_n$ in space S, where the relations of $e_1, \ldots, e_n$ are described by relational network N_R with relational values $w_{ij}, i \neq j, i, j = 1, \ldots, n$.

We make a few remarks on this definition. First, as shown by $r_{ij}, i \neq j$, $i, j = 1, \ldots, n$, we conceptually consider all possible relations between entities $e_1, \ldots, e_n$. This implies that the graph of the network $N_R = (\{e_1, \ldots, e_n\}$, $\{r_{ij} | i \neq j, i, j = 1, \ldots, n\})$ is a *complete graph*, i.e., a graph in which any pair of nodes is connected by a link (the network in Figure 7.1 can be regarded as a complete graph; the definition of the graph of a network is given in Section 2.2.1.2 in Chapter 2). In practice, depending on the subject of concern, some relations in $\{r_{ij} | i \neq j, i, j = 1, \ldots, n\}$ are insignificant and may be neglected. Even in such a case, we conceptually consider all possible relations $r_{ij}, i \neq j, i, j = 1, \ldots, n$ by assigning a specific value to the relational values w_{ij} of the insignificant relations (as will be seen, all values $w_{ij}, i \neq j, i, j = 1, \ldots, n$ are considered in Moran's I statistics). Second, in general, we may have $r_{ij} \neq r_{ji}$ or $w_{ij} \neq w_{ji}$. This implies that N_R may be a directed network (Section 4.2.1 in Chapter 4). In this chapter, however, we deal with only symmetric relations, $r_{ij} = r_{ji}$ or $w_{ij} = w_{ji}$.

Because the abstract autocorrelation is characterized by a relational network, it could be alternatively called an *abstract network autocorrelation*, which might correspond to the network autocorrelation used by the sociologists referred to in the above review. However, in this book, we will use the term *abstract autocorrelation*, because we reserve the term *network autocorrelation* for a more narrowly defined autocorrelation to be described later.

The abstract autocorrelation is embodied in many types of autocorrelation by specifying entities $e_1, \ldots, e_n$, their attribute values $x_1, \ldots, x_n$, relations $r_{ij}, i \neq j, i, j = 1, \ldots, n$, their relational values $w_{ij}, i \neq j, i, j = 1, \ldots, n$, and spaces or contexts S (Figure 7.2). The entities $e_1, \ldots, e_n$ may be, for instance, stores and districts, which are *physical entities*; or societies and cultures, which are *nonphysical entities*. The attribute values $x_1, \ldots, x_n$ of entities may be, for instance, the price of a hamburger at a fast-food shop, the number of accidents in a district, the number of languages used in a society, or whether or not a culture is teetotal. The relations $r_{ij}, i \neq j, i, j = 1, \ldots, n$ may be, for instance, that two stores are standing side by side or two districts are at least 5 km apart, which are *physical relations*; that two political groups oppose each other's income redistribution policy; or that two societies are similar in historical relatedness of languages (White, Burton, and Dow, 1981), which are *nonphysical relations*. The relational values $w_{ij}, i \neq j, i, j = 1, \ldots, n$ may be, for instance, $w_{ij} = 1$ if two entities are connected, $w_{ij} = 0$ otherwise; $w_{ij} = 2$ if two entities are within 2 km, and so forth (note that the relational values are customarily called *spatial weights* in spatial analysis, and *weights* in sociology; the latter case is discussed by Leenders (2002) in depth). The spaces (or contexts) S may be, for instance, social space and religious space, which are *nonphysical spaces*; or regional space and architectural space, which are

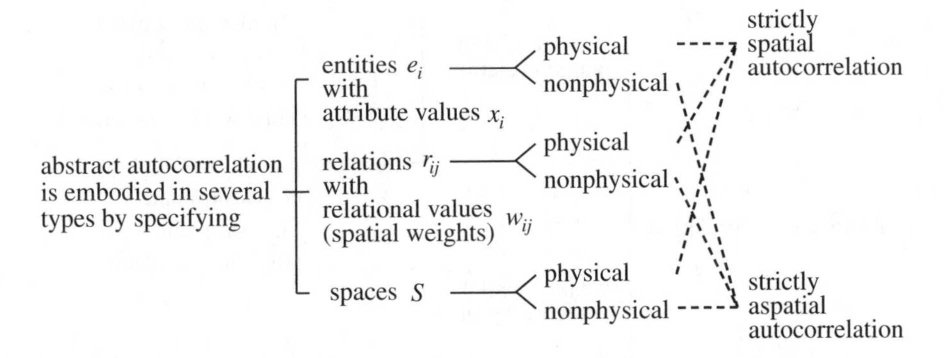

Figure 7.2 Abstract autocorrelation embodied with respect to entities, relations and spaces.

physical spaces. The physical spaces and nonphysical spaces might correspond to the *real* or *tangible space* and the *abstract* or *intangible space*, respectively, mentioned in Black (1992). Black (1992) remarked that the network autocorrelation used by sociologists might be better named *social autocorrelation*. In the above terms, *social autocorrelation* means the abstract autocorrelation embodied in a social space. The *social networks* studied in sociology (e.g., Scott, 1991; Wasserman and Faust, 1994; Knoke and Yang, 2008) are relational networks in a social space.

As noted above, there are many kinds of spaces, but in *spatial* analysis and *spatial* statistics, the space primarily means the physical space of the real world. In fact, a classic book, *Spatial Autocorrelation*, by Cliff and Ord (1973) mainly assumed that entities $e_1, \ldots, e_n$ are subareas forming an area in the real world. In what follows, we will use the term *space* for physical space without the adjective *physical*; other spaces are referred to with adjectives, such as *social* space.

We can classify the above embodied autocorrelations according to whether the entities, relations and spaces are physical or nonphysical. The number of possible combinations is 2 (physical or nonphysical entities) $\times$ 2 (physical or nonphysical relations) $\times$ 2 (physical or nonphysical spaces). The two extreme classes are the class in which entities, relations and spaces are all physical, and the class in which they are all nonphysical (indicated by the broken line segments in Figure 7.2). We call the former class *strictly spatial autocorrelations* and the latter *strictly aspatial autocorrelations*. Other classes are referred to as *weakly spatial autocorrelations* (Figure 7.3). In this volume, we focus on the strictly spatial autocorrelations and we refer to them simply as *spatial autocorrelations*.

In the literature of spatial analysis, most studies assume that the entities $e_1, \ldots, e_n$ are subareas forming a tessellation of a study area, which are represented by polygons embedded in a continuous plane with Euclidean distance. We call such spatial autocorrelations *planar spatial autocorrelations*. It should be noted that planar spatial autocorrelation seemingly deals with attribute values over a

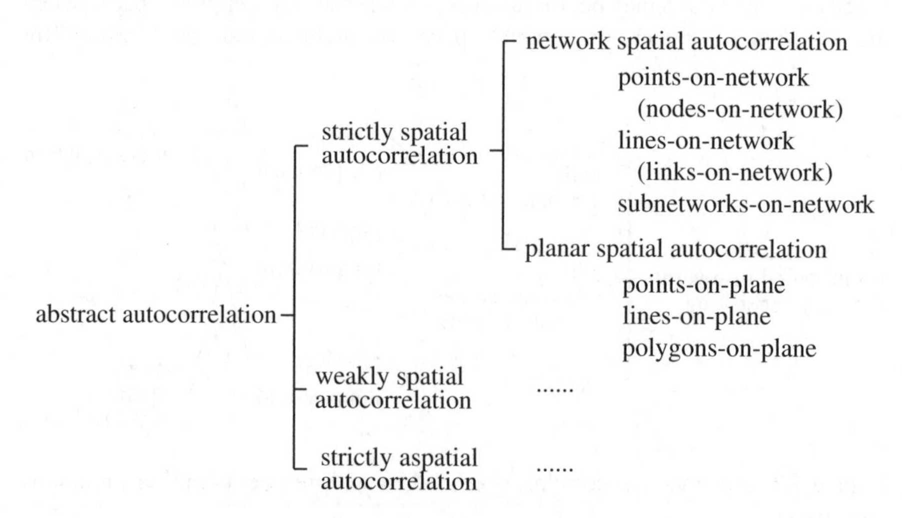

Figure 7.3 Classification of autocorrelations.

continuous plane but the plane is treated as discrete subareas; therefore, the mathematical structure for the planar spatial autocorrelation is the same as that for the (discrete) abstract autocorrelation.

Spatial autocorrelations are not restricted to the planar spatial autocorrelation. Entities $e_1, \ldots, e_n$ may be embedded in a physical network space, such as a street network, a river network or a pipeline network (other examples are shown in Chapter 1). We call such spatial autocorrelations *network spatial autocorrelations*. Note that these network spatial autocorrelations should be distinguished from the network autocorrelations used in sociology.

Network spatial autocorrelations can be classified according to the geometrical form of entities $e_1, \ldots, e_n$ on a physical network. If the entities (with attribute values $x_1, \ldots, x_n$) are points, for example, fast-food shops standing alongside a street, and x_i is the price of a hamburger (e.g., Figure 7.4a), then we refer to such a network spatial autocorrelation as a *points-on-network autocorrelation*. In the special case in which the points are the nodes of the network, we call the points-on-network autocorrelation a *nodes-on-network autocorrelation*. If the geometrical form of entity e_i is a line segment on a network, for example, a 100 m road segment as used in Black (1991), Flahaut, Mouchart, and Martin (2003), Steenberghen *et al.* (2004) and Yamada and Thill (2010) (e.g., Figure 7.4b), and x_i is the number of accidents that have occurred in each line segment, then we call such a network spatial autocorrelation a *lines-on-network autocorrelation*. In particular, if all links of a network are entities that are line segments with attribute values, we call it a *links-on-network autocorrelation*. The concept of network autocorrelation used by Black and Thomas (1998) might correspond to the links-on-network autocorrelation. More generally, if the geometrical form of entity e_i is a set of connected line segments forming a subnetwork of a physical network, for example, streets in a district (e.g., Figure 7.4c), we call such a network spatial autocorrelation a *subnetworks-on-network autocorrelation* (the collection of subnetworks may form a tessellation of a network). In the same manner, we can specify planar spatial autocorrelation as *points-on-plane*, *lines-on-plane* or *polygons-on-plane* autocorrelations, where the distances are defined in terms of Euclidean distances on a plane. In the most frequently used polygons-on-network autocorrelation, the polygons form a

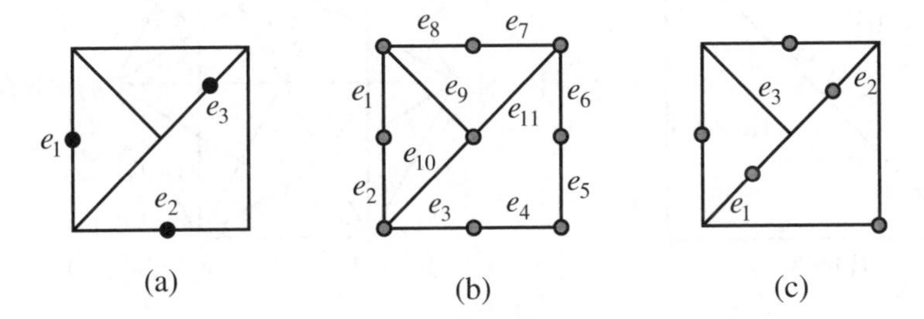

Figure 7.4 Entities for network spatial autocorrelations: (a) points (the black circles) for points-on-network autocorrelation, (b) line segments (segmented by the gray circles) for lines-on-network autocorrelation, (c) subnetworks (partitioned by the gray circles) for subnetworks-on-network autocorrelation.

tessellation of a region. Note that weakly spatial autocorrelations and strictly aspatial autocorrelations could be classified in a similar manner, but such classifications are beyond the scope of this chapter.

Black (1992) called particular attention to the autocorrelations of flow values, referred to as *flow autocorrelations*. Conceptually, flow autocorrelation looks different from other autocorrelations (Figure 7.5a), but mathematically it can be treated in the same way. Suppose that entity e_i^* is a pair of regions $e_i^* = \{e_{i1}, e_{i2}\}$, the attribute value x_i of e_i^* is the flow value between regions e_{i1} and e_{i2}, and the relation r_{ij}^* between e_i^* and e_j^* is the relation between $\{e_{i1}, e_{i2}\}$ and $\{e_{j1}, e_{j2}\}$; for instance, one example of r_{ij}^* is that a common region, say $e_{i2} = e_{j2}$, is the destination of flows; another example is that r_{ij}^* the maximum value of the distances between either of e_{i1}, e_{i2} and either of e_{j1}, e_{j2}. Then a flow autocorrelation is described by the relational network $N_R = (\{e_1^*, \ldots, e_n^*\}, \{r_{ij}^* | i \neq j, i, j = 1, \ldots, n\})$ with $\{x_1, \ldots, x_n\}$ and $\{w_{ij} | i \neq j, i, j = 1, \ldots, n\}$ (Figure 7.5b); this structure is exactly the same as that of the relational network of the abstract autocorrelation (Figure 7.1). Consequently, we do not require special treatments for flow autocorrelations. Note that flow autocorrelations may be spatial or aspatial; they correspond to links-on-network autocorrelations, such as the autocorrelation of the numbers of traffic accidents on road links (line segments between intersections).

Although there are various kinds of autocorrelations, the data for examining autocorrelation are the set of attribute values $\{x_1, \ldots, x_n\}$ and the set of relational values $\{w_{ij}, i \neq j, i, j = 1, \ldots, n\}$; these datasets are common to all the autocorrelations introduced above. No special statistical treatments are required for network spatial autocorrelations. The statistics for planar spatial autocorrelations can be directly applied to those for network spatial autocorrelations without any modifications. For this reason, this chapter focuses only on Moran's I statistics as representative examples of many statistics for network spatial autocorrelations, and provides an introduction to Moran's I statistics. Readers who wish to know how to

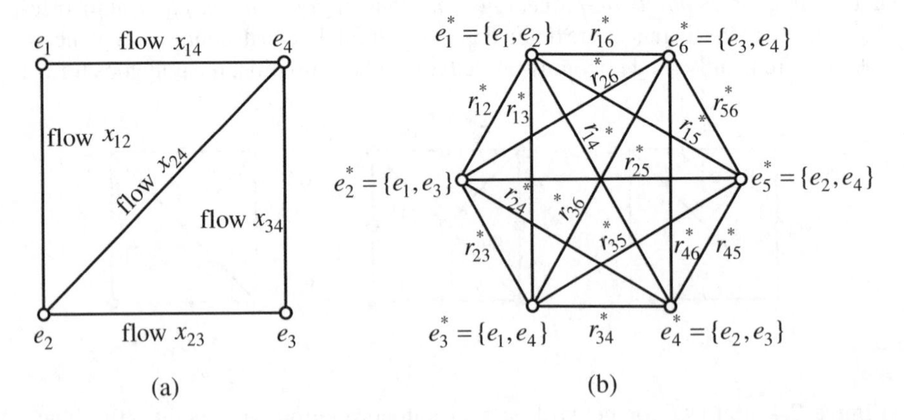

Figure 7.5 Flow autocorrelation: (a) flows between entities, (b) the relational network of the flows in (a).

derive Moran's I statistics as well as other spatial autocorrelation statistics in depth should consult textbooks devoted to spatial autocorrelation statistics, such as Cliff and Ord (1973, 1981), Goodchild (1986), Griffith (1987, 2003), Odland (1988), and Tiefelsdorf (2000).

7.2 Spatial randomness of the attribute values of network cells

In Chapter 2, we defined the concept of complete spatial randomness (CSR) for analyzing the distribution of points on a network. This definition is not applicable to the spatial randomness of the attribute values of entities on a network. Alternatively, we define two types of spatial randomness on a network, *permutation spatial randomness* and *normal variate spatial randomness*, in the subsequent two subsections.

7.2.1 Permutation spatial randomness

In network spatial autocorrelations, entities $e_1, \ldots, e_n$ are assumed to be points (including nodes), line segments (including links) or subnetworks on a network (Figure 7.4) and the data of their attribute values of interest $x_1, \ldots, x_n$ are associated with these entities. We refer to such spatial data units as *network cells*. Under 'permutation spatial randomness' (the precise definition will be given later), we consider that the attribute values $x_1, \ldots, x_n$ are fixed, but the assignment of these values to n network cells $L_1, \ldots, L_n$ is random. To illustrate this random assignment with a simple example, consider a network consisting of three network cells L_1, L_2 and L_3 (Figure 7.6). We denote the assignment in which x_1, x_2 and x_3 are placed on L_1, L_2 and L_3, respectively, by a permutation of (x_1, x_2, x_3). Then, the six possible permutations are (x_1, x_2, x_3), (x_1, x_3, x_2), (x_2, x_1, x_3), (x_2, x_3, x_1), (x_3, x_1, x_2) and (x_3, x_2, x_1) (Figure 7.6).

In these terms, spatial randomness is stated as every permutation occurring with the same probability. Under this randomness assumption, the expected value of the attribute value assigned to the first network cell L_1 is $(x_1 + x_1 + x_2 + x_2 + x_3 + x_3)/6$; that assigned to the second cell L_2 is $(x_2 + x_3 + x_1 + x_3 + x_1 + x_2)/6$; and that to the third cell L_3 is $(x_3 + x_2 + x_3 + x_1 + x_2 + x_1)/6$. All

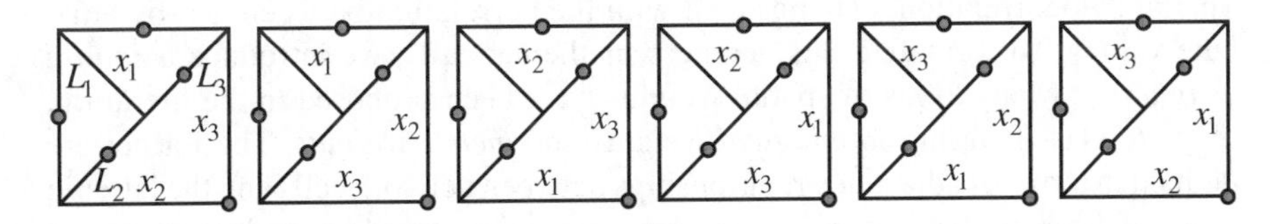

Figure 7.6 Permutation spatial randomness. The network consists of subnetwroks L_1, L_2 and L_3, to which attribute values x_1, x_2 and x_3 are assigned (the gray circles are boundary points).

these values are equal, implying that the expected attribute values are uniform across the three network cells L_1, L_2 and L_3. This uniformity corresponds to the uniform distribution assumed in the CSR for a point distribution (Section 2.4.2 in Chapter 2). In general, let $(x'_1, \ldots, x'_n)$ be a permutation of $x_1, \ldots, x_n$, indicating that attribute values $x'_1, \ldots, x'_n$ are assigned to $L_1, \ldots, L_n$, respectively, where x'_i is one of $x_1, \ldots, x_n$ and let $\mathrm{P}[(x_1, \ldots, x_n)]$ be the set of all possible permutations of $x_1, \ldots, x_n$, for example, $\mathrm{P}[(x_1, x_2, x_3)] = \{(x_1, x_2, x_3), (x_1, x_3, x_2), (x_2, x_1, x_3), (x_2, x_3, x_1), (x_3, x_1, x_2), (x_3, x_2, x_1)\}$ as in Figure 7.6. In these terms, we define the spatial randomness of attribute values across network cells as:

Permutation spatial randomness: Every permutation in $\mathrm{P}[(x_1, \ldots, x_n)]$ occurs with the same probability.

In spatial analysis, permutation spatial randomness is often used as a fundamental null hypothesis for a statistical test, like the CSR hypothesis in Chapter 2.

7.2.2 Normal variate spatial randomness

In permutation spatial randomness, the attribute values $x_1, \ldots, x_n$ are fixed and form a population. An alternative spatial randomness is that attribute values $x_1, \ldots, x_n$ are random samples from a population characterized by a discrete probability distribution or a continuous probability density function. Specifically, we define a spatial randomness of attribute values across network cells as:

Normal variate spatial randomness: Attribute values $x_1, \ldots, x_n$ assigned to network cells $L_1, \ldots, L_n$ are independently and identically generated from a normal distribution.

It follows from the above definition that the expected value of x_i is the same for all $i = 1, \ldots, n$, implying that the expected attribute value is uniform across all network cells. This uniformity, as in the uniformity of the permutation randomness, corresponds to the uniform distribution assumed in the CSR for a point distribution.

7.3 Network Moran's *I* statistics

Spatial autocorrelation statistics deal with the correlation between nearby attribute values. To formalize this nearness mathematically, we introduce a *spatial weight*, w_{ij}. Many kinds of spatial weights have been proposed in the literature; they are classified into *topological weights* and *metric weights*. The former are defined in terms of adjacency relationships between network cells and the latter in terms of distances between representative points. One possible topological weight is $w_{ij} = 1$ if network cells L_i and L_j are adjacent; otherwise, $w_{ij} = 0$. This weight may be standardized by $w'_{ij} = w_{ij} / \sum_{k=1}^{n} w_{ik}$ if network cells L_i and L_j are adjacent; otherwise, $w'_{ij} = 0$. In addition to these topological weights, many other topological weights for planar autocorrelations have been proposed in the literature, such as 'linear contiguity,' 'bishop contiguity,' 'rook contiguity,' and 'queen contiguity' (Kelejian and Robinson, 1995), which may or may not be applicable to network autocorrelations. Metric weights are written as a decreasing function of the shortest-path distance. Popular examples used in the literature are $\alpha \exp(-\beta d_S(p_i, p_j))$ and $\alpha / d_S(p_i, p_j)^\beta$, where $d_S(p_i, p_j)$ is the shortest-path distance between p_i and p_j, and α and β are positive constants.

Moran's I statistics include two types: *local* Moran's I statistic and *global* Moran's I statistic. The former examines the correlation between the attribute value of a specific network cell L_i and those of its surrounding network cells; the latter shows the average local Moran's I statistic across all network cells $L_1, \ldots, L_n$. The mathematical definitions of those two statistics are provided in the following two subsections.

7.3.1 Network local Moran's *I* statistic

The local Moran's I statistic on a network, *network local Moran's I* statistic, with respect to a network cell L_i, is exactly the same as the planar local Moran's I statistic, which was defined by Anselin (1995a) as:

$$I_i = \frac{n(x_i - \bar{x}) \sum_{j=1}^{n} w_{ij}(x_j - \bar{x})}{\sum_{j=1}^{n} (x_j - \bar{x})^2}, \tag{7.1}$$

where $\bar{x} = \sum_{i=1}^{n} x_i / n$, the average of the attribute values of all network cells. Roughly speaking, the denominator is the variance of the attribute values for all network cells and the numerator is the covariance between the attribute value of network cell L_i and those of its surrounding network cells.

The expected value and variance of I_i under the null hypothesis of the permutation spatial randomness are the same as those of the planar local Moran's I statistic, which were derived by Anselin (1995a) as:

$$\mathrm{E}(I_i) = -\frac{\sum_{j=1}^{n} w_{ij}}{n-1}, \tag{7.2}$$

$$\begin{aligned} \mathrm{Var}(I_i) &= \frac{n}{n-1} \left(1 - \frac{\sum_{i=1}^{n} z_i^4}{\left(\sum_{j=1}^{n} z_j^2 \right)^2} \right) \left(\sum_{j=1, j \neq i}^{n} w_{ij}^2 \right) \\ &+ \frac{n}{(n-1)(n-2)} \left(2 \frac{\sum_{i=1}^{n} z_i^4}{\left(\sum_{j=1}^{n} z_j^2 \right)^2} \right) \left(\sum_{k=1, k \neq i}^{n} \sum_{h=1, h \neq i}^{n} w_{ik} w_{ih} \right), \\ &- \frac{1}{(n-1)^2} \left(\sum_{j=1}^{n} w_{ij} \right)^2, \end{aligned} \tag{7.3}$$

where $z_i = x_i - \bar{x}$. Note that E(I_i) becomes $-1/(n-1)$ if the standardized weight $w'_{ij} = w_{ij}/\sum_{k=1}^{n} w_{ik}$ is used.

The exact distribution function of I_i is not yet known. Therefore, Anselin (1995b) recommended Monte Carlo simulations for testing the null hypothesis of the permutation spatial randomness. He also recommended simulation using the conditional permutation in which x_i is fixed and permutations are randomly chosen from $P[(x_1, \ldots, x_{i-1}, x_{i+1} \ldots, x_n)]$. From these simulations, we can obtain the upper and lower critical values I_i^* and I_i^{**} of the random variable I_i under the null hypothesis with significance level α: for example, the probability of the random variable I_i being less than I_i^* is α and the probability of its being greater than I_i^{**} is also α. If an observed value of I_i is less than I_i^*, then the attribute values are positively autocorrelated (i.e., nearby attribute values are similar) with confidence level $1-\alpha$; if an observed value of I_i is greater than I_i^{**}, then the attribute values are negatively autocorrelated (i.e., nearby attribute values are dissimilar) with confidence level $1-\alpha$.

The expected value and variance of I_i under the null hypothesis of the normal variate spatial randomness were explicitly written in Leung, Mei, and Zhang (2003); they are given by equation (7.2) and:

$$\mathrm{Var}(I_i) = \frac{1}{(n-1)^2(n+1)}\left(\frac{(n-1)^2}{n}\sum_{j=1}^{n} w_{ij}^2 - \frac{n^2-3n+4}{n^2}\left(\sum_{j=1}^{n} w_{ij}\right)^2\right). \quad (7.4)$$

7.3.2 Network global Moran's *I* statistic

The global Moran's *I* statistic on a network, *network global Moran's I* statistic, is exactly the same as the planar global Moran's *I* statistic, which is defined by Moran (1948) as:

$$I = \frac{n\sum_{i=1}^{n}\sum_{j=1}^{n} w_{ij}(x_i-\bar{x})(x_j-\bar{x})}{\left(\sum_{i=1}^{n}\sum_{j=1}^{n} w_{ij}\right)\sum_{i=1}^{n}(x_i-\bar{x})^2}. \quad (7.5)$$

In terms of the network local Moran's I_i statistic, this equation is written as:

$$I = \frac{\sum_{i=1}^{n} I_i}{\left(\sum_{i=1}^{n}\sum_{j=1}^{n} w_{ij}\right)}. \quad (7.6)$$

This equation shows that the global Moran's *I* statistic is proportional to the average of network local Moran's $I_1, \ldots, I_n$ statistics over all network cells.

The expected value and variance, respectively, of I under the null hypothesis of the permutation spatial randomness were derived by Cliff and Ord (1973) as:

$$\mathrm{E}(I) = -\frac{1}{n-1}, \tag{7.7}$$

and

$$\begin{aligned}\mathrm{Var}(I) = &-\frac{1}{(n-1)(n-2)(n-3)\left(\sum_{i=1}^{n}\sum_{j=1}^{n} w_{ij}\right)^2} \\ &\times\left\{\frac{1}{2}n(n^2-3n+3)\sum_{i=1}^{n}\sum_{j=1}^{n}(w_{ij}+w_{ji})^2\right. \\ &-n^2\sum_{i=1}^{n}\left(\sum_{j=1}^{n}w_{ij}+\sum_{j=1}^{n}w_{ji}\right)^2+3n\left(\sum_{i=1}^{n}\sum_{j=1}^{n}w_{ij}\right)^2 \\ &-\frac{n\sum_{i=1}^{n}z_i^4}{\left(\sum_{j=1}^{n}z_j\right)^2}\left[\frac{1}{2}(n^2-n)\sum_{i=1}^{n}\sum_{j=1}^{n}(w_{ij}+w_{ji})^2\right. \\ &\left.\left.-2n\sum_{i=1}^{n}\left(\sum_{j=1}^{n}w_{ij}+\sum_{j=1}^{n}w_{ji}\right)^2+6\left(\sum_{i=1}^{n}\sum_{j=1}^{n}w_{ij}\right)^2\right]\right\} \\ &-\frac{1}{(n-1)^2}.\end{aligned} \tag{7.8}$$

The expected value and variance of I under the null hypothesis of the normal variate spatial randomness were also derived by Cliff and Ord (1974) as equation (7.6) and:

$$\begin{aligned}\mathrm{Var}(I) = &\frac{1}{(n-1)(n+1)\left(\sum_{i=1}^{n}\sum_{j=1}^{n}w_{ij}\right)^2} \\ &\times\left\{\frac{n^2}{2}\sum_{i=1}^{n}\sum_{j=1}^{n}(w_{ij}+w_{ji})^2-n\sum_{i=1}^{n}\left(\sum_{j=1}^{n}w_{ij}+\sum_{j=1}^{n}w_{ji}\right)^2+3\left(\sum_{i=1}^{n}\sum_{j=1}^{n}w_{ij}\right)^2\right\} \\ &-\frac{1}{(n-1)^2}.\end{aligned} \tag{7.9}$$

To use the network global Moran's I statistic for testing the null hypothesis, we should know its probability distribution function. A method for computing the exact distribution was obtained by Tiefelsdorf and Boots (1995), and Tiefelsdorf (1998), and requires numerical integration. Cliff and Ord (1974) showed that as n increases, the distribution of global Moran's I statistic approaches the normal distribution under fairly mild conditions (for details, see Cliff and Ord (1974), Section 2.4). King (1981) showed that the global Moran's I statistic test is the locally best invariant. Therefore, using this property, we can test the null hypothesis in the same manner as for the network local Moran's I statistic in Section 7.3.1 (except that the conditional permutation spatial randomness is not used).

7.4 Computational methods for Moran's *I* statistics

Computing network Moran's I statistics does not require special geometrical computations but does require the basic ones described in Chapter 3. Therefore, it is straightforward to compute network local and global Moran's I statistics on a network $N = (V, L)$, where $V = \{v_1, \ldots, v_{n_V}\}$ is the set of vertices, $L = \{l_1, \ldots, l_{n_L}\}$ is the set of links and $\tilde{L} = \bigcup_{i=1}^{n_L} l_i$ is the set of points forming the network (including the nodes of V). There are three steps to prepare network data for network Moran's I statistics computation: tessellating a network into network cells, determining representative points of the network cells, and computing spatial weights between all pairs of the network cells.

The first task is to tessellate $\tilde{L} = \bigcup_{i=1}^{n_L} l_i$ into network cells $L_1, \ldots, L_n$. This tessellation can be achieved by several means; for example, we can tessellate the network $\tilde{L}$ according to postal codes, and regard the tessellated segments as network cells. In this case, we simply insert the boundary points of the network cells into N. This can be done by the method shown in Section 3.1.5.1 of Chapter 3 in time proportional to the number of the boundary points.

If the network cells are not given *a priori*, we should determine a method for tessellating a network into network cells in our own way. A typical method for this tessellation is to construct the network Voronoi diagram (Chapter 4). First, we choose a subset $P = \{p_1, p_2, \ldots, p_n\}$ of V and use these nodes as representative points for the network cells. Next, we construct the network Voronoi diagram for P and regard the Voronoi subnetworks as network cells. As shown in Section 4.3 in Chapter 4, the network Voronoi diagram can be constructed in $O(n_L \log n_V)$ time, where n_L and n_V are the number of links and that of nodes of N, respectively (note that n does not appear explicitly because $n \leq n_V$).

The second task is to determine the representative points $p'_1, \ldots, p'_n$ of $L_1, \ldots, L_n$. They may be given by arbitrary points on $L_1, \ldots, L_n$, or may be the centers of the network cells $L_1, \ldots, L_n$. The latter computational method is shown in Section 3.4.1.5 in Chapter 3; the computation of the center of a network cell L_i requires $O(m_i^2 \log m_i)$ time, where m_i denotes the number of links in the network cell L_i. On the other hand, if the network cells are defined by the Voronoi subnetworks, we regard the generator points $\{p_1, p_2, \ldots, p_n\}$ as the representative points; this requires no additional computation. In any case, we denote the resulting network as $N' = (V', L')$, where the set V' of nodes is given by $V' = V \cup P$ and the set L' of links is given by the refined links obtained by inserting the points of P in L.

The third task is to compute spatial weights w_{ij}, $i, j = 1, \ldots, n$. Because we manage the network $N' = (V', L')$ with the winged-edge data structure mentioned in Section 3.1.2 in Chapter 3, it is easy to find adjacency relationships between network cells and to compute the shortest-path distances between representative points using the method shown in Section 3.4.1. The shortest-path distances from one point in $\{p_1, p_2, \ldots, p_n\}$ to all the other points can be computed in $O(n_L \log n_L)$ time and hence all the shortest-path distances, i.e., all the weights $\{w_{ij}, i, j = 1, 2, \ldots, n\}$, can be computed in $O(n n_L \log n_L)$ time. If we define $w_{ij} = 0$ for nonadjacent network cells L_i and L_j, we must compute only the distances between adjacent network cells. However, the worst-case time complexity is still $O(n n_L \log n_L)$, because the number of adjacent network cells is of $O(n)$, and the computation of the shortest-path distance between two points requires $O(n_L \log n_L)$ time. By achieving the above tasks, we can obtain the values of $\{x_1, \ldots, x_n\}$, $\{p_1, \ldots, p_n\}$ and $\{w_{ij}, i, j = 1, \ldots, n\}$, with which we can compute the values of equations (7.1)–(7.9).

8

Network point cluster analysis and clumping method

In this chapter, we present two closely related methods for point pattern analysis on networks; namely, the *network point cluster analysis* and the *network clumping method*. Both methods are designed to provide a response to, for instance, question Q5 raised in Chapter 1, i.e., Q5: How can we locate clusters of fashionable boutiques alongside downtown streets? We generalize this question as:

Q5′: For a given set of points on a network, how can we find sets of clusters within which points are close to each other but between which the points are apart?

We describe the statistical and computational methods for responding to this general question in this chapter.

Most spatial methods for point pattern analysis are concerned with spatially clustered points. In fact, the nearest-neighbor distance method described in Chapter 5, the K function method discussed in Chapter 6 and the point density estimation method examined in Chapter 9 all generally consider spatially clustered points on networks. Consequently, we may broadly refer to analyses conducted using these methods as cluster analysis. However, the cluster analysis dealt with in this chapter does not follow such a broad definition, but is more narrowly defined; that is, it is a technique for numerical classification (or taxonomy) first proposed and formulated by Gilmour (1937), Cain (1956), Michener (1957), Michener and Sokal (1957), and Sneath (1957), among others (Sokal and Sneath (1963) integrate this early body of work). *Cluster analysis* is the collective term for the methods

Spatial Analysis along Networks: Statistical and Computational Methods, First Edition.
Atsuyuki Okabe and Kokichi Sugihara.

described in these studies, as represented in the titles of texts by Anderberg (1973), Lorr (1983), Aldenderfer and Blashfield (1984), Romesburg (1990), and Everitt, Landau, and Leese (2011).

Generally, cluster analysis is used to group objects (e.g., boutiques in Q5) according to a 'similarity' between the attribute values of the objects (e.g., the average price of clothes and the number of articles sold at each boutique). In particular, when the attribute values include spatial attribute values (e.g., distance to the nearest station and addresses; see Section 2.1.1 in Chapter 2), cluster analysis becomes *spatial cluster analysis* (Wang, 2006; Warden, 2008; Jacquez, 2008). Among the various methods available in spatial cluster analysis, this chapter focuses on a specific method in which objects are represented by points whose spatial attributes are only the locations of the objects on a network, and the 'similarity' between clusters, as measured by a function of the shortest-path distance between points on the network. We refer to this type of spatial cluster analysis as *network point cluster analysis*. Note that network point cluster analysis does not infer methods for clustering networks in terms of a similarity measure between two networks. To avoid this misunderstanding, we could alternatively employ *network-constrained point cluster analysis* or *network-based point cluster analysis*. However, for the purpose of simplicity, in this volume we use *network point cluster analysis*.

Network point cluster analysis is an extension of spatial cluster analysis for points on a plane to a network. Yiu and Mamoulis (2004) noted the importance of this extension and proposed algorithms and data structures for clustering points on a network. Likewise, Jin *et al.* (2006) proposed a statistical method for finding outliers on a network, while Clauset, Moore, and Newman (2008) formulated a method for constructing a hierarchical structure for point clusters on a network. Lastly, Sugihara, Satoh, and Okabe (2010), and Sugihara, Okabe, and Satoh (2011) developed several methods for network point cluster analysis combined with efficient computational methods for their calculation. As noted by those studies and the work reviewed in Chapter 1 of this volume, very large demand for network point cluster analysis exists.

A contrasting method to network point cluster analysis is the clumping method. In the former, points on a plane represent objects, whereas in the latter, they are disks (polygons in general) on a plane. We define a *clump* as a set of disks that overlap each other. We define the (*planar*) *clumping method* as a method for examining the composition of clumps with respect to *clump size* (defined by the number of disks forming a clump) in comparison with that obtained from randomly placed disks. Early applications include studies of aggregated laminae, dust particles, bacteria, and the overlap of circular bombed ruins, with Garwood (1947), Mack (1948, 1954), Armitage (1949), and Irwin, Armitage, and Davies (1949), among others, as the theoretical pioneers. Roach (1968) provides a comprehensive textbook account of the method, with an extension of the clumping method on a plane to a line (a very special form of network) described in Chapter 3 of his volume. Shiode and Shiode (2009a) consider an extension to a general network in a study of the spatial agglomeration of restaurants in Shibuya ward, Tokyo.

In this chapter, we first describe a general framework for network point cluster analysis and next provide several specific point clustering methods. We then formulate the clumping method on a network and show its relation to network point cluster analysis. Finally, we illustrate efficient computational methods for performing these methods.

8.1 Network point cluster analysis

We consider a planar network $N = N(V, L)$ composed of a set of nodes $V = \{v_1, \ldots, v_{n_V}\}$ and a set of links $L = \{l_1, \ldots, l_{n_L}\}$, and let $\tilde{L} = \bigcup_{i=1}^{n_L} l_i$ be the union of the links $l_1, \ldots, l_{n_L}$, i.e., a set of points forming the links including the nodes of V. Given a set of points $P = \{p_1, \ldots, p_n\}$ on $\tilde{L}$ (representing point-like entities such as boutiques alongside a street network), we wish to group the points of P into a number of nonempty subsets $P_1, \ldots, P_{n_c}$ of P, called *clusters*, that are mutually exclusive and collectively exhaustive. In mathematical terms, $P_i \subset P$, $P_i \neq \emptyset$, $P_i \cap P_j = \emptyset$ for $i \neq j$, and $\bigcup_{i=1}^{n_c} P_i = P$. We refer to the procedure adopted for this grouping as *(point) clustering*, achieved through the distance between clusters, termed the *intercluster distance* or *cluster distance*. In general, we measure the intercluster distance between P_i and P_j with a nonnegative real value function $d_C(P_i, P_j)$. As shown in Section 8.1.2, many kinds of intercluster distance functions are proposed in the literature (see also the textbooks cited above), and a variety of clustering methods is formulated corresponding to these intercluster distances.

Clustering methods are generally classified into two types: *nonhierarchical clustering* and *hierarchical clustering*. The former constructs clusters by assigning objects to predetermined centers of clusters, while the latter organizes objects into hierarchically structured clusters in a successive manner. This section focuses on hierarchical point cluster analysis. We first formulate hierarchical point cluster analysis in a general manner without specifying intercluster distances. We then describe individual point clustering methods by specifying their intercluster distances.

8.1.1 General hierarchical point cluster analysis

We start with a set of clusters, called the *initial level cluster set* C_0 (e.g., Figure 8.1), in which every point of P forms a cluster, i.e.:

$$C_0 = \{P_1^{(0)}, \ldots, P_n^{(0)}\} = \{\{p_1\}, \ldots, \{p_n\}\}.$$

To proceed to clustering, we then select an intercluster distance $d_C(P_i^{(0)}, P_j^{(0)})$ appropriate for the study of the given objects. For the initial level cluster set C_0, we then find the pair of clusters $P_*^{(0)}$ and $P_{**}^{(0)}$ in C_0 that attains the minimum value among the intercluster distances between any pair of clusters in C_0. We then delete

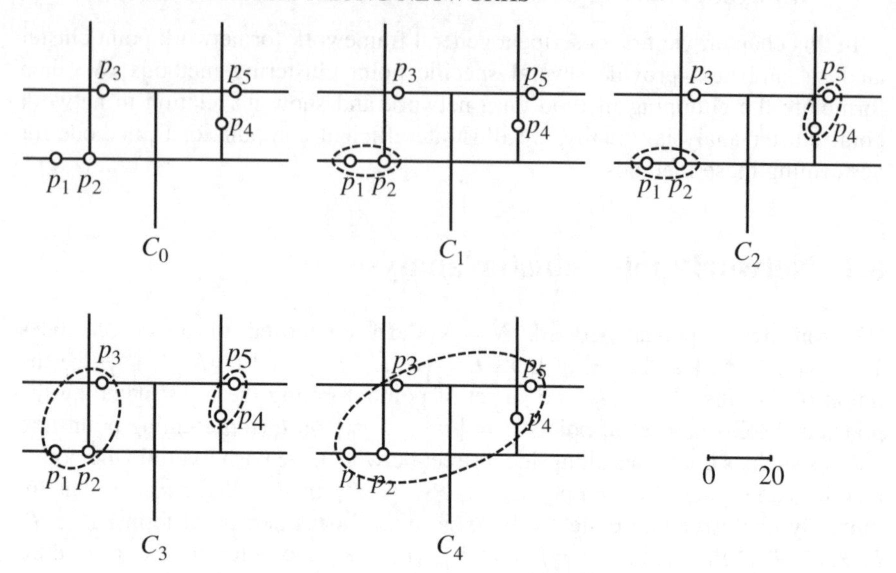

Figure 8.1 Hierarchical point clustering process.

clusters $P_*^{(0)}$ and $P_{**}^{(0)}$ from C_0, and add the union of $P_*^{(0)}$ and $P_{**}^{(0)}$ to C_0. The resulting set:

$$C_1 = \left(C_0 \backslash \{P_*^{(0)}, P_{**}^{(0)}\}\right) \cup \{P_*^{(0)} \cup P_{**}^{(0)}\}$$

is the *first-level cluster set* (Figure 8.1). For example, suppose $P_*^{(0)}$ and $P_{**}^{(0)}$ are $P_1^{(0)} = \{p_1\}$ and $P_2^{(0)} = \{p_2\}$, as in Figure 8.1. Then:

$$C_1 = \{P_1^{(1)}, \ldots, P_{n-1}^{(1)}\} = \{P_1^{(0)} \cup P_2^{(0)}, P_3^{(0)}, \ldots, P_n^{(0)}\} = \{\{p_1, p_2\}, \{p_3\}, \ldots, \{p_n\}\}.$$

For the first-level cluster set C_1, we find the pair of clusters $P_*^{(1)}$ and $P_{**}^{(1)}$ that attains the minimum value among the intercluster distances between any pair of clusters in C_1. We then merge these two clusters $P_*^{(1)}$ and $P_{**}^{(1)}$, and obtain the *second-level cluster set* (Figure 8.1) as:

$$C_2 = \left(C_1 \backslash \{P_*^{(1)}, P_{**}^{(1)}\}\right) \cup \{P_*^{(1)} \cup P_{**}^{(1)}\}.$$

We continue this process:

$$C_{k+1} = \left(C_k \backslash \{P_*^{(k)}, P_{**}^{(k)}\}\right) \cup \{P_*^{(k)} \cup P_{**}^{(k)}\} \quad \text{for } k = 0, 1, 2, 3, \ldots.$$

Eventually, at the (n–1)-th step, we obtain the *(n–1)-th level cluster set* C_{n-1}, which consists of only one set $P_1^{(n-1)}$ containing all points of P, i.e.:

$$C_{n-1} = \{P_1^{(n-1)}\} = \{P\},$$

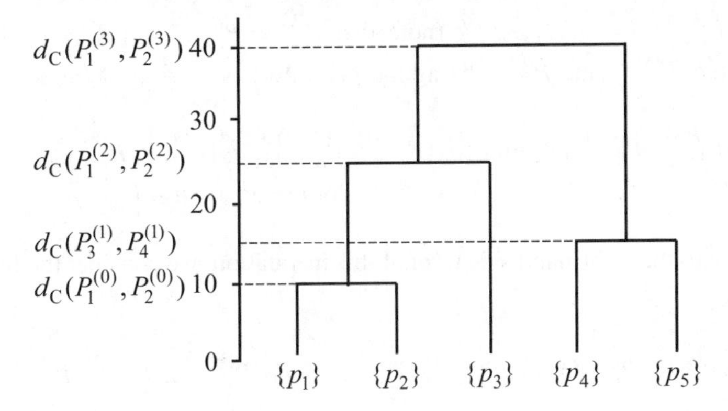

Figure 8.2 Dendrogram for the clusters in Figure 8.1.

for example, C_4 in Figure 8.1. The sequence of clusters resulting from this procedure is the *hierarchical structure of clusters*. Conventionally, the hierarchical structure is represented by the treelike diagram, as in Figure 8.2 (though upside down), called a *dendrogram*, in which the vertical axis indicates the value of the intercluster distance at which two clusters merge.

At each level of hierarchical clustering, we combine the closest clusters. This procedure, as shown in Figure 8.2, appears to imply that the intercluster distance at which two clusters merge increases with the level of hierarchical clustering, or mathematically,

Property 8.1 (monotonically increasing intercluster distance)

$$\min\{d_C(P_i^{(k-1)}, P_j^{(k-1)}),\ i \neq j;\ i, j = 1, \ldots, n-k+1\}$$
$$\leq \min\{d_C(P_{i'}^{(k)}, P_{j'}^{(k)}),\ i' \neq j';\ i', j' = 1, \ldots, n-k\} \quad \text{for } k - 1, \ldots, n-1. \quad (8.1)$$

In general, this property may or may not hold depending on the choice of intercluster distance. The latter is referred to as the *reversal phenomena* and could be acceptable in some cases (Section 6.2.5 in Anderberg (1973)). In the cluster analysis of physical points, however, Property 8.1 appears natural because the intercluster distance in this type of analysis is physical distance. In the following, we develop those clustering methods that satisfy Property 8.1.

To gain analytical tractability, we first rewrite inequation (8.1). Suppose without loss of generality, at the (k–1)-th step, $P_1^{(k-1)}$ and $P_2^{(k-1)}$ merge. Then the k-th level cluster set is written as:

$$C_k = \{P_1^{(k-1)} \cup P_2^{(k-1)}, P_3^{(k-1)}, \ldots, P_{n-k+1}^{(k-1)}\} = \{P_1^{(k)}, P_2^{(k)}, \ldots, P_{n-k}^{(k)}\}.$$

Because merger brings about change in the intercluster distances only for those clusters between $P_1^{(k-1)}$, $P_2^{(k-1)}$, $P_1^{(k-1)} \cup P_2^{(k-1)}$ and for the remaining clusters $P_i^{(k-1)}$, $i = 3, \ldots, n-k$ (note that $P_i^{(k)} = P_{i+1}^{(k-1)}$, $i = 2, \ldots, n-k$), the two clusters $P_1^{(k-1)}$ and $P_2^{(k-1)}$ being merged infer:

$$d_C(P_1^{(k-1)}, P_2^{(k-1)}) \le \min \{d_C(P_1^{(k-1)}, P_i^{(k-1)}),\ d_C(P_2^{(k-1)}, P_i^{(k-1)})\} \quad \text{for } i = 3, \ldots, n-k. \tag{8.2}$$

In relation to the right-hand side term of this inequation, we consider the following inequation:

$$\min\{d_C(P_1^{(k-1)}, P_i^{(k-1)}),\ d_C(P_2^{(k-1)}, P_i^{(k-1)})\} \le d_C(P_1^{(k-1)} \cup P_2^{(k-1)}, P_i^{(k-1)}) \quad \text{for } i = 3, \ldots, n-k. \tag{8.3}$$

This inequation implies Property 8.1, as it follows from equations (8.2) and (8.3) that:

$$d_C\,(P_1^{(k-1)}, P_2^{(k-1)}) \le \min\{d_C(P_1^{(k-1)} \cup P_2^{(k-1)}, P_i^{(k-1)}),\ i = 3, \ldots, n-k\}, \tag{8.4}$$

which is equivalent to equation (8.1). In developing the computational methods in Section 8.3, we use equation (8.3).

The objective of point cluster analysis is to find spatially agglomerated points on a network, such as the diamond rows on Manhattan Island in New York and the used (secondhand) book rows located in Kanda in Tokyo. Used book rows mean that stores in a row are spatially 'connected' along streets, but different rows are spatially 'disconnected' from each other. Because people access the stores through streets, it is then natural to represent the 'connected' stores or 'disconnected' rows in terms of the shortest-path distances between stores on a street network. To be explicit, consider a cluster $P_i = \{p_{i1}, \ldots, p_{in_i}\}$ in $C = \{P_1, \ldots, P_{n_c}\}$ and let $\tilde{L}_C(P_i)$ be a subset of $\tilde{L}$ obtained from the union of the shortest paths between any pair of points in P_i. We refer to $\tilde{L}_C(P_i)$ as the *cluster domain* of P_i. An example is shown in Figure 8.3, where the bold lines indicate the cluster domains of clusters P_1, P_2, and

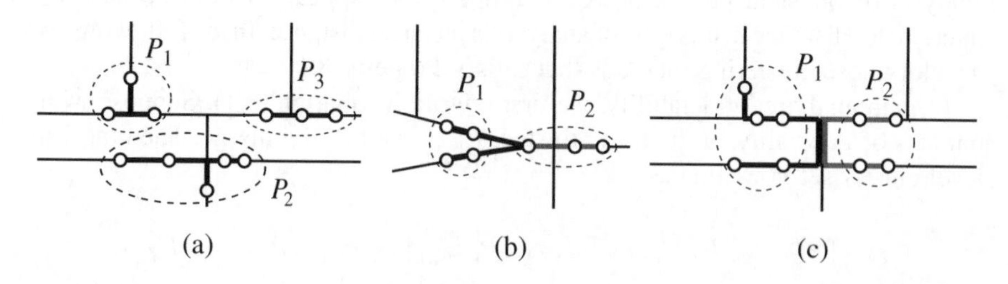

Figure 8.3 Cluster domains indicated by the bold line segments: (a) intracluster connected or intercluster disconnected; (b), (c) intracluster disconnected or intercluster connected.

P_3. If the cluster domain of P_i does not intersect with any other cluster domains, as in Figure 8.3a, we can say that cluster P_i is *intracluster connected* or, equivalently, *intercluster disconnected.* Otherwise (as in panels (b) and (c)), we can say that cluster P_i is *intracluster disconnected* or *intercluster connected.*

As mentioned in the introduction to Section 8.1, clusters $P_1, \ldots, P_{n_c}$ are defined as mutually exclusive sets. Therefore, one may consider that cluster domains are also mutually exclusive, i.e.:

Property 8.2 (intracluster connected or intercluster disconnected cluster set)

For a cluster set $C = \{P_1, \ldots, P_{n_c}\}$, equation $\tilde{L}_C(P_i) \cap \tilde{L}_C(P_j) = \phi$ holds for $i \neq j,\ i, j = 1, \ldots, n_c$.

In general, this property may or may not hold. It depends on the choice of intercluster distance. For instance, when both spatial and non spatial attributes characterize objects (Section 2.1.1 in Chapter 2), spatial clustering methods not satisfying Property 8.2 may be acceptable. However, when only their location characterizes objects, point clustering methods should satisfy Property 8.2 because the spatial form of clustered points on a network is the major concern of spatial analysis. In Section 8.2, we develop clustering methods that satisfy both Properties 8.1 and 8.2.

In conjunction with intraconnectedness, we define an additional important spatial concept. For a cluster set $C = \{P_1, \ldots, P_{n_c}\}$ (e.g., Figure 8.4a), we choose two clusters P_i and P_j, and construct a new cluster set $C\backslash\{P_i, P_j\} \cup (P_i \cup P_j)$. If the cluster $P_i \cup P_j$ is intracluster connected in $C\backslash\{P_i, P_j\} \cup (P_i \cup P_j)$, i.e., the cluster domain of $P_i \cup P_j$ does not intersect with that of P_k, $k \neq i, j$, we can say that the two clusters P_i and P_j are *adjacent*; otherwise, *not adjacent*. Figure 8.4 depicts an example, where clusters are given by P_1, P_2 and P_3 (panel (a)). Clusters P_1 and P_2 are adjacent because $\tilde{L}(P_1 \cup P_2) \cap \tilde{L}(P_3) = \varnothing$ (panel (b)), while clusters P_1 and P_3 are not adjacent because $\tilde{L}(P_1 \cup P_3) \cap \tilde{L}(P_2) \neq \varnothing$ (panel (c)). In hierarchical point cluster analysis, we execute clustering using adjacent relations between clusters.

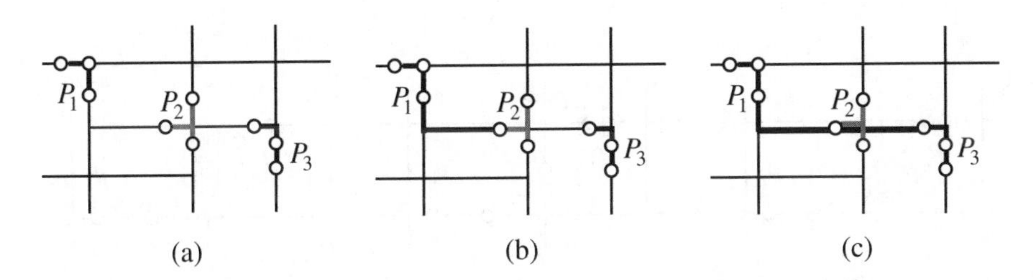

Figure 8.4 Adjacent relations (cluster domains are indicated by the bold line segments): (a) clusters P_1, P_2, P_3, (b) P_1 and P_2 are adjacent clusters because $\tilde{L}(P_1 \cup P_2) \cap \tilde{L}(P_3) = \varnothing$, (c) P_1 and P_3 are not adjacent clusters because $\tilde{L}(P_1 \cup P_3) \cap \tilde{L}(P_2) \neq \varnothing$.

8.1.2 Hierarchical point clustering methods with specific intercluster distances

In the preceding subsection, the intercluster distance was implicit. In this subsection, we describe individual clustering methods by specifying the intercluster distance. To avoid lengthy notation, we do not explicitly indicate the k-th level, and denote a cluster set at a certain level by $C = \{P_1, \ldots, P_{n_c}\}$, $P_i \subseteq P$, $i = 1, \ldots, n_c$, points belonging to a cluster P_i by $P_i = \{p_{i1}, \ldots, p_{in_i}\}$, and the shortest-path distance between points p_i and p_j on the network as $d_S(p_i, p_j)$. In terms of $d_S(p_i, p_j)$, we define several point clustering methods with the following intercluster distances.

8.1.2.1 Network closest-pair point clustering method

We first consider the network point clustering method with the intercluster distance, $d_{C1}(P_i, P_j)$, given by the minimum value among the shortest-path distances between every point in P_i and every point in P_j, or mathematically:

$$d_{C1}(P_i, P_j) = \min\{d_S(p_{ik}, p_{jl}) | \, p_{ik} \in P_i, \, p_{jl} \in P_j\}. \tag{8.5}$$

We term this intercluster distance the *closest-pair (intercluster) distance*. We refer to the network hierarchical point clustering method with $d_{C1}(P_i, P_j)$ as the *network closest-pair point clustering method*. This method corresponds to the *nearest-neighbor* or *single linkage method* in ordinary cluster analysis. The original idea was proposed by Sneath (1957) and elaborated by Gower (1967), Johnson (1967), and Zahn (1971), among others. Figure 8.5a illustrates a simple example. Suppose that at a given level, say the k-th level, the clusters are P_1, P_2, and P_3 indicated by the broken ovals in the figure. Then the closest-pair intercluster distances $d_{C1}(P_1, P_2), d_{C1}(P_1, P_3)$ and $d_{C1}(P_2, P_3)$ are given by the broken line segments. Because the relation $d_{C1}(P_2, P_3) < d_{C1}(P_1, P_2) < d_{C1}(P_1, P_3)$ holds, clusters P_2 and P_3 are merged at the $(k+1)$-th level. Section 12.2.4 in Chapter 12 shows an actual application of the network closest-pair point clustering method to sports clubhouses in Shibuya ward in Tokyo.

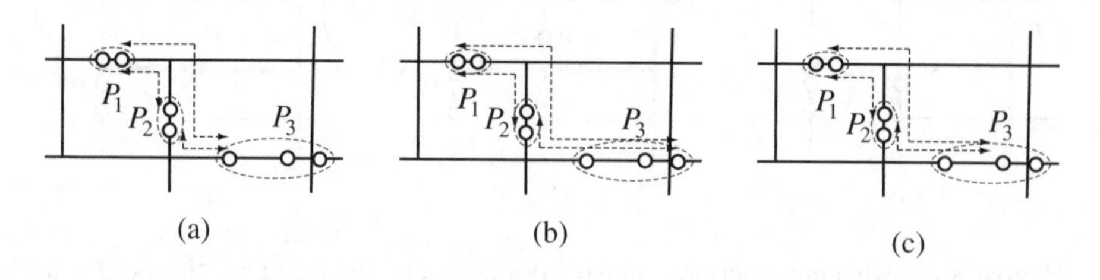

Figure 8.5 Intercluster distances: (a) closest-pair distances, (b) farthest-pair distances, (c) average-pair distances.

8.1.2.2 Network farthest-pair point clustering method

We next consider the intercluster distance, $d_{C2}(P_i, P_j)$, defined by the maximum value among the shortest-path distances between any point in P_i and any point in P_j, or mathematically:

$$d_{C2}(P_i, P_j) = \max\{d_S(p_{ik}, p_{jl}) | p_{ik} \in P_i, \; p_{jl} \in P_j\}. \tag{8.6}$$

We refer to this intercluster distance as the *farthest-pair (intercluster) distance*, and the hierarchical point cluster analysis conducted using this intercluster distance as the *network farthest-pair point clustering method*. This method is a network version of the *complete linkage method* proposed by Sorensen (1948) for ecological studies. In Figure 8.5b, the farthest-pair distances between clusters at the k-th level are indicated by the broken line segments, which show the relation $d_{C2}(P_1, P_2) < d_{C2}(P_2, P_3) < d_{C2}(P_1, P_3)$. Therefore, clusters P_1 and P_2 are merged at the $(k+1)$-th level. The comparison of panels (a) and (b) in Figure 8.5 depicts the difference between closest-pair and farthest-pair point clustering methods.

8.1.2.3 Network average-pair point clustering method

To formulate the third network point clustering method, we first define the *average intracluster distance* in cluster P_i (not intercluster), denoted by $\mathrm{AD}(P_i)$, as the average of the shortest-path distances between all pairs of points in P_i. Mathematically, $\mathrm{AD}(P_i)$ is written as:

$$\mathrm{AD}(P_i) = \frac{2}{|P_i|(|P_i| - 1)} \sum_{p_{ij}, p_{il} \in P_i} d_S(p_{ij}, p_{il}) \tag{8.7}$$

where $|P_i|$ indicates the number of points in P_i. With this intracluster distance, we define the *average-pair (intercluster) distance*, $d_{C3}(P_i, P_j)$, as:

$$d_{C3}(P_i, P_j) = \mathrm{AD}(P_i \cup P_j). \tag{8.8}$$

We term the point cluster analysis used with this distance the *network average-pair point clustering method*. This method corresponds to the *average-linkage method* in ordinary cluster analysis, with Sokal and Michener (1958) proposing the original technique. Figure 8.5c illustrates a simple example, where the broken line segments indicate the average-pair distances at the k-th level. Because the relation $d_{C3}(P_1, P_2) < d_{C3}(P_2, P_3) < d_{C3}(P_1, P_3)$ holds, clusters P_1 and P_2 are merged at the $(k+1)$-th level. Panels (a), (b), and (c) then show that the merged clusters become different according to the choice of intercluster distances. The above-cited texts provide general rules for how to choose an appropriate intercluster distance.

8.1.2.4 Network point clustering methods with other intercluster distances

In addition to the above intercluster distances, Sugihara, Okabe, and Satoh (2011) formulate network point-clustering methods with the *diameter distance*, *median distance*, *radius distance*, and *generalized radius distance*. However, as their computational requirements in terms of computational time are presently heavy, we do not discuss these methods here.

One might wonder why there is no network version of Ward's method (Ward, 1963; Ward and Hook, 1963; Whishart, 1969), one of the most frequently used methods in ordinary cluster analysis. The reason is as follows. Ward's method constructs clusters using the sum of the squared distances from the mean to the individual data points in a cluster. In network point clustering, the mean corresponds to the center of the points forming a cluster on a network. However, the center is not uniquely determined on a network. Therefore, it is not straightforward to formulate a network point clustering method corresponding to Ward's method.

8.2 Network clumping method

The second topic in this chapter is the network clumping method. We first describe the relationship between hierarchical point cluster analysis and the network clumping method. We then show how to use the network clumping method for a statistical test.

8.2.1 Relation to network point cluster analysis

As shown in the preceding section, the ordinary point cluster analysis deals with groups of clustered points on a plane. In contrast, the ordinary clumping method deals with groups of overlapped disks, termed *clumps*, on a plane. Therefore, these methods look outwardly different. However, a closer look reveals some similarity between them. Recalling the closest-pair point clustering method, we note that for a given intercluster distance d_{C}, a cluster is formed by the points whose nearest points are all less than or equal to d_{C}. This implies that the disks centered at those points with radius $d_{\mathrm{C}}/2$ overlap each other. That is, the clusters obtained by the intercluster distance d_{C} are the same as the clumps with radius $d_{\mathrm{C}}/2$. Therefore, we can regard the clumping method as the closest-pair point clustering method. However, the major concern with point cluster analysis is the detection of clusters in a point distribution, and we pay rather less attention to statistical tests with the resulting clusters. Conversely, the major concern of the clumping method is with a statistical test with respect to the observed number of clumps. For this reason, there is a separation of the literature on these two methods. The clumping method shown in this section is an extension of the clumping method for points on a plane to a network.

8.2.2 Statistical test with respect to the number of clumps

To formulate a statistical method for the network clumping method, we consider a set of points $P = \{p_1, \ldots, p_n\}$ on a network $\tilde{L}$, and let $\tilde{L}_{\mathrm{B}}(t|p_i)$ be the subnetwork

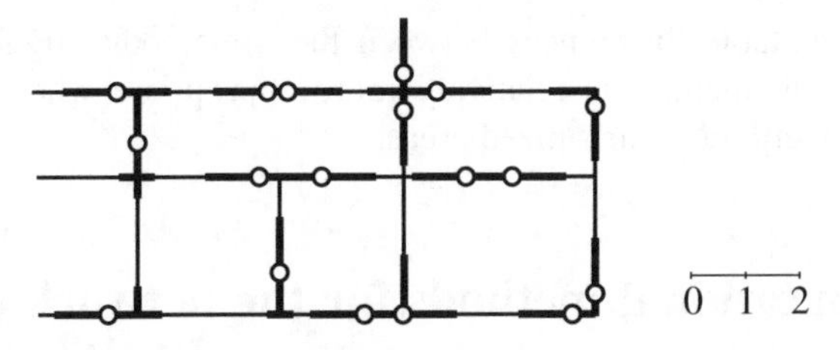

Figure 8.6 Clumps on a network (the bold line segments are buffer networks with width 1).

of $\tilde{L}$ in which the shortest-path distance from p_i is less than or equal to t. We call $\tilde{L}_{\mathrm{B}}(t|p_i)$ the *buffer network* of p_i with width t (measured by the shortest-path distance), corresponding to the *buffer zone* frequently used in the operation of geographic information systems (GIS). In Figure 8.6, the bold line segments indicate buffer networks. We then define a *clump* on a network as the set of buffer networks that overlap each other and the *size* of the clump as the number of buffer networks forming the clump. The composition of clumps on a network is described in terms of the number of clumps of size k, denoted by $n_{\mathrm{c}}(k|t)$, $k = 1, \ldots, n$. For example, in Figure 8.6, $n_{\mathrm{c}}(1|1) = 3$, $n_{\mathrm{c}}(2|1) = 6$, and $n_{\mathrm{c}}(3|1) = 1$.

Having observed these values, the reader may well consider that clumps of size two are significantly many. Here, the network clumping method tests whether an observed value of $n_{\mathrm{c}}(k|t)$ is significantly large or small, based on the distribution of a random value of $n_{\mathrm{c}}(k|t)$ that is generated by the homogeneous binomial point process of $p_1, \ldots, p_n$ on a network, that is, the points follow completely spatial randomness (CSR) (Section 2.4.2 of Chapter 2). To be explicit, for a given t, let $F_k(n_{\mathrm{c}})$ be the probability distribution function of the number $n_{\mathrm{c}}(k|t)$ of clumps of size k under the CSR hypothesis, and n_{c}^* and n_{c}^{**} be the numbers satisfying $F_k(n_{\mathrm{c}}^*) = \alpha$ and $F_k(n_{\mathrm{c}}^{**}) = 1 - \alpha$, respectively, where α is a statistical significance level (e.g., 0.05). Then, if the observed number of $n_{\mathrm{c}}(k|t)$ is smaller than n_{c}^* or larger than n_{c}^{**}, the CSR hypothesis is rejected with significance level α. If this result is obtained for $k = 2$ as in Figure 8.6, it is likely that points tend to form clumps of paired points.

To carry out this statistical test, we require the probability distribution function of $F_k(n_{\mathrm{c}})$. In the case of disks distributed on an unbounded plane, we obtain the expected value analytically using the theory of integral geometry (Mack,1948, 1954 Armitage, 1949; Kendall and Moran, 1963; Santalo, 1976). However, on a bounded network, the expected value is difficult to obtain analytically because the form of a buffer network varies from point to point on a network (Figure 8.6). Therefore, in practice, Monte Carlo simulation is employed (Shiode and Shiode, 2009a).

The ordinary clumping method fixes the radius of a disk. As an alternative, Okabe and Funamoto (2000) proposed a clumping method with a variable radius, which is equivalent to the closest-pair point clustering method, although the latter does not perform a statistical test, as in the former. Shiode and Okabe (2004), and Shiode and Shiode (2009a) extended this clumping method to a network and applied it to the study of clumps of restaurants in downtown Tokyo. Applying the planar and network clumping methods to the same data set, Shiode and Shiode (2009a) revealed some notable differences between these methods, suggesting that the network clumping method is relatively more appropriate when analyzing the distribution of facilities in urbanized areas.

8.3 Computational methods for the network point cluster analysis and clumping method

Having formulated the network hierarchical point cluster analysis and the network clumping method in the preceding two sections, in this section we develop their computational methods. We first consider a general computational framework for point cluster analysis and discuss time complexity in general terms. We then describe the computational methods for individual clustering methods characterized by specific intercluster distances. Finally, we present the counting procedure for the clumping method.

8.3.1 General computational framework

Let $N = (V, L)$ be a network and $P = \{p_1, \ldots, p_n\} \subset V$ be the set of points for which we wish to construct a hierarchical cluster structure. In this subsection, we formulate a general algorithm for hierarchical point clustering given that we choose an intercluster distance d_C. First, we formally describe the algorithm and next, its meaning.

Algorithm 8.1 (Hierarchical structure of clusters)

Input: network $N = (V, L)$, point set $P = \{p_1, p_2, \ldots, p_n\} \subset V$, and an intercluster distance d_C.
Output: hierarchical structure $(C_0, C_1, \ldots, C_{n-1})$ of clusters.
Procedure:
1. Initialization
   1.1. set $C_0 \leftarrow \{\{p_1\}, \{p_2\}, \ldots, \{p_n\}\}$.
   1.2. compute the intercluster distances $d_C(X, Y)$ for all pairs $X, Y \in C_0$.
   1.3. $i \leftarrow 0$
2. Main processing
   2.1. choose the pair X, Y of clusters from C_i such that
      (i) X and Y are adjacent,
      (ii) $d_C(X, Y)$ attains the minimum among all adjacent pairs of clusters.
   2.2. $C_{i+1} \leftarrow (C_i \backslash \{X, Y\}) \cup \{X \cup Y\}$
   2.3. compute the intercluster distances $d_C(X \cup Y, Z)$ for all $Z \in C_{i+1} \backslash \{X \cup Y\}$.
   2.4. $i \leftarrow i + 1$.
3. If $i = n - 1$ stop. Otherwise go to Step 2. □

Step 1.1 is the initialization stage, where we set an initial set C_0 of clusters. Step 1.2 computes the intercluster distances for all pairs of clusters. The main stage of the processing is Step 2, where Step 2.1 chooses the closest adjacent pair of clusters, Step 2.2 merges these clusters, and Step 2.3 computes the intercluster distances between the new cluster $X \cup Y$ and the remaining clusters. Step 3 is the iteration process in which we repeat the above procedure until all of the points form a single cluster.

To evaluate the general time complexity of the above algorithm, we assume for simplicity that $|V| = O(n)$ and $|L| = O(n)$ (for a set X, $|X|$ indicates the number of elements in the set), where n is the number of points in P. This assumption implies that most of the nodes belong to P and the network is planar. We start with n clusters, each of which comprises a single point. This initialization (Step 1.1) requires $O(n)$ time. We then compute the intercluster distances between all pairs of mutually adjacent clusters (Step 1.2).

It should be noted that even though the network is planar, the number of adjacent pairs of clusters can be as large as $O(n^2)$. To demonstrate this, we construct Figure 8.7 where the white circles represent points in P and the black circles represent other nodes, i.e., nodes in $V \backslash P$. At the initial level, each white node comprises one cluster, and a path consisting of only the black intermediate nodes connects any pair of clusters, implying that they are adjacent. Hence, the number of adjacent pairs of clusters is $O(n^2)$. Suppose that in the process of hierarchical clustering, small clusters successively merge with the larger clusters on their left, as shown in Figure 8.7. Suppose that the subset of points enclosed by the broken lines forms a cluster at some level of processing. All pairs of clusters are then also adjacent and, hence, the number of adjacent pairs of clusters is the second power of the number of current clusters.

This observation is key to evaluating the time complexity of general hierarchical clustering. Let $t(n)$ be the upper bound of the time necessary to compute the interluster distance between two clusters. At the initial level, we compute the intercluster or equivalently the shortest-path distances between all pairs of initial clusters (Step 1.2), which requires $O(n^2 t(n))$ time. In each step of merging two clusters (Step 2.2), we compute the intercluster distances from each new cluster to all adjacent previous clusters (Step 2.3). Because the number of such pairs of

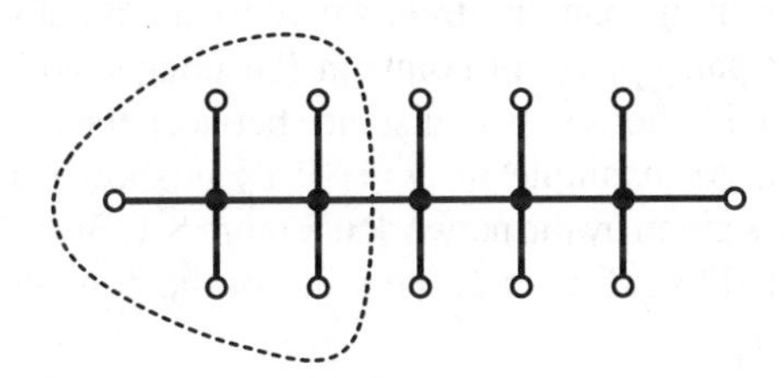

Figure 8.7 An example showing that adjacent pairs of clusters can be as large as the second power of the number of clusters.

clusters is proportional to the number of clusters, we compute the intercluster distances between $(n-1)+(n-2)+\cdots+1=\mathrm{O}(n^2)$ pairs in total. Therefore, we require $\mathrm{O}(n^2 t(n))$ time to construct the hierarchical structure of clusters. Note that we obtain this time complexity without specifying the intercluster distances. When we specify the intercluster distances, we may come across more efficient computational methods. In fact, we illustrate such examples in the following subsections.

8.3.2 Computational methods for individual intercluster distances

In the previous subsection, we assumed that $|V|=\mathrm{O}(n)$ and $|L|=\mathrm{O}(n)$ in the general computational framework. Now we remove these assumptions and consider the time complexity of specific clustering methods with three typical intercluster distances. Because we consider planar networks, we can assume that $n_L=\mathrm{O}(n_V)$, where $n_V=|V|$ and $n_L=|L|$.

8.3.2.1 Computational methods for the network closest-pair point clustering method

We first specify the general intercluster distance in the preceding subsection as the closest-pair distance. We can compute the intercluter distances, or equivalently, the shortest-path distances from one node to all of the other nodes in Step 2.1, using Dijkstra's algorithm in $\mathrm{O}(n_L \log n_V)$ time (for details, see Section 3.4.1 in Chapter 3). Importantly, because $n_L=\mathrm{O}(n_V)$ holds for planar networks, Dijkstra's algorithm runs in $\mathrm{O}(n_V \log n_V)$ time. In the main process of clustering (Step 2), each cluster can be as large as $\mathrm{O}(n)$. Consequently, we repeat the Dijkstra algorithm $\mathrm{O}(n)$ times in computing the shortest-pair distance between two clusters, and hence this task requires $t(n)=\mathrm{O}(n\, n_V \log n_V)$ time. If we straightforwardly substitute $t(n)=\mathrm{O}(n\, n_V \log n_V)$ into $\mathrm{O}(n^2 t(n))$ of the general case in Section 8.3.1, the time complexity of the network point clustering method with the closest-pair distance is given by $\mathrm{O}(n^2 t(n))=\mathrm{O}(n^3 n_V \log n_V)$. However, we can drastically reduce this time complexity by employing the following method with the minimum spanning tree.

To obtain the minimum spanning tree, we construct a network, $\bar{N}=(P,\bar{L})$, in such a way that each pair (p_i,p_j) of points in P is connected by a link in $\bar{L}$ whose length coincides with the shortest-path distance between these two points on a given network $N=(V,L)$. An example of $\bar{N}=(P,\bar{L})$ is shown in Figure 8.8 where network $N=(V,L)$ is given by the network in Figure 8.1. Because any pair of nodes in P on $\bar{N}$ is connected by a link in $\bar{L}$, the order of the number of elements in $\bar{L}$ is $|\bar{L}|=\mathrm{O}(n^2)$.

Given $\bar{N}=(P,\bar{L})$, we then construct the minimum spanning tree defined in Chapter 3. The *minimum spanning tree* for $\bar{N}=(P,\bar{L})$ is a subnetwork, $\hat{N}_{n-1}=(P,\hat{L}_{n-1})$, satisfying that $\hat{L}_{n-1}\subset\bar{L}$, the links in $\hat{L}_{n-1}$ connect all of the nodes in P by the paths consisting of the links of $\hat{L}_{n-1}$, and the total length of the

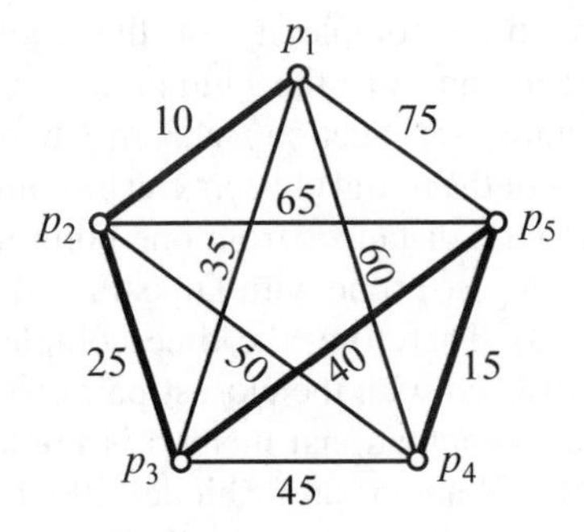

Figure 8.8 Network $\bar{N} = (P, \bar{L})$ for network $N = (V, L)$ shown in Figure 8.1 (the numbers alongside the links indicate the shortest-path distances between the nodes; the bold links form the minimum spanning tree).

links in $\hat{L}_{n-1}$ attains the minimum value among the lengths of all connected subnetworks of the form $N' = (P, L')$, $L' \subset L$.

To construct the minimum spanning tree $\hat{N}_{n-1} = (P, \hat{L}_{n-1})$, we start with the set of nodes P and an empty set of links denoted by $\hat{L}_0$, which form a subnetwork, $\hat{N}_0 = (P, \hat{L}_0)$, of $\bar{N}$. Network $\hat{N}_0 = (P, \hat{L}_0)$ consists of n isolated components, each of which consists of a single point in P. For example, in the case of $\bar{N}$ in Figure 8.8, the components are $\{p_1\}, \{p_2\}, \{p_3\}, \{p_4\}, \{p_5\}$, and $\hat{N}_0 = (\{p_1, p_2, p_3, p_4, p_5\}, \{\emptyset\})$. We then choose the link l with the minimum length among the links in $\bar{L}$, delete l from $\bar{L}$, and set $\tilde{L}_1 \leftarrow \tilde{L}_0 \cup \{l\}$. The resulting subnetwork is denoted $\tilde{N}_1 = (P, \tilde{L}_1)$, which consists of $n - 1$ components. In the example in Figure 8.8, because the minimum length among the links in $\bar{L}$ is given by the link connecting p_1 and p_2 (denoted by (p_1, p_2)), the components are $\{p_1, p_2\}, \{p_3\}, \{p_4\}, \{p_5\}$, and $\tilde{N}_1 = (\{p_1, p_2, p_3, p_4, p_5\}, \{(p_1, p_2)\})$. We proceed to a similar procedure in an iterative manner. That is, in the i-th step, we choose the link l with the minimum length from $\bar{L}$ that connects two different components of $\hat{N}_{i-1} = (P, \hat{L}_{i-1})$, delete l from $\bar{L}$, and set $\hat{L}_i \leftarrow \hat{L}_{i-1} \cup \{l\}$; the result is the i-th subnetwork, denoted by $\hat{N}_i = (P, \hat{L}_i)$. We repeat this procedure $n - 1$ times. Eventually, we obtain the subnetwork $\hat{N}_{n-1} = (P, \hat{L}_{n-1})$ with $n - 1$ links and one component in which all points are connected through the $n - 1$ links. Network $\hat{N}_{n-1} = (P, \hat{L}_{n-1})$ is the minimum spanning tree of $\bar{N}$. The bold line segments in Figure 8.8 provide an example.

Note that this construction procedure coincides with the successive constructions of the hierarchical structure of the clusters in Section 8.1.1. Indeed, the i-th-level cluster set corresponds to the set of connected components of the i-th subnetwork $\hat{N}_i = (P, \hat{L}_i)$ (observe the correspondence between the dendrogram in Figure 8.2 and $\hat{N}_i = (P, \hat{L}_i)$, $i = 1, \ldots, 4$, or $\{(p_1, p_2)\}, \{(p_1, p_2), (p_4, p_5)\}, \{(p_1, p_2), (p_4, p_5), (p_2, p_3)\}, \{(p_1, p_2), (p_4, p_5), (p_2, p_3), (p_3, p_5)\}$ in Figure 8.8). Hence, the hierarchical structure of clusters constructed by the closest-pair distance can be realized by constructing the minimum spanning tree of $\bar{N}$.

We now consider the time complexity of the above procedure. Because $|\bar{L}| = \mathrm{O}(n^2)$, the construction of the minimum spanning tree requires $\mathrm{O}(n^2 \log n)$ time (for details, see Section 3.4.3 in Chapter 3). The intercluster distances between all pairs of the initial clusters can be computed in $\mathrm{O}(n\, n_V \log n_V)$ time because the shortest-path distances from one point to all of the other points can be computed in $\mathrm{O}(n_V \log n_V)$ time with Dijkstra's algorithm (for details, see Section 3.4.1 in Chapter 3). Therefore, the time complexity of the hierarchical structure of clusters constructed with the closest-pair distance is $\mathrm{O}\,(n\, n_V \log n_V + n^2 \log n)$ time. The above computational method is implemented in the SANET software package (Okabe, Okunuki, and Shiode, 2006a, 2006b; Okabe, Satoh, and Sugihara, 2009), with an illustrative application of this software given in Section 12.2.4 of the last chapter, Chapter 12.

8.3.2.2 Computational methods for the network farthest-pair point clustering method

Secondly, we specify the general intercluster distance as the farthest-pair distance. For two clusters X and Y ($X \subset P$, $Y \subset P$, $X \cap Y = \varnothing$), the farthest-pair distance can be obtained by first computing the shortest-path distance between all pairs of points, one in X and the other in Y, and then taking their maximum. Each cluster can be as large as $\mathrm{O}(n)$, and the shortest-path distances from one point in X to all of the other points in Y can be computed in $\mathrm{O}(n_V \log n_V)$ time. Therefore, the farthest-pair distance between two clusters can be computed in $t(n) = \mathrm{O}(n\, n_V \log n_V)$ time. If we straightforwardly substitute $t(n) = \mathrm{O}(n\, n_V \log n_V)$ into $\mathrm{O}(n^2 t(n))$ of the general case in Section 8.3.1, the time complexity of the clustering method with the farthest-pair distance is given by $\mathrm{O}(n^2 t(n)) = \mathrm{O}(n^3 n_V \log n_V)$. We should note that we base this evaluation of time complexity on the assumption that the computation of the intercluster distances between clusters is mutually independent. However, we could compute the intercluster distances between larger clusters more efficiently if we were to use the intercluster distances between smaller clusters. This inspiration leads to the following computational method.

As for clustering with the shortest-path distance, we start with clusters, each containing exactly one point of P. Therefore, the intercluster or equivalently shortest-path distances between all pairs of initial clusters are computed in $\mathrm{O}(n\, n_V \log n_V)$ time. In the main stage of processing Algorithm 8.1, we merge clusters X and Y into $X \cup Y$ in Step 2.2, and then compute the farthest-pair distances d_{C2} between the new cluster $X \cup Y$ and the remaining older clusters Z. We can do this using:

$$d_{C2}(X \cup Y, Z) = \max\{d_{C2}(X, Z), d_{C2}(Y, Z)\}, \tag{8.9}$$

because the farthest pair between $X \cup Y$ and Z is either the farthest pair between X and Z or that between Y and Z. Therefore, we can compute the farthest-pair distance in constant time using equation (8.9).

Now consider the computational time of the main processing of Algorithm 8.1. In Step 2.1, we choose the pair of clusters that attains the minimum distance. This can be in $O(\log n)$ time if we use a standard data structure, as in a 2–3 tree and AVL tree (for details, see, e.g., Aho, Hopcroft, and Ullman (1974)). In Step 2.2, we merge these two clusters, and this can be in constant time. In Step 2.3, we compute the intercluster distances between a new cluster and all of the remaining clusters. This computation can be in $O(n)$ time because we can compute each intercluster distance in constant time using equation (8.9). We repeat these steps $n-1$ times in Step 2 and Step 3 and, hence, the main processing requires $O(n(n+\log n)) = O(n^2)$ time. Thus, the total time complexity of Algorithm 8.1 is $O(n\, n_V \log n_V + n^2) = O(n\, n_V \log n_V)$ because $n \leq n_V$. If most of the nodes of the network belong to P, i.e., if $n_V = O(n)$ (which may happen in many cases), the time complexity reduces to $O(n^2 \log n)$.

8.3.2.3 Computational methods for the network average-pair point clustering method

We now specify the general intercluster distance as the average-pair distance. The computation of the average-pair distance between two clusters can be in $t(n) = O(n\, n_V \log n_V)$ time because each cluster can be as large as $O(n)$, and we require computing the shortest-path distances between all pairs of points in one cluster and those in the other. Hence, the direct substitution of $t(n) = O(n\, n_V \log n_V)$ into $O(n^2 t(n))$ of the general framework presented in Section 8.3.1 gives the time complexity of the network average-pair point clustering method as $O(n^2 t(n)) = O(n^3 n_V \log n_V)$. However, we can decrease this time complexity in the following way.

At the initial level, the clusters are all one-point sets and, hence, the average-pair distances are the same as the shortest-path distances. Therefore, we can execute Step 1 in Algorithm 8.1 in $O(n\, n_V \log n_V)$ time. Next, consider the main processing of Algorithm 8.1. Suppose that we merge two clusters X and Y into $X \cup Y$. Then, the average-pair distance d_{C3} satisfies:

$$d_{C3}(X \cup Y, Z) = \frac{2}{|X \cup Y \cup Z|(|X \cup Y \cup Z| - 1)} \times \left\{ \frac{|X \cup Y|(|X \cup Y| - 1)}{2} d_{C3}(X, Y) + \frac{|Z|(|Z| - 1)}{2} \mathrm{AD}(Z) + \frac{|X \cup Z|(|X \cup Z| - 1)}{2} d_{C3}(X, Z) + \frac{|Y \cup Z|(|Y \cup Z| - 1)}{2} d_{C3}(Y, Z) \right\}, \quad (8.10)$$

where AD(Z) is the average distance between two points in Z, as defined by equation (8.7). Equation (8.10) holds because the clusters X, Y, and Z are mutually disjoint, and the average for the larger set is the weighted average of the smaller sets, where the weights are the sizes of the smaller sets. Because $d_{C3}(X, Y)$, AD(Z), $d_{C3}(X, Z)$, and $d_{C3}(Y, Z)$ were already computed when Steps 2.1–2.4 were done for X and Y, $d_{C3}(X \cup Y, Z)$ can be computed in constant time. Therefore, using the same logic as for the farthest-pair distance above, we can execute Algorithm 8.1 in $O(n\, n_V \log n_V)$ time for the average-pair distance.

8.3.3 Computational aspects of the network clumping method

In the network clumping method, we need to compute $n_c(k|t)$, i.e., the number of clumps of the size k with respect to $k = 1, 2, \ldots, n$ for a fixed buffer width t. For this purpose, we can use the result of the network closest-pair point clustering method in Section 8.3.2.1.

Let $(C_0, C_1, \ldots, C_{n-1})$ be the sequence of cluster sets obtained from the network closest-pair point clustering method, t_i $(i = 1, \ldots, n-1)$ be the closest-pair distance between the two clusters that are merged when C_i is generated from C_{i-1}, and $t_n = \infty$ for notational convention. For each given buffer width t, we find i such that $t_i \leq 2t < t_{i+1}$. Then, C_i gives the composition of clump sizes, i.e., each cluster $X \in C_i$ has the clump size $|X|$. Therefore, the value of $n_c(k|t)$, $k = 1, 2, \ldots, n$ is computed by the next procedure.

1. Find cluster set C_i such that $t_i \leq 2t < t_{i+1}$.
2. Set $n_c(k|t) \leftarrow 0$ for $k = 1, 2, \ldots, n$.
3. For each $X \in C_i$ do
   3.1. $k \leftarrow |X|$
   3.2. $n_c(k|t) \leftarrow n_c(k|t) + 1$

Using this procedure, we can easily compute $n_c\,(k|t), k = 1, \ldots, n$ for any $0 < t$. Once t is given, we can obtain the distribution of $n_c(k|t)$ using Monte Carlo simulation, with which we can use to test the CSR hypothesis, as shown in Section 8.2.2. Shiode and Shiode (2009a) provide an actual application.

9

Network point density estimation methods

The *density* of events over space is one of the key concepts in spatial analysis. In fact, and as reviewed in Chapter 1, traffic engineers, sociologists and ecologists, respectively, investigate where there are high densities of car crashes (*black zones*), street crime (*hot spots*), and roadkills. In all of these studies, frequently asked questions include Q6 in Chapter 1: How can we estimate the density of traffic accidents, roadkills, and street crime incidence; and how can we identify locations where the densities of these occurrences are high, including black zones and hot spots? More generally,

Q6′: For a given set of points on a network, how can we estimate the density of points along the network and detect significantly high-density areas on the network?

This chapter presents statistical and computational methods for responding to these kinds of questions.

To obtain statistical responses to these questions, we need to estimate the density function of points on a network. To explicitly discuss this estimation, let $N = N(V, L)$ be a network consisting of a node set $V = \{v_1, \ldots, v_{n_V}\}$ and a link set $L = \{l_1, \ldots, l_{n_L}\}$, and denote the union of the links $l_1, \ldots, l_{n_L}$ by $\tilde{L} = \bigcup_{i=1}^{n_L} l_i$, i.e., a set of points (including the nodes of V) forming the network. The problem is then how to estimate a probability density function $f(p)$, $p \in \tilde{L}$ from the observed sample points $p_1, \ldots, p_n$ on $\tilde{L}$. The methods for estimation differ

Spatial Analysis along Networks: Statistical and Computational Methods, First Edition.
Atsuyuki Okabe and Kokichi Sugihara.

depending on our knowledge of $f(p)$. For instance, if we know that the density of beauty parlors tends to exponentially decrease from the nearest subway station, or mathematically, $f(p) = \alpha \exp(-\beta d_S(p, p^*))$, where $d_S(p, p^*)$ is the shortest-path distance from a point p on $\tilde{L}$ to the nearest subway station p^*, and α and β are unknown parameters, then the estimation of $f(p)$ is equivalent to the estimation of the parameter values for α and β in $f(p)$. This type of estimation is termed *parametric estimation*. In contrast, if we know nothing about $f(p)$ except the location of the observed points, $p_1, \ldots, p_n$, the estimation of $f(p)$ without a predetermined function class is known as *nonparametric estimation*. In research processes, nonparametric estimation often precedes parametric estimation because it is an exploratory method for finding some probable parametric density function.

In this chapter, we focus on nonparametric estimation of point density on a network. The chapter consists of three sections. The first section deals with a long-established but still frequently used estimation technique, namely the *histogram*. When discussing the statistical properties of this method and its alternatives, we make the following assumption throughout.

Assumption 9.1

Sample points $p_1, \ldots, p_n$ are independently and identically generated on $\tilde{L}$ according to a continuous probability density function $f(p)$, $p \in \tilde{L}$.

Under this assumption, we prove that estimation by histogram produces bias. To avoid this shortcoming, the second section provides a class of unbiased estimators, the so-called *kernel density estimators*. Much later, Chapter 12 details an actual application of the chosen density estimation method, the equal-split continuous kernel density estimation method, as developed in Section 9.2.3. The final section describes the approaches used for computing these methods.

9.1 Network histograms

It is noteworthy that Snow (1855) was not only one of the earliest researchers employing the network Voronoi diagram (recall Section 1.2.1 in Chapter 1), but also the 'histogram on a network' by showing the number of deaths by cholera along streets using the heights of bars (Figure 1.9 in Chapter 1). In this section, we first explicitly reformulate Snow's 'histogram on a network' and then illustrate its variants.

9.1.1 Network cell histograms

In traffic accident and roadkill analyses, as reviewed in Sections 1.2.2 and 1.2.3 in Chapter 1, the most frequently used estimation method is to count the number n_i of accidents (generally points) in each segmented network L_i, referred to as a *network cell*, and divide the number by the length $|L_i|$ of the network cell. In mathematical terms, an estimator $\hat{f}(p)$ of a probability density function $f(p)$ at p is given by:

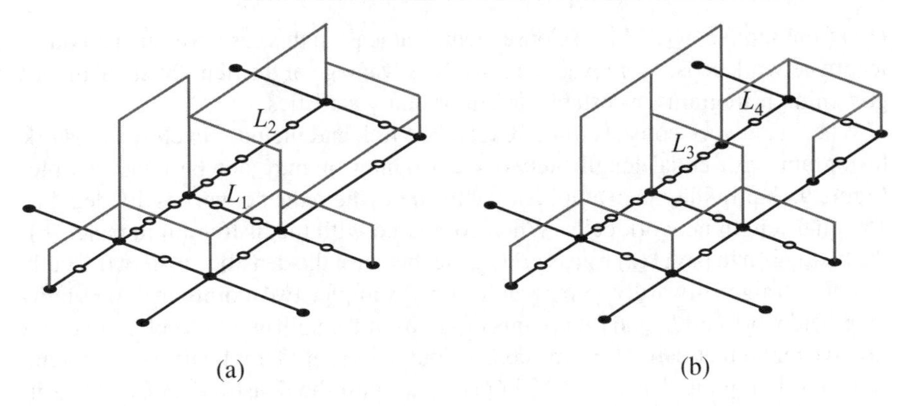

Figure 9.1 Network cell histograms of (a) unequal-length network cells, (b) equal-length network cells (the white circles are observed sample points and the black circles are the boundary points between network cells).

$$\hat{f}(p) = \frac{n_i}{n|L_i|},\ p \in L_i,\ i = 1, \ldots, m. \tag{9.1}$$

Note that $L_i \cap L_j$ includes at most finite number of points for $i \neq j$, $\tilde{L} = \bigcup_{i=1}^{m} L_i$, $n = \sum_{i=1}^{m} n_i$, and every network cell L_i is *connected* in the sense that there exists a path between two arbitrary points in L_i which is embedded in L_i. We refer to the graphic representation of the estimated function $\hat{f}(p)$ as a *network cell histogram*. Figure 9.1 provides an example, where the white and black circles, respectively, indicate the observed sample points and boundary points between the network cells.

The network histogram is an extension of the ordinary histogram in statistics used for estimating a univariate probability density function $f(x)$ with a finite number of samples, $x_1, \ldots, x_n$. The x scale is divided into equal intervals $(i-1)h \leq x < ih$, $i = 1, \ldots, m$, termed *bins*, where h is referred to as the *bin width*. An estimator for $f(x)$ is given by $\hat{f}(x) = n_i / \sum_{j=1}^{m} n_j$, $(i-1)h \leq x < ih$, $i = 1, \ldots, m$, where n_i is the number of samples in the i-th bin. While the term *histogram* was coined by Pearson (1895, p. 399), the idea itself dates back to at least the seventeenth century. The statistical theory of histograms is described in depth in texts about nonparametric density estimation, including Tapia and Thomson (1978, Chapter 1), Silverman (1986, Section 2.2), Härdle (1990, Chapter 1), Scott (1992, Chapter 3), Wand and Jones (1995, Section 1.2), and Klemela (2009, Chapter 17).

Histograms in ordinary statistics mostly adopt equal bin width (except for *adaptive histograms*, see Scott (1992, Section 3.2.8)) to make bin counts statistically comparable. In contrast, the network histogram defined by equation (9.1) may or may not adopt equal-length network cells ($|L_i| = |\tilde{L}|/m$). When a network is a very simple network, i.e., a (curved) line segment, such as an interstate expressway segment between interchanges, the density estimation of traffic accidents along the expressway is equivalent to estimating a univariate probability density function

$f(x)$ from samples $x_1, \ldots, x_n$. Consequently, at least in this case, we can use equal-length network cells. Moreover, we can take advantage of the rich statistical theory governing histograms, as established in ordinary statistics.

When a network consists of more than one link and includes cycles, a network histogram with equal-length network cells may or may not be constructible. Figure 9.1b provides an example of the former, where the network is divided into 15 equal-length network cells. When compared with the histogram in panel (b), the histogram in panel (a) is less preferable, because the densities in network cells are not straightforwardly comparable. For example, two points in a one-unit-length network cell L_1 and four points in a two-unit-length network cell L_2 in panel (a) give mathematically the same density, but their spatial implications are not the same (compare the density on L_2 in panel (a) with the densities on L_3 and L_4 in panel (b)). Although equal-length network cells are preferable, such tessellation is not always constructible for general networks. In fact, we prove in Section 9.3.1 that for a given number of network cells, networks exist that we cannot equally divide. Although Shiode and Okabe (2004), Shiode (2008), and Furuta *et al.* (2008) proposed computational methods for dividing a network into equal-length network cells, which are discussed in Section 9.3, their methods are not a guarantee of success.

9.1.2 Network Voronoi cell histograms

Standard histograms in ordinary statistics are, as noted above, histograms with equal bin width. An alternative is histograms with equal bin counts. In the network context, this implies that every network cell has the same number of sample points. However, such tessellation is generally difficult to construct except where each network cell contains a single point. Further, there are infinitely many ways to divide a network into network cells. Among many possibilities, a natural and uniquely determined way is to use 'natural neighborhoods' (proposed by Sibson (1981)), namely *network Voronoi cells*, as defined in Chapter 4.

We obtain network Voronoi cells from the network Voronoi diagram $\mathcal{V}(P) = \{V_{\mathrm{VD}}(p_1), \ldots, V_{\mathrm{VD}}(p_1)\}$ defined by equation (4.2) in Chapter 4. By definition, each Voronoi cell $V_{\mathrm{VD}}(p_i)$ includes exactly one point of P. Therefore, the density of points in $V_{\mathrm{VD}}(p_i)$ is estimated by $1/|V_{\mathrm{VD}}(p_i)|$ (where $|V_{\mathrm{VD}}(p_i)|$ is the total length of subnetwork $V_{\mathrm{VD}}(p_i)$) and an estimator $\hat{f}(p)$ for $f(x)$ is written as:

$$\hat{f}(p) = \frac{1}{n|V_{\mathrm{VD}}(p_i)|}, \; p \in V_{\mathrm{VD}}(p_i), \; i = 1, \ldots, n. \tag{9.2}$$

We refer to the graphic representation of this estimator as a *network Voronoi cell histogram*. Note that the estimator of a network Voronoi cell histogram is a special case of the estimator given by equation (9.1), where $L_i = V_{\mathrm{VD}}(p_i)$, $m = n$ and $n_i = 1$. This estimation method is a network version of the *Voronoi tessellation field estimator* (Schaap, 2007, Section 3.7), first suggested by Brown (1965) and Ord (1978).

Recalling that the ordinary Voronoi diagram is the first-nearest point Voronoi diagram, one may note that the network Voronoi cell histogram can be extended to the k-th nearest point Voronoi histogram (Section 4.2.3 in Chapter 4), where each network cell contains exactly k points of P (after allowing the network cells to overlap). In reality, there is already such an extension proposed in the literature on univariate probability density estimation (Silverman, 1986, Section 2.5). In addition, the literature also includes histogram-based methods, including frequency polygons, averaged shifted histograms and weighted averaging of rounded points (see Scott (1992, Chapters 4 and 5) and Härdle (1990, Section 1.4)). A closely related method is a network version of scan statistics developed by Shiode (2011) (for scan statistics, see Kulldorff (1997)).

In the above two subsections, we presented two types of network cell histogram. In both cases, the estimated function $\hat{f}(p)$ is a step function. Therefore, under Assumption 9.1 (i.e., $f(p)$ is assumed to be a continuous function), the discontinuous function $\hat{f}(p)$ does not coincide with the true continuous function $f(p)$. This implies that the estimator $\hat{f}(p)$ is biased. One may then consider that Assumption 9.1 does not always hold in the real world and that the true function $f(p)$ may be a step function. However, even in such a case, we cannot match the histogram breakpoints to those where $f(p)$ is discontinuous because we do not know the discontinuous points of $f(p)$ in advance. Hence, the estimator is biased except for coincidence.

9.1.3 Network cell-count method

In Section 9.1.1, we referred to equal-length network cells as ideal network cells for network histograms. In this regard, we briefly describe in this subsection the cell-count method on networks or *network cell-count method*, because the ideal network histograms and the network cell-count method use common computational techniques, i.e., those for constructing the equal-length network cells developed in Section 9.3.1. Note that the network cell-count method aims not to estimate the density of points but rather to test the complete spatial randomness (CSR) hypothesis (Section 2.4.2 in Chapter 2) with respect to the distribution of densities of network cells.

The network cell-count method is a network version of the cell-count method on a plane, referred to in the literature as the *quadrat method* or *quadrat-count method* (Haggett *et al.*, 1965; Chorley and Haggett, 1967; Diggle, 2003). According to Hopkins and Skellam (1954), the term *quadrat* originates from the Latin word *quadratum* (i.e., square) and was first used in Clements (1905). We sometimes use the term *planar cell-count method* rather than the quadrat method. A pioneering study with the quadrat method was by Matui (1932). Greg-Smith (1952) provides a review of early studies using the method. In classic spatial analysis, the quadrat method and the nearest-neighbor distance method (Chapter 5) are two of the most important methods and many textbooks on spatial analysis deal with the quadrat method at some length, for example, King (1969, Chapter 5), Pielou (1977, Chapter 9), Reply (1981, Chapter 6), and Krebs (1999, Chapter 4). Rogers (1974) is a text specializing in the quadrat method while Kolchin, Sevat'yanov, and Chistyakov (1978, Chapter 3) present theorems of the multinomial allocations upon which it rests. A simple extension of the cell-count method on a plane to a network is found in Pearson (1963), who applied it to points on a single line segment (the simplest network). One of the earliest applications of the network cell-count method on a complex network was made by Shiode and Okabe (2004), extended in Shiode (2008). These studies compared the planar cell-count method and the network cell-count method by applying the two methods to the same dataset and showed distinct differences between them.

To formulate the network cell-count method, suppose that a network $\tilde{L}$ is tessellated into m equal-length network cells, $L_1, \ldots, L_m$ (i.e., $|L_i| = |\tilde{L}|/m$, $L_i \cap L_j$ includes at most finite number of points for $i \neq j$, and $\sum_{i=1}^{m} |L_i| = |\tilde{L}|$), and that n points of P are distributed over $L_1, \ldots, L_m$ (e.g., the white circles in Figure 9.2a). Let n_i be the number of points located in L_i. Because all network cells have the same length, the value of n_i also means the density of points in L_i, where the length $|L_i|$ is standardized by $|L_i| = 1$. If the points in P follow the homogeneous binomial point process (or the CSR hypothesis defined in Section 2.4.2 in Chapter 2), then n_i is a random variable following a binomial distribution with parameters n and $1/m$, denoted by $B(j|n,\ 1/m)$, indicating the probability that $n_i = j$.

Let $n_{(j)}$ be the number of equal-length network cells containing exactly j points under the CSR hypothesis. Then, the numbers $n_{(j)}$, $j = 0, \ldots, n$ indicate the distribution of the number of network cells having j points, or equivalently, the distribution of densities j of network cells with respect to $j = 0, \ldots, n$ (e.g., Figure 9.2b for the white circles in Figures 9.1 or 9.2a). Because n_i of each network cells follows $B(j|n,\ 1/m)$ and the total number of network cells is m, the expected value of $n_{(j)}$ is approximately given by $\mathrm{E}(n_{(j)}) = m\,B(j|n,\ 1/m)$ (note that n_i and n_j $(i \neq j)$ are not completely independent, because $\sum_{i=1}^{m} n_i = n$). If the stochastic

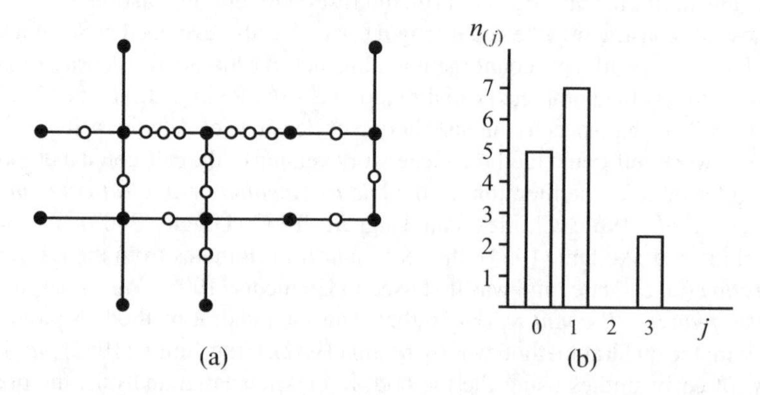

Figure 9.2 Network cell-count method: (a) the distribution of observed points (the white circles) on the equal-length network cells (the black circles indicate the boundary points), (b) the frequency distribution of the numbers $n_{(j)}$ of network cells having $j = 0, 1, 2, 3$ points.

point process of P is generated by the homogeneous binomial process, the observed distribution of $n_{(j)}$, $j = 0, \ldots, n$ may be close to the distribution of $\mathrm{E}(n_{(j)})$, $j = 0, \ldots, n$. Therefore, with these datasets, we can test the CSR hypothesis using the goodness-of-fit test (for the theory underlying this method, see Wilks (1960)). An actual application is provided by Shiode (2008, 2012).

9.2 Network kernel density estimation methods

Although network histograms are simple and easy to use in practice, they produce bias. To overcome this shortcoming, in this section we formulate a class of methods for unbiased estimation by extending the so-called the *kernel density estimation method* (*KDE method*), as established in nonparametric density estimation (see the texts cited in Section 9.1, Devroye and Gyorfi (1985), Devroye and Lugosi (2001), Eggermoni and LaRiccisa (2001)).

The seminal idea underlying the KDE method is found in Fix and Hodges (1951), with Rosenblatt (1956), Whittle (1958), Parzen (1962), Cencov (1962), Loftsgaarden and Quesenberry (1965), Schwartz (1967), and Epanechnikov (1969), among others, forming the first wave of application. To provide the intuition underlying the KDE method, Figure 9.3 depicts the simplest possible network, i.e., a (curved) line segment. We first construct 'molehills' centered at the sample points, as in Figure 9.3a. We then heap these hills vertically. As a result, we obtain a 'big mountain,' as in Figure 9.3b. The estimation method through this procedure is the KDE method. As an example, Flahaut, Mouchart, and Martin (2003) applied the KDE method to traffic accident analysis on an expressway (a line segment). However, the extension of their KDE method to general networks is not straightforward. One easy approach may be that the points on a network are regarded as those on a plane (because the network is embedded in the plane), and so we can apply the KDE method for points on a plane to these points. However, this extension produces bias (Okabe, Satoh, and Sugihara, 2009). Borruso (2003, 2005, 2008), Porta and Latora (2008), Porta *et al.* (2009), Downs and Horner (2007a, 2007b, 2008), Xie and Yan (2008), and Okabe, Satoh, and Sugihara (2009) have also

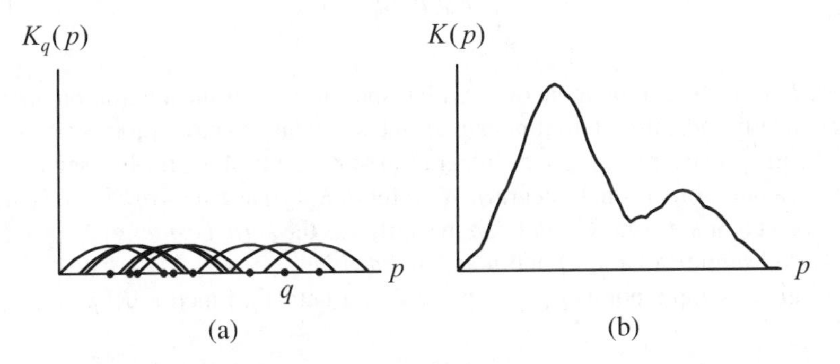

Figure 9.3 Kernel density estimation procedures on a line segment: (a) 'mole hills,' (b) a 'big mountain.'

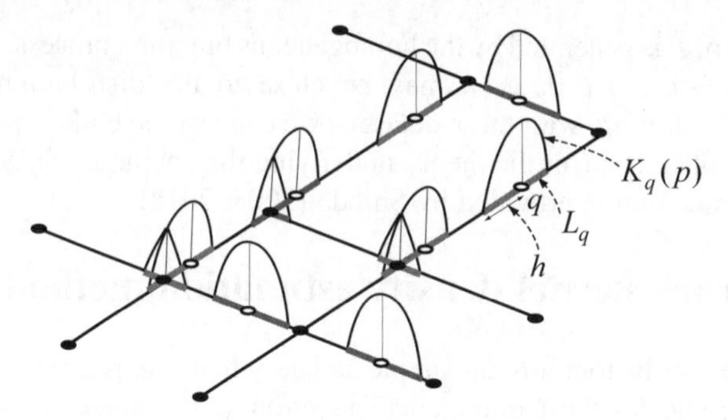

Figure 9.4 Kernel density functions on a network embedded in a plane (kernel supports indicated by the gray bold line segments).

undertaken successive attempts to formulate KDE methods for general networks. Using these studies, especially Okabe, Satoh, and Sugihara (2009), in the following two sections we discuss the general properties of kernel density functions on a network and present two classes of unbiased kernel density functions.

9.2.1 Network kernel density functions

For an arbitrary point q on a nondirected connected network $\tilde{L}$ embedded in a plane (Figure 9.4), we construct a subnetwork L_q of $\tilde{L}$ (the bold gray line segments in Figure 9.4) in such a way that the shortest-path distance between q and any point on L_q is less than or equal to h, referred to as the *buffer network* of q with width h (Section 8.2.2 in Chapter 8). For a given point q and an arbitrary point p on $\tilde{L}$, we then define a function, $K_q(p)$, satisfying:

$$K_q(p)\begin{cases} \geq 0 & \text{for} \quad p \in L_q, \\ = 0 & \text{for} \quad p \in \tilde{L}\backslash L_q, \end{cases} \tag{9.3}$$

$$\int_{p \in L_q} K_q(p)\,\mathrm{d}p = 1, \tag{9.4}$$

where $\tilde{L}\backslash L_q$ is the complement of L_q with respect to $\tilde{L}$, dp is the integration operator symbolically indicating an infinitesimal line segment around p on the network $\tilde{L}$, and the integral over $p \in L_q$ is the integration of $K_q(p)\mathrm{d}p$ along the line segments of L_q (i.e., a one-dimensional integral). We refer to $K_q(p)$ as a *network kernel density function* at q or *network KD function* in short, q as the *kernel center* of $K_q(p)$, L_q as the *kernel support* of $K_q(p)$ and h as a *bandwidth*.

For given sample points $p_1, \ldots, p_n$ on $\tilde{L}$, we define a function $K(p)$ as:

$$K(p) = \frac{1}{n}\sum_{i=1}^{n} K_{p_i}(p). \tag{9.5}$$

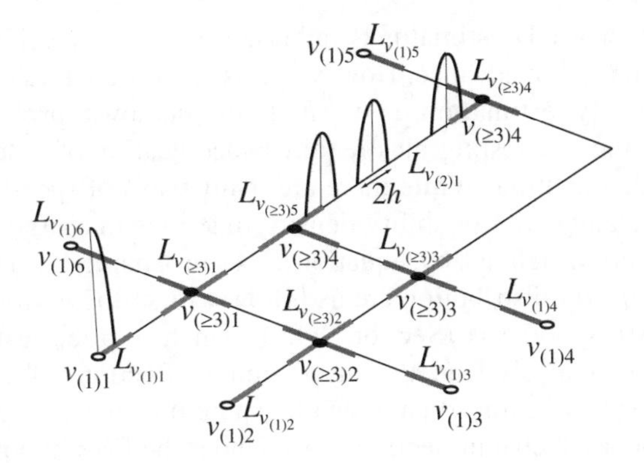

Figure 9.5 Decomposition of a network into $L_{(1)}$ (the gray straight line segments), $L_{(2)}$ (the hair straight line segments) and $L_{(\geq 3)}$(the gray T-shaped or cross-shaped line segments).

We term the value of $K(p)$ the *network kernel density estimator* for $f(p)$ at p, or the *network KD estimator*, and the method for estimating $f(p)$ with $K(p)$ as the *network kernel density estimation method* or *network KDE method.*

To discuss the properties of network KD estimators, let $V_{(1)}$ be the set of nodes of degree 1 (i.e., nodes to which one link is incident, for example, the white circles in Figure 9.5) and $V_{(\geq 3)}$ be the set of nodes of degree 3 or more (the black circles). Assume for ease of explanation that every link in L is longer than or equal to $4h$. Let $L_{v_{(1)i}}$ be the buffer network of the i-th node $v_{(1)i}$ in $V_{(1)}$ with width $2h$ (the bold gray line segments with the white circles in Figure 9.5), and $L_{(1)} = \{L_{v_{(1)i}} | v_{(1)i} \in V_{(1)}\}$; and let $L_{v_{(\geq 3)i}}$ be the buffer network of the i-th node $v_{(\geq 3)i}$ in $V_{(\geq 3)}$ with width $2h$, and $L_{(\geq 3)} = \{L_{v_{(\geq 3)i}} | v_{(\geq 3)i} \in V_{(\geq 3)}\}$ (the bold gray line segments with black circles in Figure 9.5). We delete the line segments in $L_{(1)}$ and $L_{(\geq 3)}$ from $\tilde{L}$. As a result, we obtain a set of simple line segments (every subnetwork has exactly two endpoints), denoted by $L_{(2)} = \{L_{v_{(2)i}} | v_{(1)i} \in V \backslash V_{(1)} \backslash V_{(\geq 3)}\}$ (the hairline segments in Figure 9.5).

Network KDE methods differ according to $L_{(1)}$, $L_{(2)}$ and $L_{(\geq 3)}$. The network KD estimation on simple line segments in $L_{(1)}$ and $L_{(2)}$ is equivalent to the KD estimation for univariate density estimation in ordinary statistics. Differences in the estimation of $L_{(1)}$ and $L_{(2)}$ then exist in the edge effects (Section 2.4.3 in Chapter 2). When a kernel center is on $L_{(1)i}$ of $L_{(1)}$, edge effects appear because the KD functions are always trimmed, for example, the KD function on $L_{v_{(1)1}}$ in Figure 9.5. The literature proposes several methods for adjusting the edge effects (see the texts cited earlier). In contrast, when a kernel center is on $L_{(2)i}$ of $L_{(2)}$, edge effects do not appear because its kernel support is included in $L_{v_{(1)i}}$ (e.g., the middle KD function on $L_{v_{(2)1}}$ in Figure 9.5) or is included in $L_{v_{(2)i}}$ extended by h (the right and left KD functions on $L_{v_{(2)1}}$ alongside the bold gray line segments included in $\tilde{L}$). The network KD estimation on $L_{(\geq 3)}$ is completely different from that for $L_{(1)}$ or $L_{(2)}$, and the analytical derivation appears very difficult. However, Okabe, Satoh, and Sugihara (2009) provide a few computational solutions, as shown in subsequent sections.

To test whether a KD estimator is unbiased, we must specify a probability density function $f(p)$ in advance. However, this sounds contradictory in that the purpose of density estimation is to find an unknown probability density function. Nevertheless, testing bias requires specification of a density function in theoretical examination. While there are many ways of specifying $f(p)$, one of the most fundamental probability density functions in spatial analysis is the uniform distribution defined by equation (2.4) in Chapter 2. The reasons for selecting the uniform distribution are as follows. First, the uniform distribution implies the CSR hypothesis (Section 2.4.2), which is often tested in the first phase of spatial analysis (see the discussion in Section 2.4.2). Second, all probability density functions on a network can be transformed into the uniform density function, as shown in Section 2.4.4. Under the CSR hypothesis, the bias of $K(p)$, denoted by $B(p)$, is given by the difference between the expected value of $K(p)$ and the true value of function $f(p)$ at p, or specifically, the uniform distribution, i.e.:

$$B(p) = \mathrm{E}[K(p)] - f(p) = \mathrm{E}[K(p)] - \frac{1}{|\tilde{L}|},\ p \in \tilde{L}, \tag{9.6}$$

where $\mathrm{E}[K(p)]$ is obtained from:

$$\mathrm{E}[K(p)] = \int_{q\in\tilde{L}} K_q(p)f(q)\mathrm{d}q = \frac{1}{|\tilde{L}|}\int_{q\in\tilde{L}} K_q(p)\mathrm{d}q. \tag{9.7}$$

To construct unbiased KD estimators, let $d_{\mathrm{S}}(q,p)$ be the shortest-path distance from q to p on $\tilde{L}$, and $k(d_{\mathrm{S}}(q,p))$ be a function, termed the *base kernel density function*, characterized by (for $p, q \in \tilde{L}$):

(i) $\int_{p\in\tilde{L}} k(\mathrm{d}_{\mathrm{S}}(q,p))\mathrm{d}p = 1$.

(ii) There exists $h > 0$ such that $k(d_{\mathrm{S}}(q,p)) = 0$ if $d_{\mathrm{S}}(q,p) \geq h$, and $k(d_{\mathrm{S}}(q,p)) > 0$ if $d_{\mathrm{S}}(q,p) < h$.

(iii) $k(d_{\mathrm{S}}(q,p))$ is nonincreasing with respect to $d_{\mathrm{S}}(q,p)$.

(iv) $k(d_{\mathrm{S}}(q,p))$ is continuous with respect to $d_{\mathrm{S}}(q,p)$.

In the following two subsections, we construct two unbiased KD estimators with a base KD function: the *equal-split discontinuous KD function* in Section 9.2.2 and the *equal-split continuous KD function* in Section 9.2.3. Note that the base KD function allows many forms as long as the above properties are satisfied. Consequently, both unbiased KD estimators are general in that they allow many forms.

9.2.2 Equal-split discontinuous kernel density functions

We define a network KD function $K_q(p)$ in two cases: Case 1 where the kernel center q of $K_q(p)$ does not coincide with a node in V, and Case 2 where it does. We assume for ease of explanation that the length of all cycles in a network $N(V, L)$ is greater than $2h$ (a similar result can be obtained without this assumption). In Case 1, we construct the buffer network of q with width h (i.e., kernel support), and consider the shortest path $s(q, b_i)$ from q to the boundary points b_i of the buffer network, $i = 1, 2, \ldots$ (the bold line segments in Figure 9.6). Suppose that nodes $v_{i1}, \ldots, v_{im-1}$ of V locate on the shortest path $s(q, b_i)$ from q to b_i in this order (e.g., v_{i1} and v_{i2} in Figure 9.6). We traverse the path from q to b_i visiting these nodes. For p on the first link $0 \le d_S(q,p) < d_S(q, v_{i1})$, we set $K_q(p) = k(d_S(q,p))$. When we reach v_{i1}, we split the link joining q and v_{i1} into $n_{i1} - 1$ links (e.g., $n_{i1} - 1 = 3$ in Figure 9.6). We then equally divide the value $k(d_S(q, v_{i1}))$ by $n_{i1} - 1$, where n_{i1} is the degree of node v_{i1}, and assign the divided amount to the endpoint of every link incident to v_{i1} with the exception of the traversed link, (q, v_{i1}). For p on the second link $d_S(q, v_{i1}) \le d_S(q,p) < d_S(q, v_{i2})$, we set $K_q(p) = k(d_S(q,p))/(n_{i1} - 1)$ (e.g., $k(d_S(q,p))/3$ in Figure 9.6). When we reach v_{i2}, we split the link joining v_{i1} and v_{i2} into $n_{i2} - 1$ links (e.g., $n_{i2} - 1 = 2$ in Figure 9.6). We then equally divide the value $k(d_S(q,p))/(n_{i1} - 1)$ by $n_{i2} - 1$ (e.g., $(k(d_S(q,p))/3)/2$ in Figure 9.6), where n_{i2} is the degree of node v_{i2}, and assign the divided amount to the endpoint of every link incident to v_{i2}, except for the traversed link, (v_{i1}, v_{i2}). We continue this procedure until we reach b_i. We write the KD estimator obtained through this procedure as:

$$K_q(p)=\begin{cases} \dfrac{k(d_S(q,p))}{(n_{i1}-1)(n_{i2}-1)\cdots(n_{ik-1}-1)} \text{ for } d_S(q, v_{ik-1}) \le d_S(q,p) < d_S(q, v_{ik}), \\ \hfill k = 1, \ldots, m, \\ 0 \text{ for } d_S(q,p) \ge h, \end{cases} \tag{9.8}$$

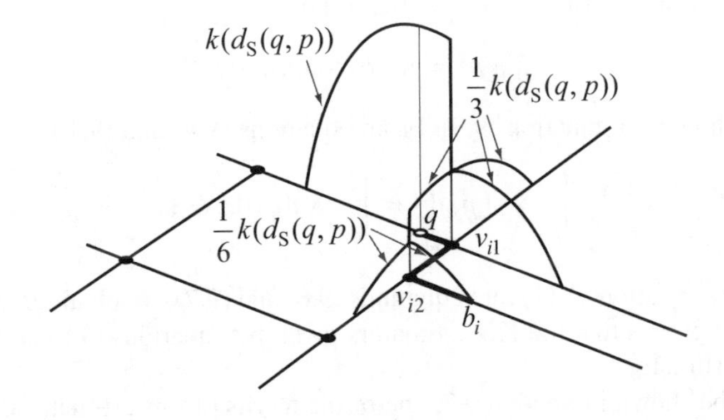

Figure 9.6 An equal-split discontinuous kernel density function centered at q, where $k(d_S(q,p))$ is a base kernel density function (the shortest path from q to b_i, $s(q, b_1)$, is indicated by the bold line segments).

where $v_{i0} = q$ and $v_{im} = b_i$. This function is a network KD function, because it satisfies equations (9.3) and (9.4).

In Case 2 ($q = v_{i1}$), noting that n_{i1} links are incident to v_{i1}, and the value of $K_q(v_{i1})$ divided by n_{i1} is assigned to these links (in Case 1 the value is assigned to the links joining v_{i1} and v_{i2}, $i = 1, 2, \ldots$, but not to the line segment joining q and v_{i1}), we obtain:

$$K_q(p) = \begin{cases} \dfrac{2k(d_S(q,p))}{n_{i1}(n_{i2}-1)\cdots(n_{ik-1}-1)} & \text{for } d_S(q,v_{ik-1}) \le d_S(q,p) < d_S(q,v_{ik}), \\ & k = 2,\ldots,m, \\ 0 \quad \text{for } d_S(q,p) \ge h. & \end{cases} \tag{9.9}$$

This function also satisfies equations (9.3) and (9.4), showing that it is a network KD function. We call the function defined by equations (9.8) and (9.9) the *equal-split discontinuous kernel density function.* We also note that this function includes many specific functions because the base KD function $k(d_S(q,p))$ allows many forms as long as they satisfy properties (i)–(iv) above.

Having observed the form of the function in Figure 9.6, one could consider the discontinuity in $K_q(p)$ is unnatural. However, this may not be the case given that space changes discontinuously at a node of degree 3 or more. Hence, the value of $K_q(p)$ may change drastically at the node. This discontinuity would not be serious if the estimator is unbiased. To examine this, for two points q and p on the link satisfying $d_S(q, v_{ik-1}) \le d_S(q,p) < d_S(q, v_{ik})$, we consider $K_q(p)$ and $K_p(q)$. As discussed in reference to the above procedure, the former is obtained by traversing from the kernel center q of $K_q(p)$ to p, while the latter is obtained by traversing reversely from the kernel center p of $K_p(q)$ to q. Because we visit the same nodes on the shortest path between p and q, we obtain $K_p(q) = k(d_S(p,q))/((n_{ik}-1)\cdots(n_{i2}-1)(n_{i1}-1))$. Comparing this with equation (9.8) and $d_S(q,p) = d_S(p,q)$, we note:

$$K_q(p) = K_p(q) \text{ for } p, q \in \tilde{L}. \tag{9.10}$$

The same holds for equation (9.9). From equations (9.4) and (9.10), we obtain:

$$\int_{p\in\tilde{L}} K_q(p)\,\mathrm{d}p = \int_{q\in\tilde{L}} K_p(q)\,\mathrm{d}q = 1. \tag{9.11}$$

Substituting equation (9.11) into equations (9.6) and (9.7), we obtain $B(p) = 0$ for $p \in \tilde{L}$. This proves that the KD estimators given by equations (9.8) and (9.9) are unbiased estimators.

As will be shown in Section 9.3, one of the merits of this estimator is the short computational time required. However, despite this, researchers may still wish to use a continuous KD function. To meet this need, we develop a continuous KD function in the following subsection.

9.2.3 Equal-split continuous kernel density functions

There appears to be no systematic method available for finding a continuous KD function on a network, and so we commonly employ heuristics. We find one solution, described as a recursive algorithm, in Okabe, Satoh, and Sugihara (2009). However, because the description does not give a clear picture of the KD function, we provide an intuitive look at the function by assuming there is one node within distance h from a kernel center q. Under this assumption, with $v_1 = v_{i1}$ and $n_1 = n_{i1}$ (for index simplicity), consider the function:

$$K_q(p) = \begin{cases} k(d_S(q,p)) \text{ for } 0 \le d_S(q,p) < 2d_S(q,v_1) - h, \\ k(d_S(q,p)) - \dfrac{n_1 - 2}{n_1} k(2d_S(q,v_1) - d_S(q,p)) \\ \qquad \text{for } 2d_S(q,v_1) - h \le d_S(q,p) < d_S(q,v_1), \\ \dfrac{2}{n_1} k(d_S(q,p)) \text{ for } d_S(q,v_1) \le d_S(q,p) < h. \end{cases} \tag{9.12}$$

An example of $K_q(p)$ is depicted in Figure 9.7. In the limit at v_1 (approaching from q), the value of $K_q(p)$ is, from the second row in equation (9.12), given by:

$$k(d_S(q, v_1)) - [(n_1 - 2)/n_1]k(2d_S(q, v_1) - d_S(q, v_1)) = (2/n_1)k(d_S(q, v_1)).$$

In the limit at v_1 (approaching from b_i), the value of $K_q(p)$ is, from the fourth row in equation (9.12), given by $(2/n_1)k(d_S(q, v_1))$. Because these two values are equal, $K_q(p)$ for equation (9.12) is continuous at node v_1. In addition, $K_q(p)$ is proved to be unbiased estimator (for the proof, see Okabe, Satoh, and Sugihara (2009, pp. 19–23); because the proof is fairly long, we omit it here). We refer to $K_q(p)$ for equation (9.12) as the *equal-split continuous kernel density function*. It should be

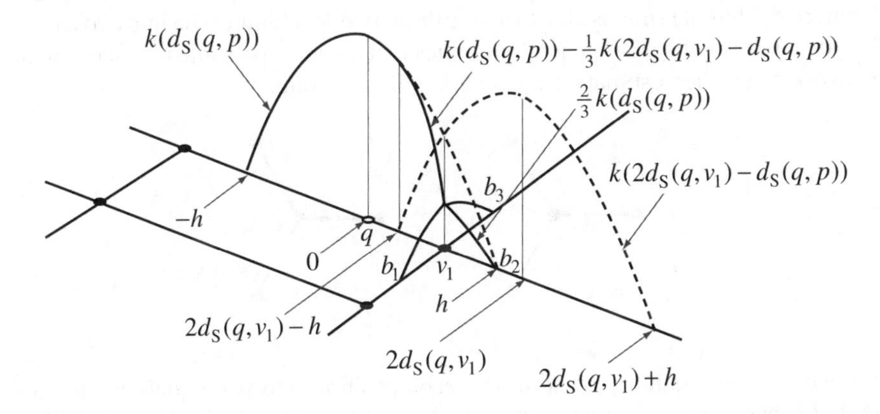

Figure 9.7 An equal-split continuous kernel density function centered at q, where $k(d_S(q, p))$ is a base kernel density function.

noted that this function includes many specific functions because we can use any base KD functions $k(d_S(q,p))$ as long as properties (i)–(iv) are satisfied. The advantages of this KD function are that it is continuous and an unbiased estimator, but one of the few disadvantages is its demanding computational time, as discussed in the following section.

9.3 Computational methods for network point density estimation

Having established the three statistical methods for nonparametric density estimation, we now develop their computational methods in this section. The section comprises three subsections dealing with the computational methods for the network cell histogram (Section 9.3.1), the equal-split discontinuous kernel density function (Section 9.3.2), and the equal-split continuous kernel density function (Section 9.3.3).

9.3.1 Computational methods for network cell histograms with equal-length network cells

When we employ network cell histograms, as noted in Section 9.1, ideal spatial data units are equal-length network cells. For a given network and a given number of network cells, such network cells may or may not be obtainable. For example, consider the simple network shown in Figure 9.8a, where three equal-length links of length one are connected at the central node v. It is clearly easy to divide this network into three equal-length network cells. However, it is equally clearly not possible to divide it into two equal-length connected subnetworks. Indeed, suppose that we select a connected subnetwork of length one-and-a-half (i.e., half of the total length) containing at least one node with degree 1. A typical possibility is then to select one link and a half of another link connected at the central node, as shown by the bold line segments in Figure 9.8b. As a result, the remaining part is not connected. Thus, it is impossible to divide the network in Figure 9.8a into two equal-length network cells. This generally implies that we must presume unequal-length network cells when estimating a network cell histogram.

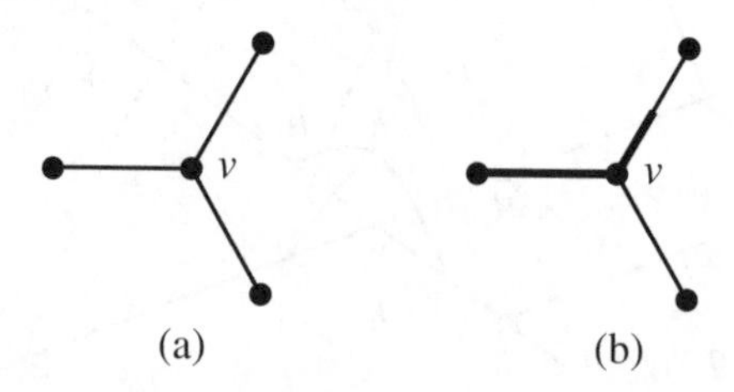

Figure 9.8 Impossibility of network decomposition into two equal-length connected subnetworks: (a) a network consisting of three equal-length links, (b) the length of the thick connected line segments is equal to the total length of thin line segments, but the thin line segments are disconnected.

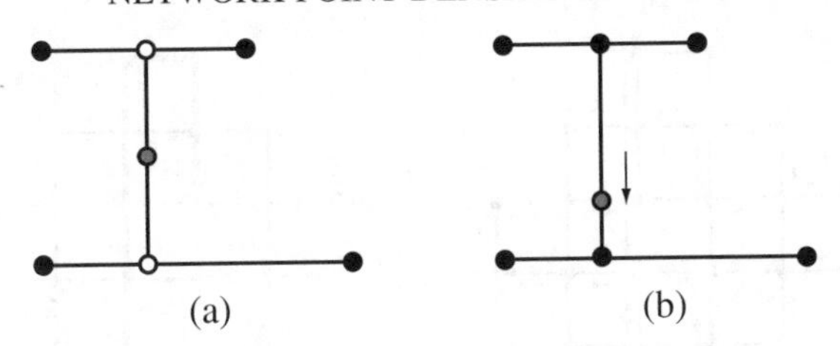

Figure 9.9 Equal-length subnetworks: (a) Voronoi subnetworks (the white circles are generator nodes), (b) subnetworks adjusted by moving the boundary point (the gray circle) using linear programming.

Nonetheless, in special cases we may still wish to use equal-length network cells if they are obtainable. The literature includes a few such attempts. For example, Shiode and Okabe (2004) and Shiode (2008) proposed a method for equal division using the shortest-path tree, but their method produces unequal-length reminder network cells at the periphery of a given network. To improve this shortcoming, Furuta, Suzuki, and Okabe (2008) proposed an alternative method using the network Voronoi diagram (Chapter 4) and linear programming. The method used comprises two phases. In the first phase, Voronoi subnetworks divide a given network by randomly generating points on the nodes of the network, where the number of chosen nodes is equal to that of network cells we wish to obtain. A simple example is depicted in Figure 9.9a where the white circles indicate the generator points of the Voronoi subnetworks (the gray circle is a boundary point). In general, the lengths of the resulting Voronoi subnetworks are not equal, as in Figure 9.9a. To improve this inequality, in the second phase, Furuta, Suzuki, and Okabe (2008) attempt to adjust the locations of the boundary points on links so that the lengths of the subnetworks become nearer to equal, as shown in Figure 9.9b. This adjustment employs linear programming. Furuta, Suzuki, and Okabe (2008) carried out numerical experiments to test the performance of their method. The results indicated that when the number of divisions was small, it tended to produce equal-length network cells. However, as the number of divisions became larger, it often failed in its objective.

As shown above, there are no general methods for dividing a network into equal-length network cells for an arbitrary number of network cells. However, in the special case where we can draw a network with a 'single-stroke path' in which two adjacent network cells are allowed to cross at nodes, equal-length cell decomposition is possible. An illustrative figure is shown in Figure 9.8, where the 'single-stroke path' for the network is indicated by the line segments (links) with the black circles (nodes) where their sequence is shown by the arrows in panel (a). In this case, we can divide the single line (the single-stroke path) into equal-length network cells for any given number of network cells. Panel (b) in Figure 9.10 provides an example of where the network divides into five equal-length network cells (where the white circles are the boundary points between the network cells).

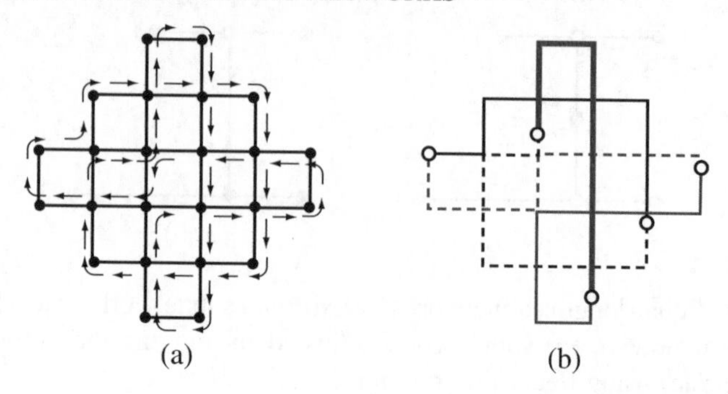

Figure 9.10 Network decomposition: (a) a single-stroke path, (b) equal-length network cells.

The 'single-stroke path' has an alternative name in graph theory, which has been studied in depth since Euler (1736) (for general discussion, see Liu (1968), Busacker and Saary (1965), Knuth (1973), and Cormen *et al.*, (2001)). To define the path precisely, for a network $N = (V, L)$ consisting of $V = \{v_1, \ldots, v_{n_V}\}$ and $L = \{l_1, \ldots, l_{n_L}\}$, consider a path $(v'_1, l'_1, v'_2, \ldots, l'_n, v'_n)$ such that links $l'_1, \ldots, l'_{n_L}$ ($l'_i \in L$, $i = 1, \ldots, n_L$) are mutually distinct (i.e., $l'_i \neq l'_j$ for any $i \neq j$), the links exhaust L (i.e., $L = \{l'_1, \ldots, l'_{n_L}\}$), $v'_i \in V$, $i = 1, \ldots, n \geq n_V$, and $v'_i = v'_j \in V$ if the path goes through the same node twice. This is termed the *Euler path*. In the specific case where the Euler path has the property of $v'_1 = v'_n$, it is termed the *Euler circuit* or *Euler tour*. To state the conditions for the existence of an Euler path, we define an *odd-degree node* as a node satisfying the condition that the number of links incident to the node is odd; otherwise, it is an *even-degree node*. In these terms, we state the following theorem (see, e.g., Liu (1968) for the proof):

Theorem 9.1

The number of odd-degree nodes of a connected network $N = (V, L)$ is even. A connected network $N = (V, L)$ has an Euler path if and only if the number of odd-degree nodes is zero or two.

Using this theorem, we can easily judge whether a given network has an Euler path. If it does, we can decompose the network into equal-length network cells. For instance, because the number of odd-degree nodes of the network in Figure 9.10a is zero, there exists an Euler path, which can be divided into an arbitrary number of equal-length line segments (forming equal-length subnetworks) as in Figure 9.10b.

9.3.2 Computational methods for equal-split discontinuous kernel density functions

Density estimation by network cell histograms is, as noted in Section 9.1, simple but biased. We can overcome this shortcoming using the equal-split discontinuous and continuous KD functions presented in preceding sections. In this section, we develop a computational method for the former, and that of the latter in the following section.

The estimation procedure for the equal-split discontinuous KD function differs according to whether a cycle (if it exists) in network $N = (V, L)$ is shorter than $2h$, i.e., twice the bandwidth. First, consider the former case as assumed in Section 9.2.2, i.e., no cycle is shorter than $2h$. In this case, the density at a point $p \in \tilde{L}$ can be computed explicitly by equations (9.8) or (9.9). More specifically, if a kernel center q is located at a node, the density is given by equation (9.9); if not, equation (9.8). In both cases, the task is to find the shortest path from q to p, compute the distance $d_S(q,p)$ between q and p, read the degrees of all of the nodes on the shortest path, and substitute these into equations (9.8) or (9.9). These tasks can be done in $O(n_L \log n_L)$ time because the most time-consuming step is to find the shortest path (this computational method is shown in Section 3.4.1 in Chapter 3).

Next, consider the general case where cycles can be shorter than $2h$ (examined in depth by Sugihara, Satoh, and Okabe (2010)), implying that there may be two or more paths with length less than h from the kernel center q to a point p on the network. Put differently, the two tails of a KD function may overlap on a cycle in $N = (V, L)$. For example, the two paths indicated by the broken lines in Figure 9.11 overlap on the line segment (v^*_{11}, v_4). In this case, the computational method for determining the value of the KD function requires the following three tasks. First, modify the given network $N(V, L)$. Second, compute the 'split ratio' at every node on each path with length h originating from each kernel center q. Third, sum the densities obtained from the 'split ratio' with respect to all of the paths originating from the kernel center q, and with respect to all kernel centers. In the following, we explain these tasks in detail and formalize the sequence of the tasks as an algorithm.

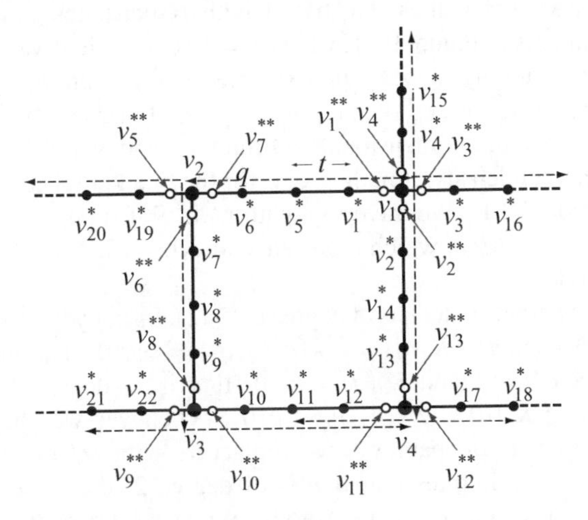

Figure 9.11 Insertion of new nodes (the small black circles) and dummy nodes (the white circles) around the nodes of degree 3 or more in a given network $N(V, L)$ (the large black circles are nodes in V). The broken lines indicate the propagation processes for computing from the kernel center q.

The first task is to modify a given network $N(V, L)$. We choose a small positive number t and insert sufficiently many new nodes with interval t on all of the links (we denote the set of new nodes as V^*). Figure 9.11 provides an illustrative example where the large black circles indicate the nodes in V (the given nodes) and the small black circles indicate the new nodes in V^*. We next insert a dummy node on every link incident to the nodes with degree 3 or more in their close neighborhoods, for example, the white circles in Figure 9.11. The dummy nodes are used for a special treatment at the nodes of degree 3 or more (recall, as shown in Figure 9.6, that the equal-split discontinuous KD function is discontinuous at these nodes). As noted at v_{i1} in Figure 9.6, the KD function has multiple values at these nodes (similarly, multiple values at v_1, v_2, v_3, v_4 in Figure 9.11). To manage these multiple values, we employ dummy nodes. The dummy nodes divide the existing links into finer links, for example, dummy node v_1^{**} divides link (v_1^*, v_1) into two links, (v_1^*, v_1^{**}) and (v_1^{**}, v_1). We assign the length of (v_1^{**}, v_1) to the length of (v_1^*, v_1^{**}) and zero to (v_1^{**}, v_1), and manage the multiple values at every node of degree 3 or more using the values at the dummy nodes (e.g., $v_1^{**}, v_2^{**}, v_3^{**}, v_4^{**}$ in Figure 9.11) around the node (e.g., v_1 in Figure 9.11). Note that we assign a value to each node in V for computational convenience, which is a kind of weighted average, but not one of the multiple values. The resulting new network consists of the set of nodes given by $V \cup V^* \cup V^{**}$ and the set of refined links by inserting $V^* \cup V^{**}$ in L of $N(V, L)$. To avoid complicated notation, in what follows we denote the resulting network simply by $N(V, L)$ (the suffixes of the resulting nodes and links are renamed).

The second task is to compute the density at the nodes of V. This computation process is indicated by the broken lines in Figure 9.11, which can be viewed as propagation from a kernel center. To execute this process, we consider a set S_q of the paths that start at a kernel center q and have length h (such paths are indicated by the broken line segments in Figure 9.11, where $h = 9t$ (t is an interval length between two consecutive dummy nodes; an example is the path s_{q1} going through $v_7^{**}, v_2, v_6^{**}, v_7^*, v_8^*, v_9^*, v_8^{**}, v_3, v_{10}^{**}, v_{10}^*, v_{11}^*, v_{12}^*, v_{11}^{**}, v_4$). Let $k(d)$ be, as denoted in Section 9.2.1, a base one-dimensional KD function and $K_q(p)$ be the equal-split discontinuous kernel density at node $p \in V$ for the one kernel center at $q \in V$ (as denoted in Section 9.2.1). We traverse a path s_{qi} in S_q from q, carrying a positive value α, called a *split ratio,* which represents how the base kernel density value is split at every node.

Suppose that a kernel center q is of degree $n_q (\geq 2)$. Initially we set $\alpha = 2/n_q$, and compute the density at q by $K_q(q) = \alpha\, k(d_S(q, q)) = \alpha\, k(0)$. This means that if q is of degree 2, then $\alpha = 1$ and $K_q(q) = k(0)$. If q is of degree 3 or more, then $\alpha = 2/n_q < 1$ and $K_q(q) = 2\, k(0)/n_q < k(0)$. Whenever we encounter a node $p \in V$ with degree n_p on path s_{qi}, we replace α with $\alpha/(n_p - 1)$, written as $\alpha \leftarrow \alpha/(n_p - 1)$. This implies that if p is of degree 2, the value of α does not change and we move on to the next node. When we encounter a node $v \in V$ of degree 3 or more, the value α of node v is split into $(n_p - 1)$ equal values, i.e., $\alpha/(n_p - 1)$; the split ratio of dummy node $v' \in V$ next to node p is given by $\alpha/(n_p - 1)$; and the density at dummy node v' is computed by $K_q(p') = \alpha\, k(d_S(q, v'))$. For example, on path s_{q1} in Figure 9.11, when we encounter node v_2 of degree $n_{v_2} = 3$, we split the value $\alpha = 1$ at v_2 into two $\alpha/(n_{v_2} - 1) = 1/(3 - 1) = 1/2$; assign 1/2 to the split ratios of dummy nodes v_5^{**} and v_6^{**}; and compute the density at dummy node v_6^{**} by $K_q(v_6^{**}) = k(d_S(q, v_6^{**}))/2$. In this way, we successively compute the densities at the nodes on path s_{qi}.

The third task is first to sum the densities at nodes across all of the paths in S_q. This procedure is necessary because the paths in S_q may overlap; for instance, node v_{12}^* in Figure 9.11 is included in the path passing through q, v_2, v_3, v_4 and the path passing through q, v_1, v_4, v_{11}^*. We then sum the densities at the nodes with respect to all kernel centers on $N(V, L)$.

We formalize the above process in the following algorithm, where d is the shortest-path distance from kernel center q to node v along the path we have traversed, α is the split ratio, v is the node we are visiting, l is the link incident to v that has been traversed, and 'nil' in Step 3 implies that this parameter conveys no information.

Algorithm 9.1 (Equal-split discontinuous kernel density function for a single kernel center)

Input: network $N(V, L)$, kernel center $q \in V$, base kernel density function $k(d)$, bandwidth
h, degree n_v at node $v \in V$.

Output: $K_q(v)$ for all $v \in V$.

Procedure:

1. $K_q(v) \leftarrow 0$ for all $v \in V$
2. $d \leftarrow 0$ and $\alpha \leftarrow 2/n_q$
3. ESD-Kernel $(d, \alpha, q, \text{nil})$
4. stop □

Here, ESD-Kernel is a procedure, the core part of Algorithm 9.1, defined as follows (ESD stands for *e*qual *s*plit *d*iscontinuous).

Procedure ESD-Kernel (d, α, v, l)

1. $K_q(v) \leftarrow K_q(v) + \alpha\, k(d)$
2. for each link l' incident to v other than $l' = l$ do
   2.1. set v' as the other node of l'
   2.2. $d' \leftarrow d + d(v, v')$
   2.3. if $h > d'$ then
      2.3.1. $\alpha' \leftarrow \alpha \frac{1}{(n_v - 1)}$
      2.3.2. ESD-Kernel(d', α', v', l')

endif

3. return □

In Step 1 of ESD-Kernel, the value $K_q(v)$ is updated by adding the base kernel density value at distance d multiplied by the split ratio α. Through this addition, the value of $K_q(v)$ is accumulated if there were two or more paths from q to v within distance h (for example, at v^*_{12} in Figure 9.11). In Step 2, we propagate the computation to all of the links from v other than the already traversed link. For each such link l', we find the other terminal node v' in Step 2.1, and compute the distance d' from q to v' via v in Step 2.2. In Step 2.3, we check whether the distance is within h. If so, we compute the split ratio α' at v in Step 2.3.1 and execute the procedure ESD-Kernel recursively at Step 2.3.2. If d' is greater than h, then the computation does not propagate in that direction, and we thus terminate the recursion. When Algorithm 9.1 terminates, the kernel density value for a single center is assigned to $K_q(v)$ for all $v \in V$. For a given set $\{q_1, q_2, \ldots, q_n\} \subset V$ of sample points (kernel centers), we compute the KD function $K(v)$ for all $v \in V$ by executing the procedure ESD-Kernel repeatedly for every sample point in $\{q_1, q_2, \ldots, q_n\}$ using the following algorithm.

Algorithm 9.2 (Equal-split discontinuous kernel density function for all sample points)

Input: network $N(V, L)$, set $\{q_1, q_2, \ldots, q_n\} \subset V$ of sample points, base kernel density function $k(d)$, bandwidth h, degree n_V at node $v \in V$.

Output: equal-split discontinuous kernel density function $K(v)$ for all v.

Procedure:

1. $K(v) \leftarrow 0$ for all $v \in V$
2. for $i \leftarrow 1$ until n do

    $d \leftarrow 0$ and $\alpha \leftarrow 2/n_{p_i}$

    ESD-Kernel $(d, \alpha, p_i,$ nil$)$
3. stop

□

The above algorithm is implemented in the SANET software package (Okabe, Okunuki, and Shiode, 2006a, 2006b; Okabe, Satoh, and Sugihara, 2009) and we illustrate its application to actual data in Section 12.2.5 in Chapter 12.

The time complexity of Algorithm 9.2 is proportional to the number of executions of procedure ESD-Kernel because Step 1 of Algorithm 9.2 can be in $O(n_v)$ time, and all other substeps are in constant time. The exact time complexity depends on the structure of a network and bandwidth h. The time complexity increases if the distances between nodes with degree 3 or more become shorter relative to bandwidth h.

9.3.3 Computational methods for equal-split continuous kernel density functions

Although estimation using the above equal-split discontinuous KD function is unbiased, users may prefer a continuous KD function. If this is the case, they can use the equal-split continuous KD function formulated in Section 9.2.3. The computational method with this function varies according to the length of bandwidth h. When no link is shorter than bandwidth h, as assumed in Section 9.2.3, the KD function

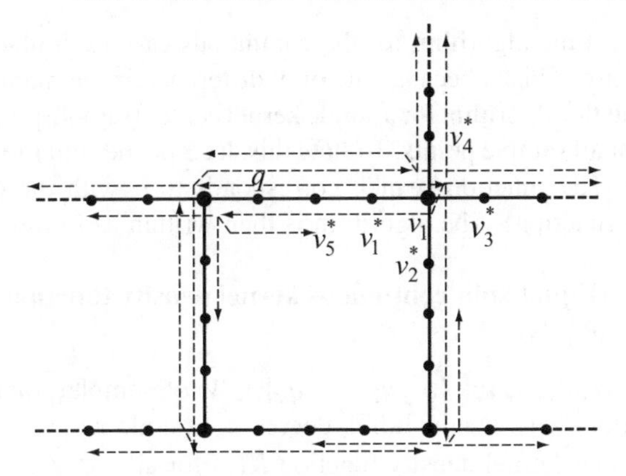

Figure 9.12 Insertion of new nodes (the small black circles) in a given network $N(V, L)$ (the large black circles are nodes in V). The broken lines indicate the propagation processes for computing from the kernel center q.

at p can be explicitly computed using equation (9.12). Next, consider the general case where the links are not necessarily longer than bandwidth h (as in Figure 9.12). Recalling that the equal-split continuous KD function has a unique value at every node, which contrasts to the case of the equal-split discontinuous KD function, we do not have to insert dummy nodes (the white circles in Figure 9.11, but there are no such circles in Figure 9.12). The splitting procedure at a node described in Section 9.2.3 applies to all relevant nodes recursively in the same manner as the equal-split discontinuous KD function in Algorithms 9.1 and 9.2. However, the manner of splitting and the directions of propagation are different (compare the broken lines in Figure 9.11 (forward propagation) with those in Figure 9.12 (not just forward but also backward propagation)).

Suppose that we carry on the splitting procedure traversing from a kernel center q toward, for example, node v_1 in Figure 9.12. In the discontinuous case in Figure 9.11, when we reach v_1, we assign the split kernel density value equally to the dummy nodes v_2^{**}, v_3^{**}, v_4^{**} around v_1 other than the dummy node v_1^{**} on the traversed path. In the continuous case in Figure 9.12, we assign the split kernel density value to not only v_2^*, v_3^*, v_4^* but also v_1^* on the traversed path (so-called backward propagation) in such a way that the kernel density value at v_1 multiplied by $-(n_v - 2)/n_v$ is assigned to v_1^*, and that multiplied by $2/n_v$ is assigned to v_2^*, v_3^*, v_4^*. Note that these two assignments correspond to the second, third and fourth lines of equation (9.12). Because $2/n_v$ is assigned to $n_v - 1$ nodes, the sum of all the split values is $-(n_v - 2)/n_v + (n_v - 1)\ (2/n_v) = 1$, implying that the total amount is preserved by the split procedure. Note that if the degree of a node is 2 (for example, v_5^*), the split value is $-(n_v - 2)/n_v = -(2 - 2)/2 = 0$ for v_6^* and $2/n_v = 2/2 = 1$ for v_1^*, implying that the value is not split at a node of degree 2. Also note that the split value for a traversed node, say v_1^* is negative if $n_v \geq 3$, which means the kernel density value at v_1^* is not increased, but rather decreased, by the split (see Figure 9.7). By this treatment, we can hold the continuity of the KD function at $v \in V$.

The structure of the algorithm for the continuous case is similar to that for the discontinuous case. This is because the only difference is the manner of splitting, whereby we omit the algorithm for a single kernel center (sample point), and present the algorithm for all sample points. To state this, let β be the split ratio for the equal-split continuous KD function (which corresponds to α with the equal-split discontinuous KD function). The algorithm is then written as follows.

Algorithm 9.3 (Equal-split continuous kernel density function for all sample points)

Input: network $N(V, L)$, set $\{q_1, q_2, \ldots, q_n\} \subset V$ of sample points, base kernel density function $k(d)$, bandwidth h, degree n_v at node $v \in V$.
Output: continuous kernel density function $K(v)$ for all $v \in V$.
Procedure:

1. $K(v) \leftarrow 0$ for all $v \in V$
2. for $i \leftarrow 1$ until n do
   $d \leftarrow 0$ and $\beta \leftarrow 2/n_{p_i}$
   ESC-Kernel $(d, \beta, p_i, \text{nil})$
3. stop □

The procedure ESC-Kernel is defined as follows (ESC stands for *e*qual *s*plit *c*ontinuous).

Procedure ESC-Kernel (d, β, v, l)

1. $K(v) \leftarrow K(v) + \beta\, k(d)$
2. for each link l' incident to v
   2.1. set v' as the other node of l'
   2.2. $d' \leftarrow d + d(v, v')$
   2.3. if $h > d$ then
   if $l = l'$ then
   2.3.1. $\beta' \leftarrow \beta \dfrac{2 - n_v}{n_v}$
else
   2.3.2. $\beta' \leftarrow \beta \dfrac{2}{n_v}$
   endif
   2.3.3. ESC-Kernel (d', β', v', l')
endif
3. return □

In Step 2, we consider all links l' incident to v, including the traversed link l. Step 2.3 contains two cases. The first is for link l (Step 2.3.1), and the second is for the other links (Step 2.3.2). Updating the split ratio β at these steps, we execute the procedure ESC-Kernel recursively at Step 2.3.3. Section 12.2.5 in Chapter 12 provides the implementation of the above algorithms in the SANET software (Okabe, Okunuki, and Shiode, 2006a, 2006b; Okabe, Satoh, and Sugihara, 2009) and its application to actual data.

The time complexity can be evaluated in the following way. Step 1 of Algorithm 9.3 can be done in $O(n_v)$ time. The computational time for Step 2 is proportional to the number of executions of the procedure ESC-Kernel, because all of the other substeps can be done in constant time. Recall that at the nodes of degree 2, the computation only propagates forward, whereas ESC-Kernel is called at many times at a node with degree 3 or more. Indeed the number of calls at the node is the same as the degree of the node because the kernel value should be split and distributed to all the links. Hence, if there are many such nodes within the distance h from the sample points, the computational time increases. The exact complexity depends on the structure of a given network, the bandwidth of a base KD function, and the distance between nodes with degree 3 or more.

Having established the computational methods for two unbiased KDE methods in the preceding two subsections, the reader may wonder which KDE method is more computationally efficient. We can intuitively understand the difference by observing the propagation processes shown by the broken lines in Figures 9.11 and 9.12. As shown, the propagation process of the equal-split discontinuous KD function is shorter than that of the equal-split continuous KD function because the former process includes only forward propagation while the latter process also includes backward propagation, implying that the former is more efficient than the latter. In conclusion, when computational time is crucial, we recommend the former method. On the other hand, when continuity in the estimated density function is important, we prefer the latter method.

10

Network spatial interpolation

This chapter extends two well-known methods for spatial interpolation, called *inverse-distance weighting* and *kriging*, to networks. A reason for choosing these two from the many interpolation methods available is that they represent two major approaches to spatial interpolation. Inverse-distance weighting is a representative of the *deterministic approach*, i.e., formulating interpolation methods without probabilistic concepts, and kriging represents the *probabilistic* or *geostatistical approach*, i.e., formulating interpolation methods with probabilistic concepts. These methods can answer, for instance, question Q7 in Chapter 1: How can we spatially interpolate an unknown NO_x (nitrogen oxides) density at a point on a road using known NO_x densities at observation points in a high-rise building district, such as Midtown Manhattan? A more generalized question is:

Q7′: Given known attribute values at a finite number of sample points on a network, how can we interpolate or predict an unknown attribute value at an arbitrary point on the network?

The key concept in these questions is *interpolation*, a class of methods for predicting an unknown value from a set of known values. According to Gasca and Sauer (2000), the first researcher to develop a mathematical method for interpolation was John Wallis in the mid-seventeenth century; he computed π by successive interpolations (Section 15.07 in Jeffreys and Jeffreys (1988)). Interpolation with several variables was developed in the second half of the nineteenth century, although geographical spaces were not explicitly treated at that time. In the twentieth century, Steffensen (1927) organized the methods of interpolation and

Spatial Analysis along Networks: Statistical and Computational Methods, First Edition.
Atsuyuki Okabe and Kokichi Sugihara.

published a now-classic book, entitled *Interpolation*. The book, however, does not explicitly deal with spatial interpolation. The extensive developments of spatial interpolation methods began in the latter half of the twentieth century. Those studies were reviewed by Lam (1983), Myers (1994), and Mitas and Mitasova (1999).

It could be said that spatial interpolation is empirically based on the *first law of geography* stated by Tobler (1970) (as in Chapter 7): everything is related to everything else, but near things are more related than distant things. In the context of spatial interpolation, this law may be stated as: an attribute value at a point is more similar to those at near points than those at distant points. The nearness may be measured by several types of distance metric. Most spatial interpolation methods, as noticed in the above reviews, assume Euclidean distances on a plane. However, this assumption may not be acceptable in some situations. A clear example is depicted in Figure 10.1, where the winding curve indicates the road named the I-ro-ha Slope in Nikko, Japan. Suppose that we wish to predict the unknown altitude at the point p_0 using the known altitudes at the sample points p_1, p_2, p_3, p_4. If we measure the nearness by Euclidean distances between p_1, p_2, p_3, p_4 and p_0, then the points p_1 and p_4 are closer to p_0 than are the points p_2 and p_3. However, the altitude at p_0 is more similar to those at p_2 and p_3 than those at p_1 and p_4. This example suggests that the shortest-path distance should be used for interpolating the altitude of roads. In fact, Shiode and Shiode (2009b, 2011) tested an interpolation method for predicting road altitudes and concluded that spatial interpolation using shortest-path distances was more precise than that using Euclidean distances.

Spatial interpolation methods have many varieties, including inverse-distance weighting, kriging, natural neighbor interpolation, splines, trend surfaces, and triangulated irregular network (TIN)-based interpolation. These methods can be classified into *distance-based methods* and *fitted-function methods*. The former methods predict the value at a point in terms of the distances between sample points and a prediction point; and the latter methods by fitting a continuous function to the values at the sample points. The former include inverse-distance weighting, kriging,

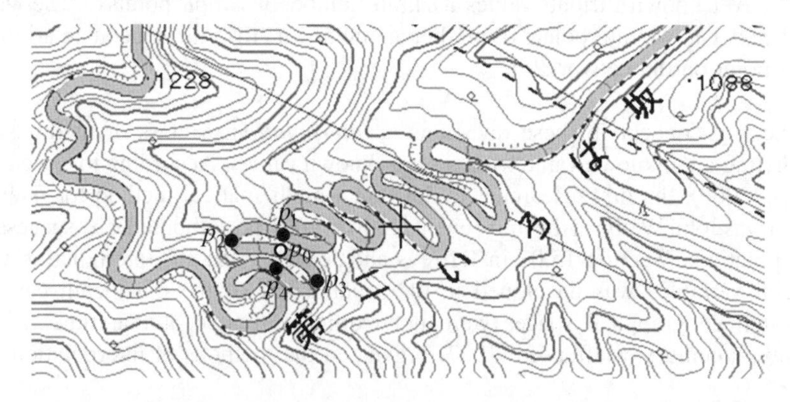

Figure 10.1 Interpolation of the altitude at p_0 using known altitudes at p_1, p_2, p_3 and p_4 on the I-ro-ha Slope in Nikko, Japan (the map is part of the 1/25 000 map published by the National Land Survey Agency, Japan).

and natural neighbor interpolation; the latter include splines, trend surfaces and TIN-based interpolation. It should be noted that most of those methods deal with interpolation on a plane. The extension of the distance-based methods on a plane to networks began around 2000 (Cressie and Majure, 1997; Shiode and Okabe, 2003). The extension of the fitted-function methods to a network lagged behind, because of difficulties in treating continuity or smoothness at nodes of a network; initial attempts began around 2010 (Hiyoshi, 2009, 2010). Because the latter extension looks premature at present, this chapter focuses on the former.

This chapter consists of three sections. The first section describes inverse-distance weighting on a network, and the second section illustrates kriging on a network. The last section explains efficient computational methods for carrying out these two methods. The chapter provides an introductory explanation of these spatial interpolation methods. It deals with only nondirected networks. Directed networks, such as interpolation on a river network with flows, are beyond the scope of this chapter. The reader who wishes to thoroughly understand those two methods as well as the other methods mentioned above should consult the textbooks cited in each section.

10.1 Network inverse-distance weighting

We start discussion on spatial interpolation with inverse-distance weighting. An original idea of inverse-distance weighting is found in Horton (1923), who applied it to rainfall interpolation in a region (he used the term, *weighting interpolated values*). The computer implementation of this idea was developed by Shepard (1968), followed by Gordon and Wixom (1978), Franke (1982), Barnhill and Stead (1984), Tobler and Kennedy (1985), Watson and Philip (1985), and Watson (1992, Section 2.5 (6)), among others. The method on a plane was extended to networks by Shiode and Okabe (2003), and Shiode and Shiode (2009b, 2011) empirically tested the method.

10.1.1 Concepts of neighborhoods on a network

To formulate an inverse-distance weighting method mathematically, we suppose a nondirected network $N = (V, L)$, consisting of a set of nodes $V = \{v_1, \ldots, v_{n_V}\}$ and a set of links $L = \{l_1, \ldots, l_{n_L}\}$. Let $\tilde{L} = \bigcup_{i=1}^{n_L} l_i$, (i.e., the set of points forming the network including the nodes), and let $z_1, \ldots, z_n$ be the observed attribute values at sample points $p_1, \ldots, p_n$ on $\tilde{L}$. We wish to predict an unknown value z_0 at an arbitrary point p_0 ($p_0 \neq p_1, \ldots, p_n$) on $\tilde{L}$ using known values in a neighborhood of p_0, denoted by $P_{\mathrm{N}}(p_0)$. The neighborhood $P_{\mathrm{N}}(p_0)$ can be specified in several ways. A first one is defined using the network Voronoi diagram generated by points $p_0, p_1, \ldots, p_n$, referred to as *generator points* (Section 4.1.1 in Chapter 4). With this diagram, $P_{\mathrm{N}}(p_0)$ is defined by the generator points having Voronoi cells adjacent to the Voronoi cell of p_0. This concept is an extension of the *natural neighborhood* on a plane defined by Sibson (1981) to a network. A second one, called the *k-th nearest neighborhood*, is defined by the first, second, ... , k-th nearest points from p_0, where k is predetermined (all the distances are shortest-path distances). A third one, called the *distance-cut-off neighborhood*, is defined by the set of points p_i such that the shortest-path distance $d_S(p_0, p_i)$ from p_0 to p_i is less than or equal to a predetermined value. Last, in an extreme case, $P_N(p_0)$ may be given by all sample points $p_1, \ldots, p_n$.

10.1.2 Network inverse-distance weighting predictor

We predict the value at p_0 as the weighted average of the known attribute values at the points of a neighborhood $P_N(p_0)$. To be explicit, let:

$$w_i = \frac{d_S(p_0, p_i)^{-\alpha}}{\sum_{p_j \in P_N} d_S(p_0, p_j)^{-\alpha}}, \tag{10.1}$$

where α is a positive predetermined parameter. With this weight, we interpolate the value at p_0 as:

$$\hat{z}_0 = \sum_{p_i \in P_N(p_0)} w_i z_i. \tag{10.2}$$

An actual example is shown in Section 12.2.6 in the last chapter of this volume.

We make three remarks on this interpolator. First, there are no definite rules for determining the value of α. Horton (1923) used $\alpha = 1$. The most popular choice is $\alpha = 2$, so that the data are inversely weighted as the squared distance (Webster and Oliver, 2001, Section 3.1.4). Watson (1992, Section 2.5) discussed the general properties of $0 < \alpha < 1$, $\alpha = 1$ and $\alpha > 1$. Lu and Wong (2008) reported that values between one and five are used in the related literature; they also proposed a variable α in accordance with the pattern of points in $P_N(p_0)$.

Second, the choice of the number k of elements in $P_N(p_0)$ is rather *ad hoc*. Zimmerman *et al.* (1999) reported that many empirical studies used values between 3 and 9. It should be noted that these empirical numbers were obtained for spatial interpolation on a plane; they are not guaranteed for that on a network. Shiode and Shiode (2009b, 2011) obtained an appropriate combination of α and k for interpolating altitudes on a street network using *cross-validation*, i.e., a procedure for finding appropriate parameter values by iterative point removal and prediction (see, e.g., the outline of the cross-validation procedure in De Smith, Goodchild, and Longley (2007, Section 6.6); for details, see Isaaks and Srivastava (1989, Chapter 15)).

Third, the inverse-distance weighting method has advantages as well as disadvantages, which are discussed in Shepard (1964), Watson (1992, Section 2.5), and Burrough and McDonnell (1998, Chapter 6). Here we briefly refer to one advantage and one disadvantage. The advantage is that this weighting is fast and

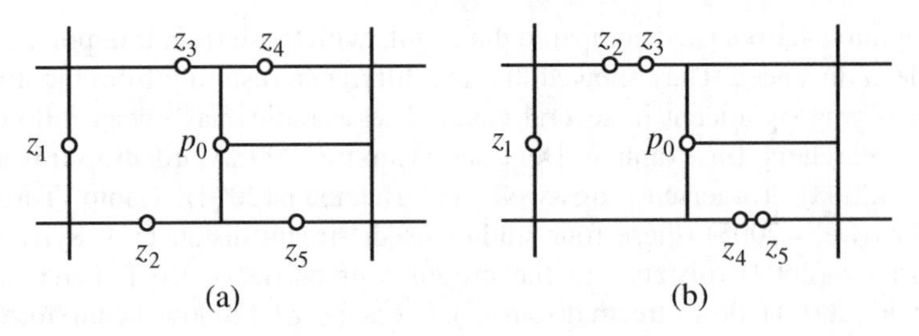

Figure 10.2 Two configurations of the attribute values z_1, z_2, z_3, z_4, and z_5 that predict the same value at p_0 with the inverse-distance weighting method.

easy to compute and straightforward to interpret. In fact, the main task is merely to compute the distances between sample points and the point at which the value is to be predicted. The disadvantage is the method's insensitivity to the configuration of observed sample points. For example, the configurations of the observed sample points in panels (a) and (b) in Figure 10.2 produce the same prediction value at p_0, because in both cases the values z_1, z_2, z_3, z_4, and z_5 are at the same shortest-path distances from p_0. We shall discuss this disadvantage in comparison with kriging in the subsequent section.

10.2 Network kriging

Contrasted with inverse-distance weighting – a deterministic method, is kriging – a probabilistic method. Kriging is very frequently used in *geostatistics*, i.e., statistics for earth sciences (including geology, hydrology and meteorology). The term *kriging* was named after Krige (1951) by Matheron (1963). Krige's research life was recounted by Camisani-Calzolari (2003) and the origin of kriging was discussed by Cressie (1990) in depth. There are a number of textbooks dealing with kriging in detail, for example, Journel and Huijbregts (1991, Chapter 5), Cressie (1991, Chapter 3), Wackernagel (1995, Chapters 10–13, 15, 16), Armstrong (1998), Chiles and Delfiner (1999, Chapter 3), Stein (1999), Webster and Oliver (2001, Chapters 8–10), and Mase and Takeda (2001, Chapter 6). It should be noted that the kriging methods in those textbooks mostly assume a plane with Euclidean distances, referred to as *planar kriging*. In contrast, this chapter is concerned with kriging on a network using shortest-path distances, referred to as *network kriging*.

Two of the earliest studies on network kriging are by Cressie and Majure (1997), and Little, Edwards, and Porter (1997). Cressie and Majure (1997) called attention to kriging on a river network, and discussed the predictability of pollution (nitrate) caused by livestock wastes in streams. Little, Edwards, and Porter (1997) predicted contaminant and water-quality variables in estuaries using 'minimum in-water distances' and compared this with interpolation with Euclidean distances. The comparison showed that the former distance improved prediction accuracy on the order of 10–30%. Rathbun (1998) interpolated salinity using 'water distances,' defined as the distance constrained by irregularly shaped regions (three rivers flowing into a harbor) and compared the results with those from interpolation with Euclidean distances. They showed that the difference resulting from the different distances was significant in several cases. These studies have been followed by many researchers, for example, Dent and Grimm (1999), Gardner, Sullivan, and Lembo (2003), Torgersen, Gresswell, and Bateman (2004), Ganio, Torgersen, and Gresswell (2005) (these four studies used 'stream distances'), Krivoruchko and Gribov (2004) (distance in the presence of barriers), Hoef, Peterson, and Theobald (2006) ('flow-stream distance'), Cressie *et al.* (2006) (the mixture of the Euclidean distance and 'river distance'), and Curriero (2006) ('water metric').

Conceptually, kriging and inverse-distance weighting are the same in that both predict an unknown value using a weighted average of known observed values. The difference is in the derivation of the weights. As seen in the preceding section, the derivation of the inverse-distance weighting predictor is straightforward. In contrast, the deviation of kriging predictors, which are to be illustrated in the subsequent subsections, is fairly complicated. To avoid getting lost, we outline the steps of the derivation.

Section 10.2.1 introduces three kriging models: *simple kriging*, *ordinary kriging* and *universal kriging*. Section 10.2.2 specifies these kriging methods by adding assumptions about stationary processes, so-called *second-order stationary processes*, *intrinsic stationary processes* and *network isotropy*, and introduces the concept of a *variogram*. Section 10.2.3 specifies the variogram as a continuous function with parameters, referred to as a *variogram model*. To estimate the parameter values of the variogram model from data, the same section shows the procedure for constructing a discrete function obtained from observed data, called an *experimental variogram*. The parameters are estimated by fitting the variogram model to the observed experimental variogram. The final section, Section 10.2.4, formulates kriging predictors using the estimated variogram model.

10.2.1 Network kriging models

To describe a class of kriging models, we consider a stochastic process (Chapter 3) in which a random variable $Z(p)$ is generated at every point p on a network $\tilde{L}$ (for example, the density of NO_X along roads), and suppose that the observed values $z(p_1), \ldots, z(p_n)$ are the random realizations of the variables $Z(p_1), \ldots, Z(p_n)$ at the points $p_1, \ldots, p_n$ on $\tilde{L}$.

Kriging is modeled in several ways according to our prior knowledge about $Z(p)$. When we know that the expected value $\mathrm{E}(Z(p))$ is written as:

$$\mathrm{E}(Z(p)) = \beta, \tag{10.3}$$

and we know the value of parameter β exactly, then the kriging is termed a *simple kriging model*.

In practice, however, we often do not know the value of β. In this case, the kriging is termed an *ordinary kriging model*. If we do not have any information about the spatial structure of $Z(p)$ at all, we must assume the ordinary kriging model.

We usually have some knowledge about the structural factors of the random variable $Z(p)$. If we know that the structure is written as:

$$\mathrm{E}(Z(p)) = \sum_{i=1}^{m} \beta_i f_i(p), \tag{10.4}$$

and we know the explicit form of the functions $f_1(p), \ldots, f_m(p)$ but do not know the values of their coefficients $\beta_1, \ldots, \beta_m$, then the kriging is termed a *universal kriging model*. The simplest form of equation (10.4) is:

$$\mathrm{E}(Z(p)) = \beta_1 f_1(p) + \beta_2 f_2(p) = \beta_1 + \beta_2 f_2(p). \tag{10.5}$$

An actual example is that the value $Z(p)$ is the roadside land price at p, and $f_2(p)$ is the log of the shortest-path distance from p to the central station at p_s in a city, or mathematically, $f_2(p) = \log d_S(p_s, p)$. In addition to the above three models, the class of kriging models includes block kriging, cokriging and disjunctive kriging. The reader who wants to understand these kriging models should consult the textbooks cited above.

10.2.2 Concepts of stationary processes on a network

To predict an unknown value by a kriging method, we must make several assumptions about 'stationary' processes. First, we refer to the concepts of the stationary process defined on a plane, and next modify them on a network.

Second-order stationary processes on a plane

For arbitrary points p_i and p_j in a domain S in the Cartesian space, if the relations:

$$\mathrm{E}[Z(p_i)] = \mathrm{E}[Z(p_j)] = \beta, \tag{10.6}$$

$$\mathrm{Cov}[Z(p_i), Z(p_j)] = C(\overrightarrow{p_j p_i}), \tag{10.7}$$

hold, where $C(\overrightarrow{p_j p_i})$ is a function of the vector $\overrightarrow{p_j p_i}$ from p_j to p_i and β is a constant, then the stochastic process $Z(p), p \in S$ is termed the second-order stationary process.

Note that the term 'second' derives from the second moment (the covariance) of the variable $Z(p)$.

Equation (10.6) implies that the expected attribute value of $Z(p)$ is the same over S. Equation (10.7) implies that the covariances are the same if their vectors are the same. For example, in Figure 10.3a, $\mathrm{Cov}[Z(p_3), Z(p_4)] = \mathrm{Cov}[Z(p_5), Z(p_6)]$ holds, because $\overrightarrow{p_4 p_3} = \overrightarrow{p_6 p_5}$, whereas $\mathrm{Cov}[Z(p_3), Z(p_4)] = \mathrm{Cov}[Z(p_1), Z(p_2)]$ does not necessarily hold, because $\overrightarrow{p_4 p_3} \neq \overrightarrow{p_2 p_1}$. However, these relations are arguable when we analyze events occurring on a network. Generally, distances along a network (the shortest-path distances) are more effective for stochastic processes on network events (imagine that the locations of stores are affected by consumers' trip behavior on

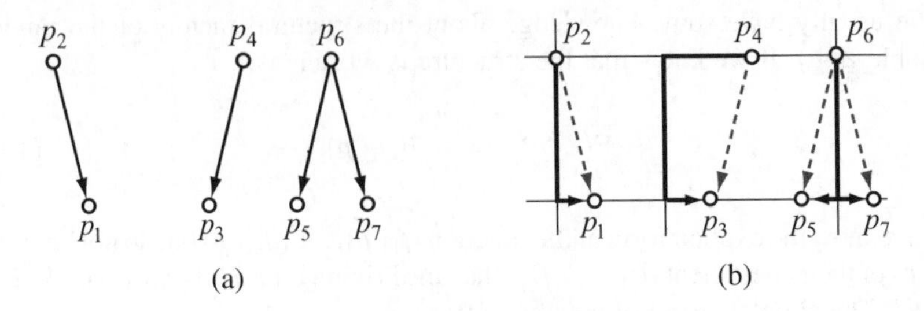

Figure 10.3 The configuration of points on a plane (a) and that on a network (b) for discussing the second-order stationary process.

streets). Therefore, directed crow-flight distances are not always meaningful for analyzing network events. An example is shown in Figure 10.3b (the configuration of the points is the same as that on the plane in Figure 10.3a). The shortest path from p_4 to p_3 is distinctively different from that from p_6 to p_5, although $\overrightarrow{p_4p_3} = \overrightarrow{p_6p_5}$; therefore, on a network, $\mathrm{Cov}[Z(p_3), Z(p_4)] = \mathrm{Cov}[Z(p_5), Z(p_6)]$ does not seem to hold. In contrast, $\mathrm{Cov}[Z(p_1), Z(p_2)] = \mathrm{Cov}[Z(p_5), Z(p_6)]$ seems to hold, because their shortest-path distances are the same, $d_S(p_6, p_5) = d_S(p_2, p_1)$.

For the above reasons, we assume in the following derivation that the stochastic process $Z(p)$, $p \in \tilde{L}$ satisfies:

Network isotropy

For any pair of point pairs (p_i, p_j) and $(p_{i'}, p_{j'})$ on $\tilde{L}$ such that $d_S(p_i, p_j) = d_S(p_{i'}, p_{j'})$, if the equation:

$$\mathrm{Cov}[Z(p_i),\ Z(p_j)] = \mathrm{Cov}[Z(p_{i'}), Z(p_{j'})] \tag{10.8}$$

holds, then the stochastic process $Z(p)$, $p \in \tilde{L}$ satisfies *network isotropy.*

The term 'isotropy' might not sound natural on a network, because directions on a network are always restricted (e.g., only two directions are allowed from p_1 on the network in Figure 10.3b). However, isotropy holds for allowable directions. Under the network isotropy assumption, we specify the second-order stationary processes on a network as follows.

Second-order stationary processes under the network isotropy assumption

For arbitrary points p_i and p_j on $\tilde{L}$, if equation (10.6) and:

$$\mathrm{Cov}[Z(p_i), Z(p_j)] = C(d_S(p_i, p_j)) \tag{10.9}$$

hold, where $C(d_S(p_i, p_j))$ is a function of the shortest-path distance $d_S(p_i, p_j)$, then the stochastic process $Z(p)$, $p \in \tilde{L}$ satisfies the *second-order stationary process* under the network isotropy assumption. The function $C(d_S(p_i, p_j))$ is a covariance in ordinary statistics, but in geostatistics, it is customarily called a *covariogram.*

Similarly, paralleling the intrinsic stationary processes defined on a plane, we modify them on a network as follows.

Intrinsic stationary processes under the network isotropy assumption

For arbitrary p_i and p_j on $\tilde{L}$, if the equations:

$$\mathrm{E}[Z(p_i)] - \mathrm{E}[Z(p_j)] = 0, \tag{10.10}$$

$$\mathrm{Var}[Z(p_i) - Z(p_j)] = 2\gamma(d_S(p_i, p_j)) \tag{10.11}$$

hold, where $\gamma(d_S(p_i, p_j))$ is a function of the shortest-path distance $d_S(p_i, p_j)$, then the stochastic process $Z(p), p \in \tilde{L}$ satisfies the *intrinsic stationary process* under the network isotropy assumption. The function $2\gamma(d_S(p_i, p_j))$ is termed a *variogram*, and $\gamma(d_S(p_i, p_j))$ a *semivariogram* or simply *variogram*.

Note that to avoid lengthy terms, we call 'second-order or intrinsic stationary processes under the network isotropy assumption' simply 'second-order or intrinsic stationary processes.' Note also that the covariogram, variogram and semivariogram on a plane are defined in terms of the vector $\overrightarrow{p_i p_j}$, but they are defined in terms of $d_S(p_i, p_j)$ on a network, because we assume network isotropy.

Equation (10.10) is equivalent to equation (10.6), but β is not explicitly written in equation (10.10). Therefore, we can avoid the errors resulting from estimating β in equation (10.6). Second-order stationary processes imply intrinsic stationary processes but the converse is not always true. If $Z(p)$ is a second-order stationary process, then $\mathrm{Cov}[Z(p_i), Z(p_i)] = \mathrm{Var}\,[Z(p_i)] = C(d_S(p_i, p_i)) = C(0)$. This implies that not only the expected value but also the variance of $Z(p)$ is constant over $\tilde{L}$. Therefore, if $Z(p)$ is a second-order stationary process, then equation (10.11) is written as:

$$\begin{aligned}\gamma(d_S(p_i, p_j)) &= \mathrm{Var}[Z(p_i)]/2 + \mathrm{Var}[Z(p_j)]/2 - \mathrm{Cov}[Z(p_i), Z(p_j)] \\ &= C(0) - C(d_S(p_i, p_j)).\end{aligned} \tag{10.12}$$

This equation is informative in that it shows the relationship between a variogram and its corresponding covariogram.

Recalling equation (10.4) in relation to equations (10.6) and (10.10), one might wonder why second-order and intrinsic stationary processes cannot be applied to universal kriging; in fact, equation (10.4) of universal kriging shows that $\mathrm{E}(Z(p))$ is not constant over $\tilde{L}$. This is true, but if we consider the variable $Z'(p) = Z(p) - \sum_{i=1}^{m} \beta_i f_i(p)$ in place of $Z(p)$, then $\mathrm{E}[Z'(p)] = 0$ (i.e., $\beta = 0$), to which the second-order and intrinsic stationary processes are applicable.

10.2.3 Network variogram models

To predict an unknown value $Z(p_0)$ at an arbitrary point p_0 on $\tilde{L}$ by a kriging method, we should know the covariogram $C(d_S(p_i, p_j))$ or variogram $\gamma(d_S(p_i, p_j))$ for all possible pairs (p_i, p_j), $i, j = 1, \ldots, n$. If we have a large number of datasets

$\{z(p_1), \ldots, z(p_n)\}$ of observed attribute values (say, NO_X values at $p_1, \ldots, p_n$ for 30 days), we can directly estimate $\gamma(d_S(p_i, p_j))$. However, such abundant data are rarely available in practice; we usually have only one set of data (for example, social data are observed only once in most cases). To utilize a small amount of data efficiently, we make the following five assumptions.

Assumption 10.1

The stochastic process $Z(p)$, $p \in \tilde{L}$ satisfies network isotropy.

Assumption 10.2

The stochastic process $Z(p)$, $p \in \tilde{L}$ satisfies the intrinsic stationary process assumption.

Assumption 10.3

The variogram function $\gamma(d)$ is nondecreasing with respect to the shortest-path distance d.

Assumption 10.4

If $d_S(p_i, p_j)$ is sufficiently large, then $\gamma(d)$ is close to $C(0)$. In the limit, $\lim_{d \to \infty} \gamma(d) = C(0)$.

Assumption 10.5

$\gamma(d)$ is discontinuous at the origin and $\lim_{d \to 0} \gamma(d) = \gamma_0 > 0$.

In geostatistical terms, $C(0)$ is called a *sill*, γ_0 a *nugget* and $d^* = \min\{d | \gamma(d) = C(0)\}$ a *range* (see Figure 10.4). The former two terms are geological ones. In geology, a sill is a tabular body of intrusive igneous rock,

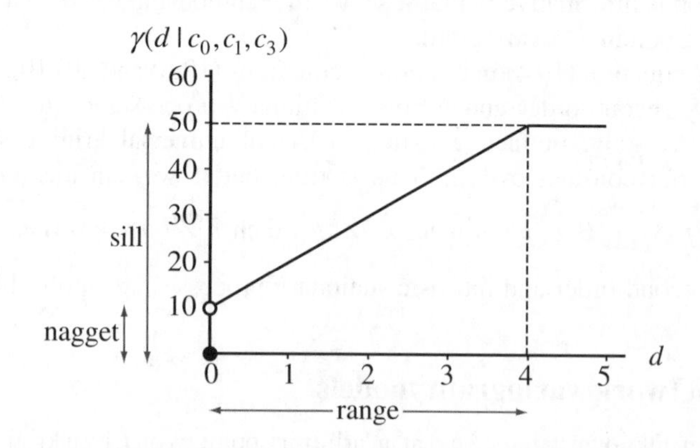

Figure 10.4 A bounded linear variogram model $\gamma(d|c_0, c_1, c_2)$, where $c_0 = 10$ (the nugget), $c_1 = 10$, $c_2 = 4$, the sill is $c_0 + c_1 c_2 = 50$, and the range is 4.

which seemingly appears as a horizontal line of a variogram function (Figure 10.4). A nugget is a gold nugget, which is larger than the gold dust in a gold deposit. The estimation of gold content changes drastically according to whether or not gold nuggets are included in a sample (for details, see IIIA in Journel and Huijbregts, 1991). This drastic change corresponds to the discontinuous behavior of a variogram around the origin (the black and white circles at $d = 0$ in Figure 10.4).

It follows from the above assumptions and equation (10.12) that the positive covariance $\mathrm{Cov}[Z(p_i), Z(p_j)]$ (weakly) decreases as the distance $d_S(p_i, p_j)$ increases; eventually, there is no correlation between $Z(p_i)$ and $Z(p_j)$. This property is a restatement of the first law of geography (mentioned in the introduction to this chapter) in terms of covariances.

The variogram $\gamma(d)$ is a continuous function with respect to the shortest-path distance d, but observable values of $\gamma(d)$ are at finite points. Therefore, we must estimate $\gamma(d)$ by modeling a form of the function $\gamma(d)$, referred to as a *variogram model*. Many forms of variogram model are proposed in the literature, including the bounded linear model, power model, spherical model, exponential model, and Gaussian model (Chapter 6 in Webster and Oliver (2001)). Here we refer to the simplest model, a *bounded linear model*, defined by:

$$\gamma(d|c_0, c_1, c_2) = \begin{cases} 0, & d = 0, \\ c_0 + c_1 d, & 0 < d \leq c_2, \\ c_0 + c_1 c_2, & d > c_2, \end{cases} \tag{10.13}$$

which is depicted in Figure 10.4 when $c_0 = 10$, $c_1 = 10$, and $c_2 = 4$. Note that most variogram models have three parameters.

The next task is to estimate the parameter values of a variogram function, for example, c_0, c_1 and c_2 in equation (10.13). To this end, we should construct a dataset for estimating the parameter values. This set is given by an *experimental variogram*, which is obtained from the following procedure. For ease of explanation, we use a very simple example in Figure 10.5.

First, we construct the set D_k consisting of the shortest-path distances $d_S(p_i, p_j)$ satisfying $d_{k-1} < d_S(p_i, p_j) \leq d_k$, $k = 1, \ldots, m$, where $d_0 = 0$ and d_m is the maximum shortest-path distance among $\{d_S(p_i, p_j),\ i, j = 1, \ldots, n\}$. In the example in

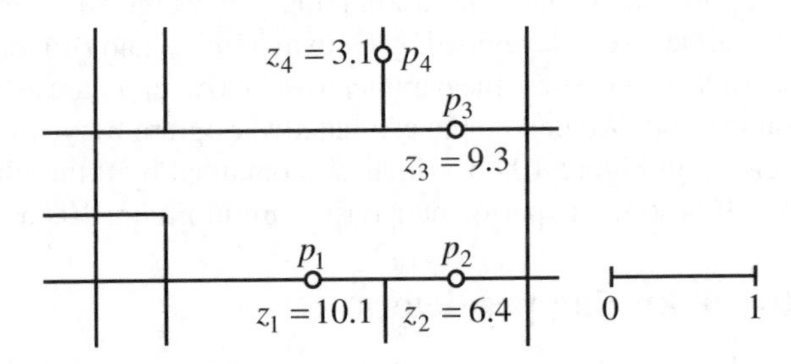

Figure 10.5 The configuration of points $p_1, \ldots, p_4$ on a network and the attribute values $z_1, \ldots, z_4$ at those points.

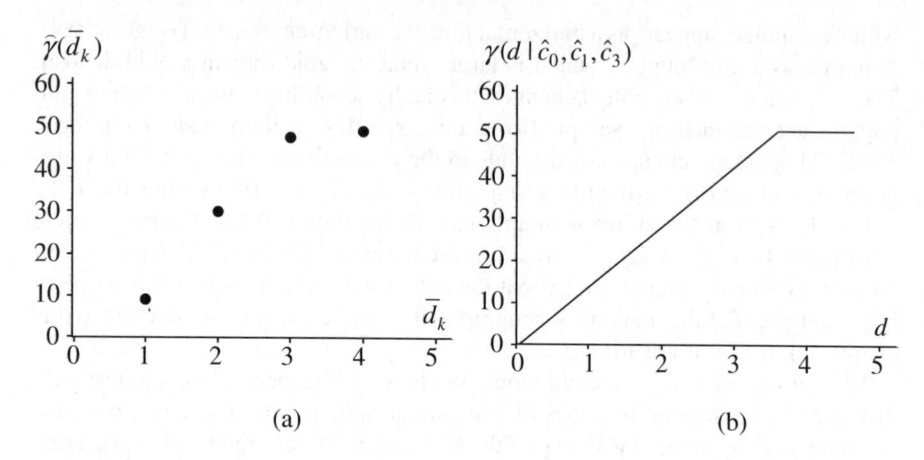

Figure 10.6 (a) The experimental variogram obtained from the attribute values at the points in Figure 10.5, and (b) the estimated variogram for the experimental variogram in (a).

Figure 10.5, we fix $d_1 = 1$, $d_2 = 2$, $d_3 = 3$, and $d_4 = 4$; then $D_1 = \{d_S(p_1, p_2), d_S(p_3, p_4)\}$, $D_2 = \{d_S(p_2, p_3)\}$, $D_3 = \{d_S(p_1, p_3), d_S(p_2, p_4)\}$, $D_4 = \{d_S(p_1, p_4)\}$. (Note that in practice, it is recommended that there should be more than 30 elements in each D_k.)

Second, we compute the average $\bar{d}_k$ of the shortest-path distances included in D_k and regard $\bar{d}_k$ as the representative value of the interval $d_{k-1} < d_S(p_i, p_j) \leq d_k$. In the example in Figure 10.5, $\bar{d}_1 = 1$, $\bar{d}_2 = 2$, $\bar{d}_3 = 3$, and $\bar{d}_4 = 4$. Third, we compute the average of $(z(p_i) - z(p_j))^2$ across all pairs (p_i, p_j) that satisfy $d_S(p_i, p_j) \in D_k$, and regard the resulting average as the value of a function at $\bar{d}_k$, denoted by $\hat{\gamma}(\bar{d}_k)$. The resulting discrete function $\hat{\gamma}(\bar{d}_k)$, $k = 1, \ldots m$ is termed an *experimental variogram*. The experimental variogram of the example in Figure 10.5 is shown in Figure 10.6a.

Given an experimental variogram, we estimate the parameter values c_0, c_1, c_2 of a variogram model, $\gamma(d|c_0, c_1, c_2)$. A simple and naïve method is to minimize the value of $\sum_{k=1}^{m}\{\gamma(\bar{d}_k|c_0, c_1, c_3) - \hat{\gamma}(\bar{d}_k)\}^2$ with respect to c_0, c_1, c_2. Cressie (1985) alternatively proposed a method of minimizing the weighted average, and its computational methods were developed by Gotway (1991), and Brunell and Squibb (1993). Once a criterion is chosen, the parameter values c_0, c_1, c_2 are estimated using an optimization method. We denote the estimated variogram by $\gamma(d|\hat{c}_0, \hat{c}_1, \hat{c}_3)$. An example is depicted in Figure 10.6b, which was obtained by fitting the variogram model in Figure 10.4 to the experimental variogram in Figure 10.6a.

10.2.4 Network kriging predictors

Having estimated a variogram, we now predict an unknown value $Z(p_0)$ at p_0 with the 'best linear unbiased predictor.' The *best predictor* means that the average squared prediction error is the minimum, i.e., the predictor $\hat{Z}(p_0)$ attains the minimum value of $\mathrm{E}[(\hat{Z}(p_0) - Z(p_0))^2]$ among all possible predictors. The *linear predictor* means that the predictor is written as:

$$\hat{Z}(p_0) = w_1 Z(p_1) + \cdots + w_n Z(p_n) = \boldsymbol{w}'\boldsymbol{Z}, \tag{10.14}$$

where

$$\boldsymbol{w} = \begin{pmatrix} w_1 \\ \vdots \\ w_n \end{pmatrix}, \ \boldsymbol{w}' = (w_1, \ldots, w_n), \ \boldsymbol{Z} = \begin{pmatrix} Z(p_1) \\ \vdots \\ Z(p_n) \end{pmatrix}.$$

The *unbiased predictor* means that $\mathrm{E}[\hat{Z}(p_0)] = \beta$ for ordinary kriging and $\mathrm{E}[\hat{Z}(p_0)] = \sum_{i=1}^{m} \beta_i f_i(p_0)$ for universal kriging (if the reader is not familiar with vector and matrix operations, see the derivations in Isaaks and Srivastava (1989, pp. 278–90)).

We first derive the best linear unbiased predictor for ordinary kriging. It follows from $\mathrm{E}[\hat{Z}(p_0)] = \beta$, and equations (10.3) and (10.14) that $\mathrm{E}[\hat{Z}(p_0)] = w_1\beta_1 + \cdots + w_n\beta_1 = \beta_1$. Therefore, $w_1 + \cdots + w_n = 1$ or $\boldsymbol{w}'\boldsymbol{1} = 1$ (where $\boldsymbol{1}$ is the column vector in which each element is 1). The minimization problem is to minimize the value of $\mathrm{E}[(\hat{Z}(p_0) - Z(p_0))^2]$ subject to $\boldsymbol{w}'\boldsymbol{1} = 1$. After several steps of calculation (for details, see Isaaks and Srivastava (1989, pp. 281–4)), $\mathrm{E}[(\hat{Z}(p_0) - Z(p_0))^2]$ is written as $-\boldsymbol{w}'\Lambda\boldsymbol{w} + 2\boldsymbol{w}'\gamma'_0$, where $\gamma'_0 = (\gamma(d_S(p_1, p_0)|\hat{c}_0, \hat{c}_1, \hat{c}_2), \ldots, \gamma(d_S(p_n, p_0)|\hat{c}_0, \hat{c}_1, \hat{c}_2))$ and Λ is the matrix having as its *ij*-th element $\gamma(d_S(p_i, p_j)|\hat{c}_0, \hat{c}_1, \hat{c}_2)$. To solve the minimization problem, we use the Lagrange multiplier method. The objective function is written as $\varphi(\boldsymbol{w}, \mu) = -\boldsymbol{w}'\Lambda\boldsymbol{w} + 2\boldsymbol{w}'\gamma_0 - \mu(\boldsymbol{w}'\boldsymbol{1} - 1)$, where μ is a parameter. Partially differentiating $\varphi(\boldsymbol{w}, \mu)$ with respect to $(\boldsymbol{w}, \mu)$, equating the resulting equations to zero, and solving the resulting simultaneous equations with respect to $(\boldsymbol{w}, \mu)$ (for a detailed derivation, see Isaaks and Srivastava (1989, pp. 284–7)), we obtain:

$$\begin{pmatrix} \boldsymbol{w} \\ \mu \end{pmatrix} = \begin{pmatrix} \Lambda & \boldsymbol{1} \\ \boldsymbol{1}' & 0 \end{pmatrix}^{-1} \begin{pmatrix} \gamma_0 \\ 1 \end{pmatrix}. \tag{10.15}$$

Using Cramer's formula, we can explicitly write $\boldsymbol{w}$ as:

$$\boldsymbol{w} = \Lambda^{-1} \left(\gamma_0 - \frac{\det\begin{pmatrix} \Lambda & \gamma_0 \\ \boldsymbol{1} & 1 \end{pmatrix}}{\det\begin{pmatrix} \Lambda & \boldsymbol{1} \\ \boldsymbol{1} & 0 \end{pmatrix}} \boldsymbol{1} \right), \tag{10.16}$$

where 'det' in the equation means the determinant operator. The best linear unbiased predictor for ordinary kriging is $\hat{Z}(p_0) = \boldsymbol{w}'\boldsymbol{Z}$, where $\boldsymbol{w}$ is given by equation (10.16).

Similarly, we can obtain the best linear unbiased predictor for universal kriging as $\hat{Z}(p_0) = \boldsymbol{w}'\boldsymbol{Z}$, where:

$$\boldsymbol{w} = \Gamma^{-1}\boldsymbol{C}_0 + \Gamma^{-1}\boldsymbol{X}(\boldsymbol{X}^{-1}\Gamma^{-1}\boldsymbol{X})^{-1}(\boldsymbol{x}_0 - \boldsymbol{X}\Gamma^{-1}\boldsymbol{C}_0), \tag{10.17}$$

where X is the matrix having as its ij-th element $f_j(p_i)$, $x_0 = (f_1(p_0), \ldots, f_1(p_n))'$, Γ is the matrix having $C(d_S(p_i, p_j))$ as its ij-th element, and $C_0 = (C(d_S(p_1, p_0)), \ldots, C(d_S(p_n, p_0)))'$. Note that $(X^{-1}\Gamma^{-1}X)^{-1}X'\Gamma^{-1}Z$ is the generalized least square estimator (GLS) for $(\beta_1, \ldots, \beta_m)$ (for GLS, see, e.g., Wonnacott and Wonnacott (1970, Chapter 16)).

As noted in the introduction to Section 10.2, kriging and inverse-distance weighting are conceptually the same in that both predict an unknown value as the weighted average of observed values. However, the weights are different between them. The implications of the difference are discussed by Weber and Englund (1992, 1994), Gotway *et al.* (1996), Zimmerman *et al.* (1999), Malvic and Durekovic (2003), and Lloyd (2005), among others (note that their discussions are about the differences on a plane). To discuss the difference on a network, we present Figure 10.7, where $z_1 = 7.5$, $z_2 = 7.0$, $z_3 = 5.0$ and $z_4 = 4.0$, which are common to the panels (a), (b) and (c).

As we have already noted with Figure 10.2, a disadvantage of the inverse-distance method is that it is insensitive to the configuration of sample points. In fact, if the inverse-distance weighting method is applied to the configurations of sample points in (a) and (c) of Figure 10.7, the weights will not change because the shortest-path distances from the sample points to the prediction point are the same in both cases. In contrast, if kriging is applied, the weights are different between the configurations in (b) and (c) of Figure 10.7. This implies that the kriging method is sensitive to the configuration of sample points. It is suggested that the difference be noted in

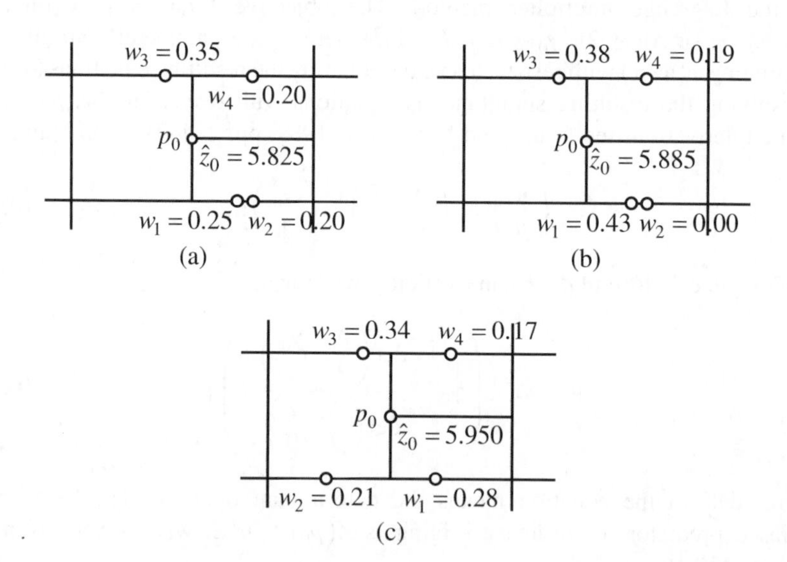

Figure 10.7 Weights and predicted values: (a) the weights obtained from the inverse-distance weighing method of equation (10.1) with $\alpha = 2$; (b), (c) those obtained from the ordinary kriging method of equation (10.16) with the bounded linear experimental variogram of equation (10.13) (this example is personally provided by Y. Li).

weights between the inverse-distance weighting method in (a) and those by the kriging method in (b). In (b), surprisingly, the weight w_2 is almost zero. The reason is that the kriging method considers a positive correlation between the nearby attribute values, for example, $z_1 = 7.5$ and $z_2 = 7.0$, and regards z_1 as the representative value of z_1 and z_2; as a result, the weights are given by $w_1 = 0.43$ and $w_2 = 0.00$. However, this difference should not be overemphasized. The value produced by the weights $w_1 = 0.25$ and $w_2 = 0.20$ in (a) is $0.25 \times 7.5 + 0.20 \times 7.0 = 3.275$, whereas that by $w_1 = 0.43$ and $w_2 = 0.00$ in (b) is $0.43 \times 7.5 + 0.00 \times 7.00 = 3.225$; the difference in the grouped contributions (i.e., $w_1 z_1 + w_2 z_2$) to the prediction value $\hat{z}_0$ is 2%.

In conclusion, there are no all-round spatial interpolation methods. We should carefully choose an appropriate method by considering the characteristics of events in each subject of research, as an appropriate method for one subject is not always appropriate for another subject. In particular, methods may be different between interpolation on a plane and that on a network. In practice, the comparison of interpolation methods in Table 6.3 of Burrough and McDonnell (1998) is helpful for the consideration of an appropriate choice.

10.3 Computational methods for network spatial interpolation

The computations for performing inverse-distance weighting, ordinary kriging and universal kriging are almost algebraic computations except for a few geometric ones, which are the basic computations shown in Chapter 3.

The initial setup is to place the points $p_0, p_1, \ldots, p_n$ on a network $\tilde{L}$. This task is done by inserting those points as new nodes, followed by dividing the associated links into two shorter links (see Section 3.1.5.1 in Chapter 3). The insertion of a new node into a link can be done in constant time, and hence the insertion of all $n + 1$ points is done in $O(n)$ time.

In many cases, points $p_1, p_2, \ldots, p_n$ are fixed because they are given data points, while the point p_0 is free because p_0 is an arbitrary point at which we want to estimate the value. In other words, $p_1, p_2, \ldots, p_n$ are inserted to a given network only once, but p_0 is given repeatedly at different locations. In such a case, $p_1, p_2, \ldots, p_n$ are inserted in a preprocessing stage (a computational method for this task is shown in Section 3.1.5.1 in Chapter 3), and p_0 is inserted when we want to interpolate the value. Once this task is done, the remaining parts of the computations for inverse-distance weighting and kriging are achieved by the methods shown in the following two subsections respectively.

10.3.1 Computational methods for network inverse-distance weighting

The major task is to compute the shortest-path distance from p_0 to the points in a neighborhood $P_N(p_0)$ of p_0. This is done by constructing the shortest-path tree rooted at p_0 by Algorithm 3.5 in Section 3.4.1 of Chapter 3 in

$O((n_V + n)\log(n_V + n))$ time for a planar network $N = (V, L)$, where $n_V = |V|$. Once the distances are obtained, the interpolation of the value at p_0 is done by equations (10.1) and (10.2) in $O(n)$ time. We can achieve this computation by a tool in SANET software (Okabe, Okunuki, and Shiode, 2006a,b; Okabe and Satoh, 2009), as shown in Figure 12.21 in the last chapter of this volume.

10.3.2 Computational methods for network kriging

The major task is to compute the shortest-path distances between $p_0, p_1, \ldots, p_n$. In the preprocessing stage where p_0 is not included, we should compute the shortest-path distances between points $p_1, p_2, \ldots, p_n$. A simple, straightforward method for this computation is to construct the shortest-path trees rooted at p_i for $i = 1, 2, \ldots, n$. This construction can be done in $O(n(n_V + n)\log(n_V + n))$ time, because each shortest-path tree can be constructed in $O((n_V + n)\log(n_V + n))$ time (Section 3.4.1.2 in Chapter 3).

Readers might feel that this way of computing the shortest-path distances between all pairs of points $p_1, p_2, \ldots, p_n$ is not efficient. Indeed, the distance between p_i and p_j would be computed twice: once for the shortest-path tree rooted at p_i and again for that rooted at p_j. To avoid this duplication, we generate $n(n-1)/2$ boxes $A(i, j)$, $1 \leq i < j \leq n$, initialize them empty, and put the value $d_S(p_i, p_j)$ in $A(i, j)$ whenever $d_S(p_i, p_j)$ is computed for the first time, until all the boxes are filled.

To fill the boxes efficiently, the following property is useful.

Property 10.1

Let T_i be the shortest-path tree of network $N = (V, L)$ rooted at p_i. If the path from a leaf node (i.e., a node to which only one link is incident, for example, p_3 in T_1 in Figure 10.8b) to the root p_i in T_i contains points p_j and p_k (e.g., p_2 and p_3), the distance between p_j and p_k measured along this path is the shortest-path distance between them.

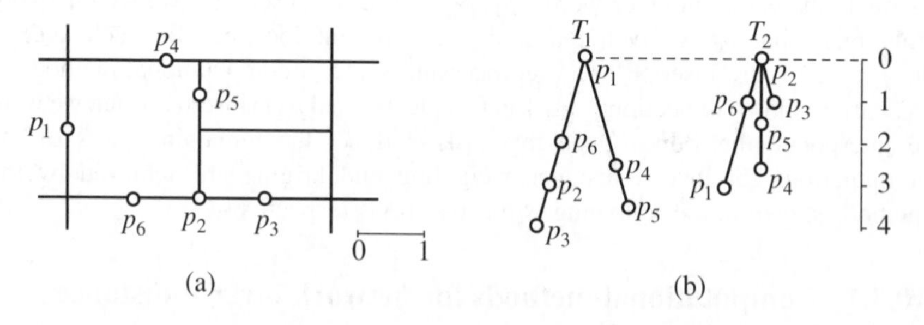

Figure 10.8 Shortest-path trees spanning $p_1, \ldots, p_6$ placed on a network: (a) a network, (b) the shortest-path trees rooted at p_1 and p_2 (the vertical distance indicates the shortest-path distance from the roots).

This property can be proved by contradiction. Suppose that p_j and p_k are contained in this order in the path from a leaf node to the root in T_i, but that there is a shorter path, say s, between p_j and p_k that is not in T_i. Then, the path from p_i to p_j given by the path from p_i to p_k in T_i followed by the inverse path of s that is not in T_{p_i}, is shorter than the path from p_i to p_j in T_i. This contradicts the fact that T_i is the shortest-path tree. This completes the proof.

Because of Property 10.1, we can expect that once the shortest-path tree rooted at p_i is constructed, the shortest-path distance $d_S(p_j, p_k)$ for some pair $j \neq i, k \neq i$ can be obtained in addition to all the shortest-path distances from p_i to other points. Using this property, we can fill the boxes $A(i,j)$, $1 \leq i < j \leq n$, in the following way.

First, we construct the shortest-path tree T_1 rooted at p_1 (e.g., Figure 10.8b). Then we compute the shortest-path distances $d_S(p_1, p_j)$ for all $j = 2, 3, \ldots, n$, and we put them in $A(1,j)$. In the case of Figure 10.8, $A(1,2) = 3$, $A(1,3) = 4$, $A(1,4) = 2.5$, $A(1,5) = 3.5$, $A(1,6) = 2$. Next, for each leaf node p_j in T_1, we list all the points, say, $q_1 = p_j, q_2, \ldots, q_m, q_{m+1} = p_1$ on the path from p_j to the root p_1 in T_1. For instance, $q_1 = p_3, q_2 = p_2,\ q_3 = p_6, q_4 = p_1$ in Figure 10.8. For each pair (q_k, q_l), $1 \leq k, l \leq m$, we compute the shortest-path distance $d_S(q_k, q_l)$ along the path and put it in the associated box if the box is empty (e.g., $A(2,3) = 1$, $A(2,6) = 1$, $A(3,6) = 2$, $A(4,5) = 1$) because Property 10.1 holds. Next we check whether the boxes $A(2,i), i = 3, 4, \ldots, n$ are all filled. If they are, it is not necessary to construct the shortest-path tree rooted at p_2. Otherwise (in the example, $A(2,4)$ and $A(2,5)$ are empty), we construct the shortest-path tree T_2 rooted at p_2 and fill the distances in the boxes $A(2,i)$, $i = 3, 4, \ldots, n$. Furthermore, according to Property 10.1, there may be some pairs, say (q_k, q_l), included in the shortest-path tree T_2. In this case, we put them in the associated boxes $A(k,l)$. In this manner, we construct the shortest-path trees and fill the boxes $A(i,j)$ until no boxes are left empty.

The time complexity of this method depends on the structure of a given network and on the order in which we choose the roots of the shortest-path trees. In a lucky case, it might happen that the shortest-path tree rooted at p_1 is just a single path starting at p_1. Then, the shortest-path distances between all pairs of points can be computed as the path distance along this tree. In the worst case, on the other hand, we must construct n shortest-path trees rooted at all n points, and hence the worst-case time complexity is still $O(n(n_V + n) \log (n_V + n))$. However, the coefficient that should be multiplied to this order to find the time actually required can be expected to be much smaller than that of the simple method. For this reason, we recommend the above box-filling method for practical computations.

Once we know the shortest-path distances for all pairs of points, we can estimate the variogram by the method in Section 10.2.3, and can predict the value at p_0 by equations (10.14) and (10.16) for ordinary kriging, and by equations (10.14) and (10.17) for universal kriging.

This property can be proved by contradiction. Suppose that p_j and p_k are contained in this order in the path from a leaf node to the root of T_i, but that there is another path q_k between p_i and p_k that is shorter than the path from p_k through p_j to p_i. Then the path from p_j to p_i is shorter if [illegible] followed by the [illegible] of a tree vertex in T_i is shorter if the path from p_j to p_i in T_i. This contradicts the fact that T_i is the shortest-path tree. This completes the proof.

Because of Property 10.1, we can expect that, once the shortest-path tree rooted at p_i is constructed, the shortest-path distances $d_s(p_i, p_j)$ for some pairs of p_j and p_k can be obtained in addition to the distances between p_i and p_j, for other points. Using this property, we can fill the boxes of (p_j, p_k) of T_i [illegible] in the following way.

First, we establish the shortest-path tree T_i from p_i as shown in Figure 10.6(a). Then we compute the shortest-path distance $d_s(p_i, p_j)$ for all p_j [illegible] and we put them in $A(1, j)$. In the case of Figure 10.6, $A(1, 2) = 1$, $A(1, 3) = 2$, $A(1, 4) = 3$, [illegible]. Next, for each leaf node p_j in T_i, we trace all the vertices on the path from p_j to p_i in the path from p_j to the root in T_i [illegible]. In Figure 10.6, for example, [illegible]. Then we compute the shortest-path distance $d_s(p_j, p_k)$ to obtain $A(2, 3)$, and put them in the boxes we visited [illegible]. In the example [illegible] $A(3, 4) = 1$, $A(2, 4) = 2$, [illegible]. [illegible]. Similarly for the [illegible] [illegible] consider p_3 and [illegible]. Then the [illegible] are empty. We construct the shortest-path tree rooted at p_j and fill the [illegible] boxes [illegible]. Furthermore, according to Property 10.1, these [illegible] are [illegible] in the shortest-path tree T_j. In this case, we put them into the associated boxes [illegible]. In this manner, we construct the shortest-path trees and fill the boxes [illegible] until no empty boxes are left.

The time complexity of this method depends on the structure of a given network and on the order in which we choose the roots of the shortest-path trees. In the best case, it could happen that the [illegible] are [illegible]. In this case, the total time for computing the shortest-path distances between all pairs of points is [illegible]. In the worst case, [illegible] [illegible] [illegible] trees, and hence the worst-case time complexity is still $O(n \log n)$ [illegible]. However, the actual time should be small when [illegible] because the [illegible] usually required can be expected to be much smaller than [illegible]. For [illegible], we recommend that we should use this method more often in actual computation.

Once we know the shortest-path distances [illegible] all pairs of [illegible], we can estimate the values [illegible] the method in Section 10.[illegible] and compute [illegible] the [illegible] (10.[illegible]) and (10.[illegible]) for [illegible] (10.[illegible]) and (10.7) for the universal kriging.

11

Network Huff model

This chapter presents a probabilistic model of a user choosing a facility from a number of alternative facilities alongside a network. The model can answer, for example, question Q8 in Chapter 1: How can we estimate the probability of a consumer choosing a specific fast-food shop among alternative shops located alongside streets in a downtown area? Closely related questions are as follows:

Q8′ : How can we estimate the demand for a specific fast-food shop alongside downtown streets?

Q8″: To locate a new fast-food shop alongside a street in a downtown area, which site is most profitable?

These questions have been the central subject of market area analysis for over a hundred years. Pioneering work was done by Launhardt (1885), but extensive studies started in the first half of the twentieth century. Early contributors were Fetter (1924), Reilly (1931), Palander (1935), Converse (1945), Stewart (1948), Hyson and Hyson (1950), Reynolds (1953), Greenhut (1952), Carrothers (1956), and Jung (1959), among others. It should be noted that these models did not explicitly consider the probabilistic nature of market areas. In the 1960s, Huff (1963, 1964, 1966) reformulated those models as a probabilistic model, called the *Huff model*, by applying behavioral choice theory (Luce, 1959; Ben-Akiva and Lerman, 1985; an axiomatic formulation of the Huff model was developed by Smith (1975)). Since then, the Huff model has been empirically tested and modified by many researchers; for instance, Bucklin (1971), Nakanishi and Cooper (1974), Mahajan,

Spatial Analysis along Networks: Statistical and Computational Methods, First Edition.
Atsuyuki Okabe and Kokichi Sugihara.

Jain, and Bergier (1977), Hodgson (1981), Weisbrod, Parcells, and Kern (1984), Wee and Pearce (1985), Berry and Parr (1988), Birkin *et al.* (1996), Diplock and Openshaw (1996), Plastria (2001), Brodsky (2003), Fernandez *et al.* (2007), Fernandez *et al.* (2007), and Toth *et al.* (2008). Today, the Huff model is still one of the most frequently used models in practical market area analysis (see, e.g., Wang (2006, Section 4.2.2)).

The original Huff model assumes that a market area is a continuous plane with Euclidean distances, referred to as the *planar Huff model*. This assumption is acceptable when market areas are large, but it becomes very restrictive when retailers analyze their detailed location strategies in an urbanized area, such as locating fast-food shops in a downtown area. The retailers assess which street and which site on the street should be the most profitable, considering consumers' trip behavior on streets, which often include one-way streets. Therefore, retailers wish to use the Huff model formulated on a street network with shortest-path distances. This model is referred to, in contrast to the planar Huff model, as the *network-constrained Huff model* or the *network Huff model* in short. The network Huff model became feasible with progress in geographic information systems (GIS) and the increase in digital street data in the 1990s. Reflecting this feasibility and data availability, a new concept, *micromarketing*, appeared in the marketing literature (Buxton, 1992). In response to this marketing, several network Huff models were proposed by Miller (1994), Okabe, Yomono, and Kitamura (1995), Okabe and Kitamura (1996), Okabe and Okunuki (2001), and Okunuki and Okabe (2002). Based on these studies, this chapter describes the network Huff model.

The chapter is organized into three sections. The first section conceptually formulates the network Huff model and its associated applications. The second and third sections develop computational methods for implementing the conceptual models formulated in the first section. Specifically, the second section shows how to compute the demand for a store, assuming that consumers' choice behavior follows the Huff model. The last section illustrates how to search for an optimal location at which the demand for a newly entering store is the largest.

11.1 Concepts of the network Huff model

In the following subsections, we mathematically describe the concepts of the network Huff model and their associated applications: the general Huff model, the planar Huff model, the network Huff model, dominant market areas (subnetworks), Huff-based demand estimation and Huff-based locational optimization.

11.1.1 Huff models

To state a general form of the Huff model, we consider a space S (which may be a plane or a network), in which n stores are located at $p_1, \ldots, p_n$. Let a_i be the attractiveness of store i, which may be a function of its floor area, the number of items sold, its parking area and so forth; let $d(p, p_i)$ be the distance between a point

p on S and the store at p_i, which may be the Euclidean distance or the shortest-path distance; and let $F(d(p,p_i))$ be a monotonically decreasing function of $d(p,p_i)$, referred to as a *distance decay function* or *distance deterrence function.* In these terms, the Huff model showing the probability of a consumer at p choosing the store at p_i is generally written as:

$$P_i(p) = \frac{a_i F(d(p,p_i))}{\sum_{k=1}^{n} a_k F(d(p,p_k))}. \tag{11.1}$$

The original Huff model, namely the *planar Huff model*, assumes that the space S is a plane, the distance $d(p,p_i)$ is the Euclidean distance and the distance decay function $F(d(p,p_i))$ is given by $d(p,p_i)^{-\alpha}$, $\alpha > 0$.

The extension of the planar Huff model to the network Huff model is conceptually straightforward. To describe it explicitly, we consider a network $N = (V,L)$ consisting of a set of nodes $V = \{v_1,\ldots,v_{n_V}\}$ and a set of links $L = \{l_1,\ldots,l_{n_L}\}$, and suppose that n stores are located at $p_1,\ldots,p_n$ on $\tilde{L} = \bigcup_{i=1}^{n_L} l_i$ (i.e., the points forming the network including the nodes of V). We measure the distance between a point p on $\tilde{L}$ and a store i at p_i by the shortest-path distance $d_S(p,p_i)$ on the network N, and we use the distance decay function $F(d_S(p,p_i)) = \exp(-\alpha\, d_S(p,p_i))$ in place of $F(d_S(p,p_i)) = d_S(p,p_i)^{-\alpha}$. One reason for adopting the former function is that the value of the latter function becomes extremely large for a consumer living next door to a store, and this extreme is not acceptable in practical marketing. In addition, the former function is more frequently used in transportation studies (e.g., Wilson, 1970; Ben-Akiva and Lerman, 1985). In these terms, the *network Huff model* is written as:

$$P_i(p) = \frac{a_i \exp(-\alpha\, d_S(p,p_i))}{\sum_{k=1}^{n} a_k \exp(-\alpha\, d_S(p,p_k))}. \tag{11.2}$$

A consumer at p chooses stores 1, ..., n with probabilities $P_1(p)$, ..., $P_n(p)$ given by equation (11.2).

11.1.2 Dominant market subnetworks

The market area of the planar Huff model is often indicated by *equal-probability lines* (*isolines*) on a plane (e.g., Figure 2.16 in Berry (1967)). In the case of the network Huff model, the isolines correspond to *equal-probability points* on $\tilde{L}$, which are derived from solving the following equation with respect to p for a given probability ρ (say, $\rho = 0.80$):

$$\frac{a_i \exp(-\alpha\, d_S(p,p_i))}{\sum_{k=1}^{n} a_k \exp(-\alpha\, d_S(p,p_k))} = \rho. \tag{11.3}$$

Note that the equal-probability lines of the planar Huff model are difficult to obtain analytically, but the equal-probability points of the network Huff model can be derived analytically. The derivation will be shown in Section 11.2.

The market areas of the Huff model overlap each other, because the consumers at p choose more than one store with the probabilities given by equation (11.2). For every point p on $\tilde{L}$, there exists a store that attains the maximum probability. Mathematically, if the relation $P_i(p) \geq P_j(p)$ holds for $j \neq i$, $j = 1, \ldots, n$, we call store i the *dominant store* at p, and the subnetwork consisting of the points at which store i is the dominant store the *dominant market subnetwork*, $L_{\mathrm{D}i}$, of store i. In mathematical terms, the dominant market subnetwork $L_{\mathrm{D}i}$ is written as:

$$L_{\mathrm{D}i} = \{p | P_i(p) \geq P_j(p), p \in \tilde{L}, j \neq i, j = 1, \ldots, n\}. \tag{11.4}$$

Substituting equation (11.2) into equation (11.4), we can alternatively write equation (11.4) as:

$$L_{\mathrm{D}i} = \{p | d_{\mathrm{S}}(p, p_i) - (\log a_i)/\alpha \leq d_{\mathrm{S}}(p, p_j) - (\log a_j)/\alpha, j \neq i, j = 1, \ldots, n\}. \tag{11.5}$$

Recalling the weighted Voronoi diagrams in Chapter 4, we notice that the dominant market subnetwork of store i is equivalent to the additively weighted network Voronoi diagram defined in Section 4.2.2 of Chapter 4.

11.1.3 Huff-based demand estimation

The Huff model provides only choice probabilities. To estimate the demand for store i, we should consider the density of consumers. Let $w(p)$ be the consumer density at p. Then, the demand, $D(p_i)$, for store i is obtained by integrating $P_i(p)w(p)$ across the domain $\tilde{L}$, or mathematically:

$$D(p_i) = \int_{p \in \tilde{L}} P_i(p) w(p) \mathrm{d}p = \int_{p \in \tilde{L}} \frac{a_i \exp(-\alpha\, d_{\mathrm{S}}(p, p_i))}{\sum_{k=1}^{n} a_k \exp(-\alpha\, d_{\mathrm{S}}(p, p_k))} w(p) \mathrm{d}p, \tag{11.6}$$

where dp is the integration operator, symbolically indicating an infinitesimal line segment including p on $\tilde{L}$ (note that dp is one dimensional). For the planar Huff model, even if the consumer density is uniform over the whole market area, i.e., $w(p) = w$, $p \in S$, the analytical derivation of $D(p_i)$ is intractable. Clearly, the uniformity assumption is unacceptable in practice. An alternative practical assumption is that the consumer density on a link l_h is uniform, $w(x) = w_h$, but it varies from link to link, l_h, $h = 1, \ldots, n_L$. Equation (11.6) is then written as:

$$D(p_i) = \sum_{h=1}^{n_L} \int_{p \in l_h} \frac{a_i \exp(-\alpha\, d_{\mathrm{S}}(p, p_i))}{\sum_{k=1}^{n} a_k \exp(-\alpha\, d_{\mathrm{S}}(p, p_k))} w_h \, \mathrm{d}p. \tag{11.7}$$

This integration still looks difficult to calculate, but it becomes tractable if equation (11.7) is written in terms of the 'shortest-path tree distance' to be defined in Section 11.2.

11.1.4 Huff-based locational optimization

Retailers wish to find which street, and which site on the street, is the most profitable. This type of locational optimization problem has been extensively studied in operations research (OR) since Hakimi (1965) (reviews on these studies are provided by Tansel, Francis, and Lowe (1983), Brandeau and Chiu (1989), Mirchandani and Francis (1990), Hale and Moberg (2003), and Revelle and Eiselt (2005)). The early models assumed that consumers deterministically choose only the nearest store, and demands are placed at nodes. Noticing that this assumption is unrealistic in micromarketing, Okunuki and Okabe (2002) developed the Huff-based competitive location model on a network in which demand is continuously distributed on the network. Similar models on a plane were formulated by Drezner (1994), Drezner, Drezner, and Salhi (2002), Drezner and Drezner (2004), Fernandez *et al.* (2007), and Toth *et al.* (2008).

Given that n stores are located on $\tilde{L}$, we wish to find the location of a new store, referred to as store 0, that attains the maximum demand. Let p_0 be a temporarily fixed location of store 0. Then, it follows from equation (11.7) that the demand for store 0 at p_0 is written as:

$$D(p_0) = \sum_{h=1}^{n_L} \int_{p \in l_h} \frac{a_0 \exp(-\alpha\, d_S(p, p_0))}{a_0 \exp(-\alpha\, d_S(p, p_0)) + \sum_{k=1}^{n} a_k \exp(-\alpha\, d_S(p, p_k))} w_h \, dp. \tag{11.8}$$

The optimization problem is to maximize $D(p_0)$ with respect to p_0 on $\tilde{L}$, or mathematically:

$$\max_{p_0 \in \tilde{L}} D(p_0) = \max_{p_0 \in \tilde{L}} \sum_{h=1}^{n_L} \int_{p \in l_h} \frac{a_0 \exp(-\alpha\, d_S(p, p_0))}{a_0 \exp(-\alpha\, d_S(p, p_0)) + \sum_{k=1}^{n} a_k \exp(-\alpha\, d_S(p, p_k))} w_h \, dp. \tag{11.9}$$

A computational method for solving this problem will be discussed in Section 11.3.

11.2 Computational methods for the Huff-based demand estimation

Having formulated the demand function and the locational optimization problem above, we now wish to develop computational methods for them. This section describes a method for computing the demand function, and the next section illustrates a method for solving the locational optimization problem.

The demand for store i is conceptually given by equation (11.7), but if we want actually to compute the demand, we must compute the integration in equation (11.7).

To make this integration analytically solvable, we rewrite the shortest-path distances in equation (11.7) in terms of the shortest-path tree defined in Chapter 3. With this distance, we can rewrite equation (11.7) and derive the analytical solution to its integration.

11.2.1 Shortest-path tree distance

We first show how to compute the shortest-path distance in equation (11.7) with the aid of the shortest-path trees rooted at stores. The shortest-path tree and its extension have already been described in Section 3.4.1 in Chapters 3, but they are explained here again to introduce notation to be used only in this chapter. For illustrative purposes, we depict a simple network in Figure 11.1a, where the network $N = (V, L)$ consists of $V = \{v_1, v_2\}$ (the black circles) and $L = \{l_1, l_2, l_3\}$; store 1 is located at p_1 and store 2 at p_2 (the white circles), respectively. We construct the shortest-path trees rooted at p_1 and p_2, which are indicated by the heavy arrowed lines in Figure 11.1b and c, respectively. These trees do not completely cover the links of the network (e.g., they include l_2 and l_3 in Figure 11.1b, and l_1 and l_3 in Figure 11.1c). On each uncovered link, there exists a point, b_{ij}, at which the shortest-path distance from its root p_i to the point b_{ij} through one end of the link is equal to that through the other end of the link (for example, in Figure 11.2a, $d_S(p_1, v_1) + d_S(v_1, b_{11}) = d_S(p_1, v_2) + d_S(v_2, b_{11})$). We call such a point a *breakpoint*. The breakpoints for the shortest-path trees in Figure 11.1b and c are shown by the gray circles in Figure 11.2a and b, respectively.

Let $P = \{p_1, \ldots, p_n\}$ be the locations of the n stores, and let B_P be the set of breakpoints for the shortest-path trees rooted at the n points (stores) of P. We insert the points of P and B_P on the links of the original network $N = (V, L)$ as nodes. The resulting node set is denoted by $V^+ = V \cup P \cup B_P$ (the white, gray and black circles in Figure 11.2c). The insertion of the nodes $P \cup B_P$ refines the original links, as in Figure 11.2c. The resulting refined link set is denoted by L^+. Through this procedure, we obtain the refined network $N^+ = (V^+, L^+)$. For this network, we again construct the shortest-path trees rooted at $p_1, \ldots, p_n$. The resulting shortest-path trees are called *extended shortest-path trees* (for precise definition, see

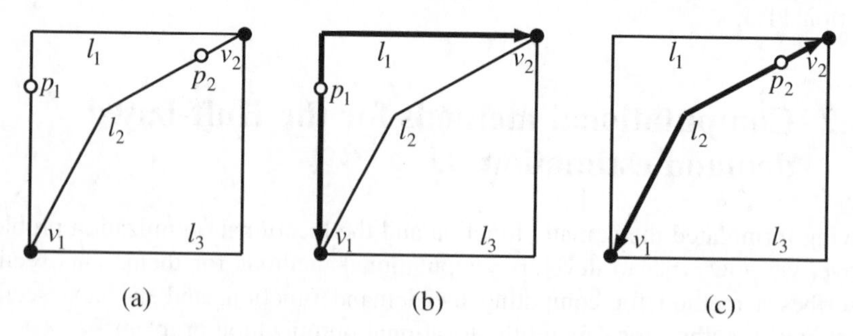

Figure 11.1 (a) A network on which two stores are located at p_1 and p_2; (b) the shortest-path tree rooted at p_1; and (c) the shortest-path tree rooted at p_2.

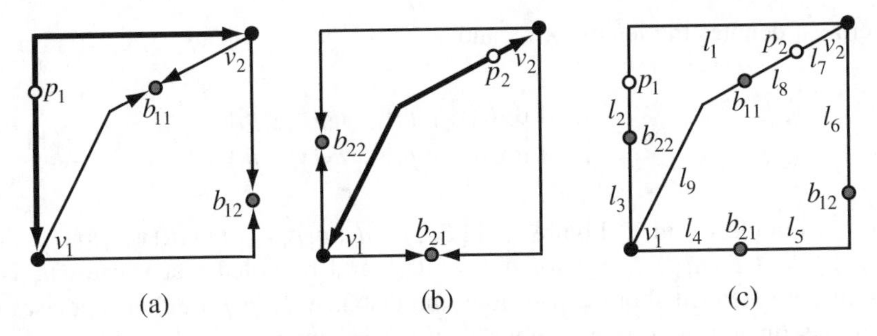

Figure 11.2 (a) The extended shortest-path tree rooted at p_1, (b) that at p_2, and (c) the refined network $N^+ = (V^+, L^+)$ of the network $N = (V, L)$ in Figure 11.1a.

Section 3.4.1.3 in Chapter 3). Examples are shown in Figure 11.3a and b (the arrowed line segments). Note that for convenience, we treat the links incident to a breakpoint as part of the shortest-path tree (e.g., the links l_8 and l_9 are incident to b_{11}) and regard the breakpoints as end nodes of both incident links which are virtually separated at the same location (this treatment is described in Section 3.4.1.3 in Chapter 3).

Because we can easily compute the shortest-path distance from p_i to every node using the extended shortest-path tree rooted at p_i, the shortest-path distance from p_i to a point p on a link $l_h \in L^+$ (the square in Figure 11.3c) is the sum of the shortest-path distance from p_i to a node of the link l_h and the shortest-path distance from that node to p. Let $v_{l_h 1}$ and $v_{l_h 2}$ be the end nodes of l_h, and $d_S(p, v_{l_h 1}) = t$ (e.g., in Figure 11.3c, $v_{l_5 1} = b_{21}$ and $v_{l_5 2} = b_{12}$ for l_5). Then, the shortest-path distance from a point p on l_h to the store at p_i, i.e., $d_S(p, p_i)$, can be written as:

$$d_S(p, p_i) = d_S(p, v_{l_h 1}) + d_S(v_{l_h 1}, p_i) = \delta_{hi} t + d_S(v_{l_h 1}, p_i) \text{ for } 0 \leq t \leq |l_h|, p \in l_h, \tag{11.10}$$

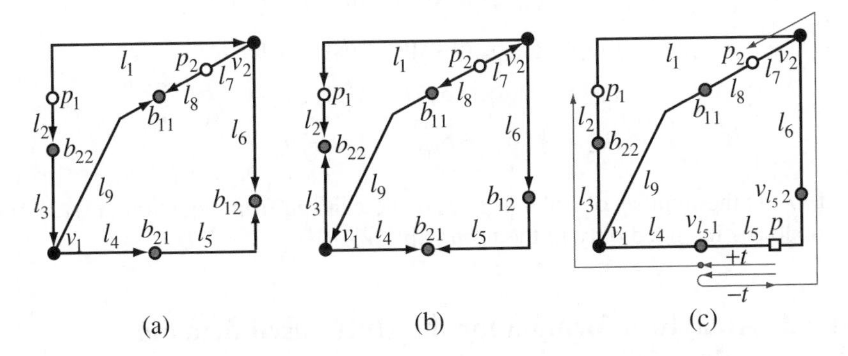

Figure 11.3 (a) The extended shortest-path tree rooted at p_1, (b) that rooted at p_2, and (c) the shortest-path tree distance $d_S(p, p_i) = \delta_{hi} t + d(v_{l_h 1}, p_i)$ of equation (11.10).

where $|l_h|$ denotes the length of l_h, and:

$$\delta_{hi} = \begin{cases} +1 \text{ if } d_S(v_{l_h1}, p_i) \leq d_S(v_{l_h2}, p_i), \\ -1 \text{ if } d_S(v_{l_h1}, p_i) > d_S(v_{l_h2}, p_i). \end{cases} \tag{11.11}$$

For example, in Figure 11.3c, $d_S(p, p_1) = t + d_S(v_{l_h1}, p_1)$ and $d_S(p, p_2) = -t + d_S(v_{l_h1}, p_2)$ for $p \in l_5$. It should be noted that because l_h is a link of the extended shortest-path tree, the distance $d_S(p, p_i)$ either increases or decreases linearly as p moves along l_h; it never happens that $d_S(p, p_i)$ increases (decreases) and then decreases (increases) within l_h. The shortest-path distance defined by equation (11.10) is obtained from the (extended) shortest-path trees, and so we term it the *shortest-path tree distance*.

11.2.2 Choice probabilities in terms of shortest-path tree distances

We now wish to rewrite equation (11.2) by substituting the shortest-path tree distance of equation (11.10) into equation (11.2). After several steps of calculation, the probability $P_i(p) = P_i(t)$ that consumers at p on the link l_h or t from the end node v_{l_h1} of l_h will choose store i at p_i, is written as:

$$P_i(t) = \frac{c_{hi1}}{c_{hi2} + c_{hi3}\exp(2\alpha\delta_{hi}t)}, 0 \leq t \leq |l_h|, h = 1, \ldots, n_{L^+}, \tag{11.12}$$

where

$$\begin{aligned} c_{hi1} &= a_i \exp(-\alpha d_S(v_{l_h1}, p_i)), \\ c_{hi2} &= \textstyle\sum_{k \in J_{hi1}} a_k \exp(-\alpha d_S(v_{l_h1}, p_k)), \\ c_{hi3} &= \textstyle\sum_{k \in J_{hi2}} a_k \exp(-\alpha d_S(v_{l_h1}, p_k)), \\ J_{hi1} &= \{j | \delta_{hj} = \delta_{hi}, j = 1, \ldots, n_{L^+}\}, \\ J_{hi2} &= \{j | \delta_{hj} \neq \delta_{hi}, j = 1, \ldots, n_{L^+}\} \end{aligned}$$

and n_{L^+} is the number of links in L^+. As we can see from equation (11.12), the variable t is included only in the term $\exp(2\alpha\delta_{hi}t)$.

11.2.3 Analytical formula for the Huff-based demand estimation

Because the market areas of the Huff model are probabilistic, the boundaries of the market areas are not distinct but are indicated by a set of the equal-probability points defined in Section 11.1.2. These points are obtained from equation (11.3), which is alternatively written with equation (11.12) as:

$$\frac{c_{hi1}}{c_{hi2} + c_{hi3}\exp(2\alpha\delta_{hi}t)} = \rho. \tag{11.13}$$

Solving this equation with respect to t, we obtain:

$$t = \frac{1}{2\alpha\delta_{hi}}\log\left(\frac{c_{hi1}}{\rho c_{hi3}} - \frac{c_{hi2}}{c_{hi3}}\right). \tag{11.14}$$

If t satisfies $0 \leq t \leq |l_h|$, then the consumer at distance t from v_{l_h1} (the endpoint of l_h) on link l_h chooses store i with probability ρ. The set of equal-probability points for $P_i(t) = \rho$ is given by the set of t values satisfying equation (11.14) and $0 \leq t \leq |l_h|$, $h = 1, \ldots, n_{L^+}$.

The demand $D(p_i)$ for store i at p_i is, from equations (11.7) and (11.12), written as:

$$D(p_i) = \sum_{h=1}^{n_{L+}} \int_0^{|l_h|} \frac{c_{hi1}w_h}{c_{hi2} + c_{hi3}\exp(2\alpha\delta_{hi}t)}\mathrm{d}t. \tag{11.15}$$

In the planar Huff model, the integration corresponding to equation (11.15) is not analytically solvable, but in the network Huff model, noticing that the integration in equation (11.15) is analytically solvable, we can obtain the analytical solution as:

$$D(p_i) = \sum_{h=1}^{n_{L+}} \frac{c_{hi1}w_h}{2\alpha\delta_{hi}c_{hi2}}\{2\alpha\delta_{hi}|l_h| - \log[c_{hi2} + c_{hi3}\exp(2\alpha\delta_{hi}|l_h|)] + \log(c_{hi2} + c_{hi3})\}. \tag{11.16}$$

Using this equation, we can exactly estimate the demand for store i at p_i.

11.2.4 Computational tasks and their time complexities for the Huff-based demand estimation

Computational tasks to achieve the above computations are:

(i) inserting the breakpoints in the original network,

(ii) generating the extended shortest-path trees of the resulting network,

(iii) computing the demand according to equation (11.16).

First, consider task (i), or the number of breakpoints to be added. To add breakpoints, we consider two cases of the shortest-path tree, T_{p_i}, rooted at p_i on the network $N = (V, L)$. The first case, referred to as the *nondegenerate case*, is that the shortest-path tree is unique; that is, there are not two shortest paths from p_i to any node v of

V in the original network $N = (V, L)$. In this case, for each link $l = (v_1, v_2)$ that does not belong to T_{p_i}, we can locate a breakpoint b on l, satisfying $d_S(p_i, v_1) + d_S(v_1, b) = d_S(p_i, v_2) + d_S(v_2, b)$ (e.g., for p_1, b_{11} and b_{12} in Figure 11.2a). Such a point always exists uniquely for each link l that does not belong to T_{p_i} (the hair lines in Figure 11.1a and b). The second case, referred to as the *degenerate case*, is that the shortest-path tree is not unique; i.e., there exist two shortest paths from p_i to a node v of V, say path $(p_i, \ldots, v_1, v)$ and path $(p_i, \ldots, v_2, v)$, $v_1 \neq v_2$, satisfying $d_S(p_i, v_1) + d_S(v_1, v) = d_S(p_i, v_2) + d_S(v_2, v)$. In this case, we arbitrarily choose one path, say $(p_i, \ldots, v_1, v)$, of the two shortest paths and include path $(p_i, \ldots, v_1, v)$ and $(p_i, \ldots, v_2)$ in T_{p_i} (link (v_2, v) is deleted). It should be noted that the breakpoint b coincides with v ($b = v$); consequently, a new node b does not appear.

To evaluate the time complexity, we consider the worst case; i.e., when the number of nodes becomes the largest by the addition of the breakpoints. The worst case occurs for the nondegenerate case, and we assume that case in the following. Under this assumption, T_{p_i} has exactly $n_V - 1$ links. Because there is exactly one breakpoint on each link that does not belong to T_{p_i}, the number of breakpoints is equal to $n_L - (n_V - 1) = n_L - n_V + 1$ for one shortest-path tree rooted at p_i. We locate the breakpoints for all roots $p_1, p_2, \ldots, p_n$. As a result, the total number of breakpoints amounts to $n(n_L - n_V + 1) = O(n\, n_L)$. The time required to construct the shortest-path tree for each p_i is of order $O(n_L \log n_L)$. Therefore, the total time complexity for the task (i) is of order $O(n\, n_L \log n_L)$.

Second, consider task (ii). The total number of breakpoints is $n(n_L - n_V + 1)$, so the total number of links of the refined network N^+ is equal to $|L^+| = n_L + n(n_L - n_V + 1) = O(n\, n_L)$ (where $|L^+|$ indicates the number of elements in the set L^+). We already constructed the shortest-path trees rooted at $p_i, i = 1, \ldots, n$ in task (i). Hence, what we have to do is to copy the tree structure information to the refined links. We do this n times, and so we require $O(n^2\, n_L)$ time in total.

Last, consider task (iii). To compute the demand $D(p_i)$ for each link of N^+ on which p (or t in equation (11.12)) lies, we must construct the index sets J_{hi1} and J_{hi2}, and the three values c_{hi1}, c_{hi2} and c_{hi3}. They require $O(n)$ time because they are operations over the n stores. We execute these computations $|L^+|$ times to obtain the summation in equation (11.16), and hence the task (iii) requires $|L^+|O(n) = O(n^2\, n_L)$ time.

In summary, the total time for computing the demand $D(p_i)$ is the sum of the times for tasks (i), (ii), and (iii). Therefore, the time complexity is of order $O(n\, n_L \log n_L) + O(n^2\, n_L) + O(n^2\, n_L) = O(n^2\, n_L)$.

11.3 Computational methods for the Huff-based locational optimization

We now wish to develop computational methods for solving the second problem referred to in the introduction, namely, the locational optimization problem given by equation (11.9); i.e., to maximize $D(p_0)$ with respect to p_0 given that n stores are located at $p_1, \ldots, p_n$ on $\tilde{L}$. The computational procedure for solving this problem takes the following four steps.

First, we formulate the demand function $D(p_0)$ with respect to its temporal location p_0 on $\tilde{L}$. For this formulation, we will define the refined link set L^{+0}. With this demand function, we try to find p_0 at which $dD(p_0)/dp_0 = 0$ holds, but this method is only effective when $\mathrm{d}D(p_0)/\mathrm{d}p_0$ exists.

In the second step, therefore, we try to find the points at which $\mathrm{d}D(p_0)/\mathrm{d}p_0$ exist (or the points at which $\mathrm{d}D(p_0)/\mathrm{d}p_0$ do not exist). To find such points, we will define the link set L^-.

The third step is to compute $\mathrm{d}D(p_0)/\mathrm{d}p_0$ by moving p_0. The algebraic form of $\mathrm{d}D(p_0)/\mathrm{d}p_0$ is written in terms of the lengths of the links of L^{+0} (not L^-). Therefore, we wish to keep the topology of the link set L^{+0} the same as p_0 moves. For this purpose, we will define the link set L^*, in such a way that the graph of the network $N^{+0} = (V^{+0}, L^{+0})$ does not change as p_0 moves along every link of the link set L^* (for graph, see Section 2.2.1.2 in Chapter 2).

The fourth step is to solve $\mathrm{d}D(p_0)/\mathrm{d}p_0 = 0$ on each link of L^*, except at the points where $\mathrm{d}D(p_0)/\mathrm{d}p_0$ does not exist, and to find an optimal location on each link of L^*. Comparing the demand values of the optimal locations on the links of L^* and the demand values at the points where $\mathrm{d}D(p_0)/\mathrm{d}p_0$ does not exist, we obtain the location for the new store that captures the maximum demand, or mathematically, the solution to equation (11.9).

The details of the four steps are described in the following four subsections. As mentioned in the above outline of the procedure, we will use four types of link set: L^+, L^{+0}, L^- and L^* in the following derivations. The reader should carefully distinguish these sets with the aid of the examples in Figure 11.4.

11.3.1 Demand function for a newly entering store

The derivation of the demand function for a newly entering store, store 0, is almost the same as that for store i in Section 11.2. Let B_0 be the breakpoints of p_0 (an illustrative example is shown in Figure 11.4a, where $B_0 = \{b_{01}, b_{02}\}$), and $V^{+0} = V^+ \cup \{p_0\} \cup B_0 = V \cup P \cup B_P \cup \{p_0\} \cup B_0$. The insertion of the point p_0 and the nodes of B_0 refines the links of L^+ (an example of L^+ is shown in Figure 11.2c). The resulting refined link set is denoted by L^{+0} (the links in L^{+0} are reindexed in Figure 11.4a).

On the network $N^{+0} = (V^{+0}, L^{+0})$, the demand for store 0 at a temporally fixed location p_0 is, from equation (11.16), written as:

$$D(p_0) = \sum_{h=1}^{n_{L^{+0}}} \frac{w_h a_0 \exp(-\alpha\, d_S(v_{l_h 1}, p_0))}{2\alpha\delta_{h0}(a_0 \exp(-\alpha\, d_S(v_{l_h 1}, p_0)) + c_{h02})}$$
$$\times\{2\alpha\delta_{h0}|l_h(p_0)| - \log[a_0 \exp(-\alpha\, d_S(v_{l_h 1}, p_0)) + c_{h02}$$
$$+c_{h03} \exp(2\alpha\delta_{h0}|l_h(p_0)|)] + \log[a_0 \exp(-\alpha\, d_S(v_{l_h 1}, p_0)) + c_{hi2} + c_{hi3}]\}. \quad (11.17)$$

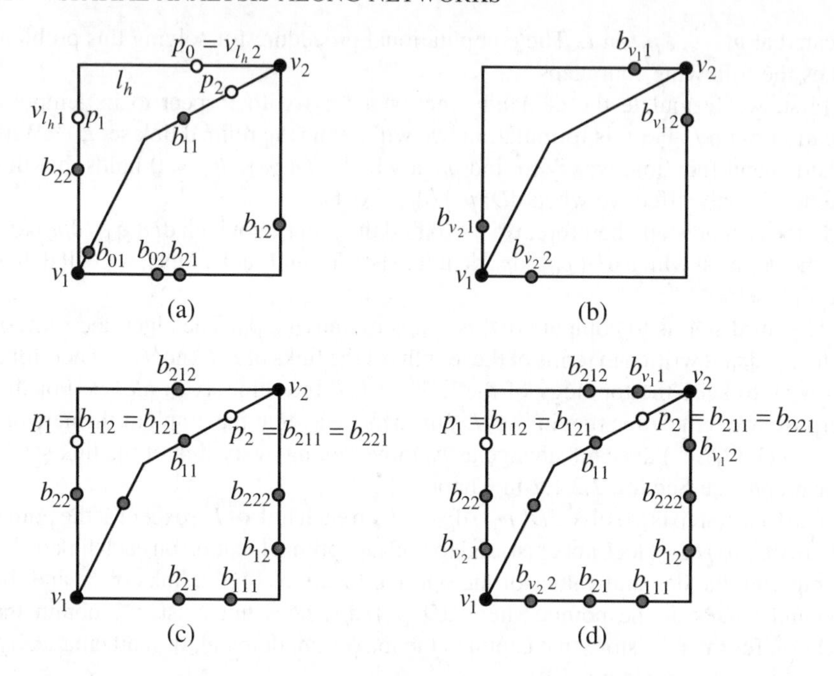

Figure 11.4 Refined networks: (a) $N^{+0} = (V^{+0}, L^{+0})$, B_0 and B_p, (b) $N^- = (V^-, L^-)$ and $B_{\geq 3}$, (c) the breakpoints B_b (three digit suffixes) of the breakpoints B_P, (d) $N^* = (V^*, L^*)$ (the beakpoints are indicated by the gray circles).

It should be noted that $|l_h|$ changes as p_0 moves if the link l_h is incident to p_0 or to the breakpoints of B_0 (e.g., link (p_0, p_1) or link (v_1, b_{02}) in Figure 11.4a); otherwise, it is a constant (e.g., link (p_1, b_{22}) in Figure 11.4a). To indicate this property explicitly, $|l_h|$ is written as $|l_h(p_0)|$. It should also be noted that c_{h02} and c_{h03} include neither p_0 nor $|l_h(p_0)|$; therefore, they are constants with respect to p_0.

11.3.2 Topologically invariant shortest-path trees

An analytical method for finding the largest-demand site of store 0 is to derive $\mathrm{d}D(p_0)/\mathrm{d}p_0$, and solve $\mathrm{d}D(p_0)/\mathrm{d}p_0 = 0$ with respect to p_0. This method assumes that $\mathrm{d}D(p_0)/\mathrm{d}p_0$ exists at p_0. Recalling that $D(p_0)$ is written in terms of the lengths of the links forming the extended shortest-path tree rooted at p_0 (Section 11.2), we wish to obtain the link set such that the derivative $\mathrm{d}D(p_0)/\mathrm{d}p_0$ exists when the extended shortest-path tree does not topologically change as p_0 moves in the interior of a link of the link set or the graph of the extended shortest-path tree does not change as p_0 moves in the interior of a link of the link set. To obtain such a link set, we modify the network $N^{+0} = (V^{+0}, L^{+0})$ by deleting the nodes of degree two and joining the links incident to the nodes of degree two. The resulting network consists of the nodes of degree one, and those of degree three or more, denoted by $V_{\geq 3}$. We next generate the breakpoints, $B_{\geq 3}$, of the nodes of $V_{\geq 3}$ and insert these points as nodes. The resulting network is denoted by $N^- = (V^-, L^-)$. An example of $N^- = (V^-, L^-)$ for the network $N^{+0} = (V^{+0}, L^{+0})$ is shown in Figure 11.4b. The resulting network N^{+0} has the following property.

Property 11.1 (Topologically invariant shortest-path tree)

For $N^- = (V^-, L^-)$, the graph of the extended shortest-path tree rooted at a point p_0 does not change if p_0 moves in the interior of a link l_h of L^-. The graph changes as p_0 crosses a point in $V_{\geq 3} \cup B_{\geq 3}$.

This property can be proved in the following way. Suppose that p_0 moves along a link $l_h \in L^-$. Let $b(p_0)$ be any breakpoint with respect to the shortest-path tree rooted at p_0. If $b(p_0)$ remains in the interior of a link in L^- as p_0 moves in the interior of l_h from one terminal node to the other, the graph structure of the extended shortest-path tree does not change. We now employ the proof by contradiction. Contrary to the property, let us assume that $b(p_0)$ passes through a node $v \in V^-$ as p_0 moves in the interior of $l_h \in L^-$. Then, at that location of p_0, there exist two distinct shortest paths from v to p_0, which in turn means that p_0 is a breakpoint with respect to v. This is a contradiction because all such breakpoints are included in V^-. This completes the proof of Property 11.1.

Property 11.1 implies that $\mathrm{d}D(p_0)/\mathrm{d}p_0$ exists at any point on $\tilde{L}$ except at the nodes of $V_{\geq 3} \cup B_{\geq 3}$. We use only this fact in the following derivations. Note that $\tilde{L}$ is the set of points forming the links of L (including their end nodes), which are the same as those forming the links of L^+, those of L^-, and those of L^{+0} (also those of L^* to be defined below).

11.3.3 Topologically invariant link sets

As p_0 moves along a link of L^{+0}, its breakpoint also moves, and it may cross a node of N^{+0} (for example, in Figure 11.4a, as p_0 moves from v_2 to p_1, b_{02} also moves and crosses b_{21}). If this occurs, the link set L^{+0} changes (the links (v_1, b_{02}) and (b_{02}, b) change to links (v_1, b_{21}) and (b_{21}, b_{02})). As a result, we must use another network, $N^{+0'} = (V^{+0'}, L^{+0'})$, in place of $N^{+0} = (V^{+0}, L^{+0})$ to compute $D(p_0)$. Because this change makes computation for finding the largest demand site of a new store complicated, we wish to use the set of links with a topology that does not change as p_0 moves along a link. To construct such a link set, we consider the breakpoints B_b of the breakpoints B_P (in Figure 11.4c, the former breakpoints are indicated by b with three-digit suffixes and the latter by two-digit suffixes). Note that by definition of the breakpoints B_P of the n points of P, the breakpoints B_b include P (e.g., $b_{112} = b_{121} = p_1$ in Figure 11.4c). We insert the breakpoints $B_{\geq 3}$ and those B_b in the links L^+ of the network N^+ (Figure 11.2c) as nodes. As a result, the network N^+ is refined as $N^* = (V^*, L^*) = (V^+ \cup B_{\geq 3} \cup B_b, L^*)$ (e.g., Figure 11.4d;

note that p_0 and B_0 are not included in V^*). The resulting network $N^* = (V^*, L^*)$ has the following property.

Property 11.2 (Topologically invariant link set)

The graph of $N^{+0} = (V^{+0}, L^{+0})$ does not change as p_0 moves in the interior of a link of the link set L^* of $N^* = (V^*, L^*)$.

Note that the network corresponding to the graph referred to in Property 11.2 and the network in which p_0 moves along its links are different; the former is $N^{+0} = (V^{+0}, L^{+0})$ and the latter is $N^* = (V^*, L^*)$, although the set of points $\tilde{L}^{+0}$ forming the links of L^{+0} and that $\tilde{L}^*$ of L^* are the same as $\tilde{L} = \bigcup_{i=1}^{n_L} l_i$ of the original network $N = (V, L)$ where $L = \{l_1, \ldots, l_{n_L}\}$.

The proof of Property 11.2 is given by Okunuki and Okabe (2002), and is too long to repeat here (for details, see pp. 247–249 of their paper). Instead, we provide an intuitively understandable example in Figure 11.5. Let v_{l_m1} and v_{l_m2} be the end nodes of a link $l_m \in L^*$, i.e., $l_m = (v_{l_m1}, v_{l_m2})$, for example, $l_m = (v_{l_m1}, v_{l_m2}) = (b_{212}, b_{v_11})$ in Figure 11.4d, which is superimposed on $\tilde{L}^{+0}$ in Figure 11.5 (the superimposed v_{l_m1} and v_{l_m2} are indicated by the broken circles). As p_0 moves from v_{l_m1} to v_{l_m2} (or from b_{212} to b_{v_11}), the breakpoints b_{01} and b_{02} for p_0 move as in panels (b) and (c) of Figure 11.5. We observe in these panels that any breakpoints for p_0 do not go across the nodes of $N^* = (V^*, L^*)$. This confirms that the graph of the shortest-path tree rooted at p_0 does not change as p_0 moves in the interior of the link $l_m = (b_{212}, b_{v_11}) = (v_{l_m1}, v_{l_m2})$ of $N^* = (V^*, L^*)$. Note that the graph of $N^{+0} = (V^{+0}, L^{+0})$ changes at the end nodes of the links of L^*, but the value of $D(p_0)$ is continuous at the end nodes of the links of L^*.

So far, we have defined four types of link set: L^+, L^{+0}, L^- and L^*, which might have caused the reader confusion. To avoid misconception, we summarize those

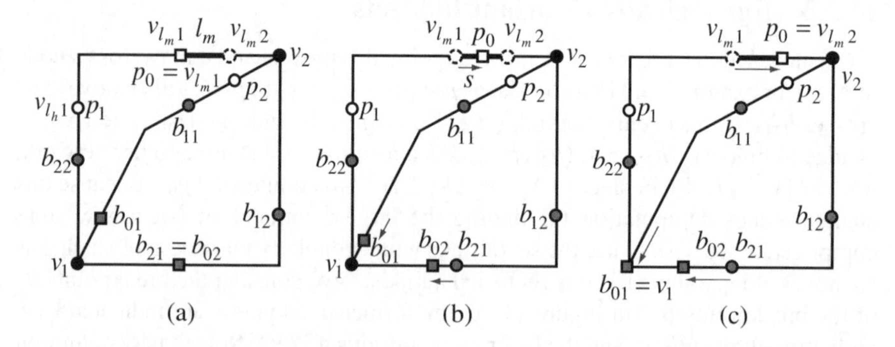

Figure 11.5 The graph of (V^{+0}, L^{+0}) does not change as p_0 moves in the interior of the link $l_m = (v_{l_m1} v_{l_m2}) = (b_{212} b_{v_11})$ of L^* in Figure 11.4d (note that the link $l_m = (v_{l_m1} v_{l_m2}) \in L^*$ of $N^* = (V^*, L^*)$ is superimposed on the network $N^{+0} = (V^{+0}, L^{+0})$ of Figure 11.4a; the result is Figure 11.5): (a) p_0 is at v_{l_m1}, (b) p_0 is at a point in the interior of $(v_{l_m1} v_{l_m2})$, (c) p_0 is at v_{l_m2}.

definitions here. First, we defined the set L^+ by inserting the existing stores P and their breakpoints B_P in the links of L of the original network $N = (V, L)$ (Figure 11.2c); the resulting network is $N^+ = (V^+, L^+)$, where $V^+ = V \cup P \cup B_P$. This refined link set L^+ is used in computing the demand function $D(p_i)$ of an existing store at $p_i \in P$ (a new store is not included). Second, we defined the link set L^{+0} by inserting a new store at a temporal location p_0 and its breakpoints B_0 in the links of L^+ (Figure 11.4a); the resulting network is $N^{+0} = (V^{+0}, L^{+0})$, where $V^{+0} = V^+ \cup \{p_0\} \cup B_0$. This refined link set L^{+0} is used in computing the demand function $D(p_0)$ of the new store at a temporarily fixed location p_0, and its derivative $\mathrm{d}D(p_0)/\mathrm{d}p_0$. This derivative is meaningful when it exists. Third, to examine this existence, we defined the link set L^- by deleting the nodes of degree two from the refined link set L^{+0} and joining the links incident to the deleted nodes and adding the breakpoints $B_{\geq 3}$ of the nodes $V_{\geq 3}$ of degree three or more (Figure 11.4b). This link set L^- is used in finding the points at which $\mathrm{d}D(p_0)/\mathrm{d}p_0$ exists or those at which $\mathrm{d}D(p_0)/\mathrm{d}p_0$ does not exist. Property 11.1 shows that $\mathrm{d}D(p_0)/\mathrm{d}p_0$ exists at any points on $\tilde{L}$ except at the points of $V_{\geq 3} \cup B_{\geq 3}$. Last, we defined the set L^* by inserting the breakpoints $B_{\geq 3}$ of $V_{\geq 3}$ and the breakpoints B_b of the breakpoints B_P in the links of L^+ (Figure 11.4d); the resulting network is $N^* = (V^*, L^*)$, where $V^* = V^+ \cup B_{\geq 3} \cup B_b$ (note that the new store at p_0 and its breakpoints B_0 are not inserted in the links of L^*, and that $V_{\geq 3}$ is included in V^+). This link set L^* is used in finding the largest demand site of the new store (i.e., the point p_0 at which $\mathrm{d}D(p_0)/\mathrm{d}p_0 = 0$ holds) by moving the new store p_0 along a link of L^*.

As p_0 moves, the lengths of some links in the link set L^{+0} vary; consequently, the network $N^{+0} = (V^{+0}, L^{+0})$ varies quantitatively. However, the algebraic form of $\mathrm{d}D(p_0)/\mathrm{d}p_0$ does not change as long as the graph of $N^{+0} = (V^{+0}, L^{+0})$ remains the same. Property 11.2 shows that the graph of $N^{+0} = (V^{+0}, L^{+0})$ does not change as long as p_0 moves on a link of L^* except at $V_{\geq 3} \cup B_{\geq 3}$ (where $\mathrm{d}D(p_0)/\mathrm{d}p_0$ does not exist according to Property 11.1). Therefore, in the subsequent section, we will find the points at which $\mathrm{d}D(p_0)/\mathrm{d}p_0 = 0$ (the left-hand side of which is written in terms of the lengths of the links of L^{+0}, not those of L^*) by moving p_0 on each link of L^* (not L^{+0}).

Summing up, the solution to the Huff-based locational optimization problem, formulated by equation (11.9), will be obtained by finding the maximum value among the demand values at the points where $\mathrm{d}D(p_0)/\mathrm{d}p_0$ exists and $\mathrm{d}D(p_0)/\mathrm{d}p_0 = 0$, and those at the nodes $V_{\geq 3} \cup B_{\geq 3}$ (where $\mathrm{d}D(p_0)/\mathrm{d}p_0$ does not exist). This procedure is explained in the subsequent section in detail.

11.3.4 Numerical method for the Huff-based locational optimization

Having established the preliminaries in the preceding subsections, we are now ready to compute $\mathrm{d}D(p_0)/\mathrm{d}p_0$. For any point p_0 on every link of L^* excluding the nodes of $V_{\geq 3} \cup B_{\geq 3}$, we differentiate $D(p_0)$ of equation (11.17) with respect to p_0.

The derivation of $\mathrm{d}D(p_0)/\mathrm{d}p_0$ is fairly lengthy, but we can obtain the analytical solution from the following equations. Let $\sum_{h=1}^{n_{L+0}} G(p_0)G_{h2}(p_0)$ be the corresponding terms in the right-hand side of equation (11.17). Then:

$$\begin{aligned}\frac{\mathrm{d}}{\mathrm{d}p_0}D(p_0) &= \frac{\mathrm{d}}{\mathrm{d}p_0}\sum_{h=1}^{n_{L+0}} G_{h1}(p_0)G_{h2}(p_0) \\ &= \sum_{h=1}^{n_{L+0}}\left\{G_{h2}(p_0)\frac{\mathrm{d}}{\mathrm{d}p_0}G_{h1}(p_0) + G_{h1}(p_0)\frac{\mathrm{d}}{\mathrm{d}p_0}G_{h2}(p_0)\right\}, \end{aligned} \quad (11.18)$$

where:

$$\frac{\mathrm{d}}{\mathrm{d}p_0}G_{h1}(p_0) = -\frac{a_0 w_h \alpha c_{h02}}{2\alpha\delta_{h0}}\frac{\exp(-\alpha d_S(v_{l_h1}, p_0))}{(a_0\exp(-\alpha\, d_S(v_{l_h1}, p_0)) + c_{h02})^2}\frac{\mathrm{d}\, d_S(v_{l_h1}, p_0)}{\mathrm{d}p_0}, \quad (11.19)$$

$$\begin{aligned}\frac{\mathrm{d}}{\mathrm{d}p_0}G_{h2}(p_0) &= 2\alpha\delta_{h0}\frac{\mathrm{d}}{\mathrm{d}p_0}|l_h(p_0)| \\ &+ \frac{a_0\alpha\exp(-\alpha d(v_{l_h1}, p_0))\dfrac{\mathrm{d}d_S(v_{l_h1}, p_0)}{\mathrm{d}p_0} - 2c_{h03}\alpha\delta_{h0}\exp(2\alpha\delta_{h0}|l_h(p_0)|)\dfrac{\mathrm{d}}{\mathrm{d}p_0}|l_h(p_0)|}{a_0\exp(-\alpha d_S(v_{l_h1}, p_0)) + c_{h02} + c_{h03}\exp(2\alpha\delta_{h0}|l_h(p_0)|)} \\ &- \frac{a_0\alpha\exp(-\alpha\, d_S(v_{l_h1}, p_0))}{a_0\exp(-\alpha\, d_S(v_{l_h1}, p_0)) + c_{hi2} + c_{hi3}}\frac{\mathrm{d}\, d_S(v_{l_h1}, p_0)}{\mathrm{d}p_0}.\end{aligned} \quad (11.20)$$

As noted in Section 11.4.1, $|l_h(p_0)|$ changes if a link $l_h \in L^{+0}$ is incident to p_0 or the breakpoints b of p_0 (i.e., $b \in B_0$); otherwise, it does not change. Let v_{l_m1} and v_{l_m2} be the end nodes of a link $l_m \in L^*$ (e.g., link (b_{212}, b_{v_11}) in Figure 11.4d or link (v_{l_m1}, v_{l_m2}) in Figure 11.5), and suppose that p_0 moves from v_{l_m1} and v_{l_m2} as in panels (a), (b) and (c) of Figure 11.5. Let v_{l_h1} and v_{l_h2} be the end nodes of a link $l_h \in L^{+0}$ (e.g., v_{l_h1} and v_{l_h2} in Figure 11.4a). Then:

$$\frac{\mathrm{d}}{\mathrm{d}p_0}|l_h(p_0)| = \begin{cases} +1 \text{ if } l_h \text{ is incident to } p_0 \text{ or a breakpoint } b \in B_0 \\ \quad \text{and } d_S(v_{l_h1}, v_{l_m1}) \leq d_S(v_{l_h1}, v_{l_m2}), \\ -1 \text{ if } l_h \text{ is incident to } p_0 \text{ or a breakpoint } b \in B_0 \\ \quad \text{and } d_S(v_{l_h1}, v_{l_m1}) > d_S(v_{l_h1}, v_{l_m2}), \\ 0 \text{ if } l_h \text{ is neither incident to } p_0 \text{ nor a breakpoint } b \in B_0. \end{cases} \quad (11.21)$$

$$\frac{\mathrm{d}}{\mathrm{d}p_0}d_S(v_{l_h1}, p_0) = \begin{cases} +1 \text{ if } d_S(v_{l_h1}, v_{l_m1}) \leq d_S(v_{l_h1}, v_{l_m2}), \\ -1 \text{ if } d_S(v_{l_h1}, v_{l_m1}) > d_S(v_{l_h1}, v_{l_m2}). \end{cases} \quad (11.22)$$

For example, as Figure 11.5 shows, for $l_h = (v_{l_h1}, v_{l_h2}) = (p_1, p_0) \in L^{+0}$ (which is incident to p_0), $\mathrm{d}|l_h(p_0)|/\mathrm{d}p_0 = \mathrm{d}|(p_1, p_0)|/\mathrm{d}p_0 = 1$; for $l_h = (v_{l_h1}, v_{l_h2}) = (v_2, p_0) \in L^{+0}$ (which is incident to p_0), $\mathrm{d}|l_h(p_0)|/\mathrm{d}p_0 = \mathrm{d}|(v_2, p_0)|/\mathrm{d}p_0 = -1$; for $l_h = (v_{l_h1}, v_{l_h2}) = (b_{22}, p_1) \in L^{+0}$ (which is neither incident to p_0 nor $b \in B_0$), $\mathrm{d}|l_h(p_0)|/\mathrm{d}p_0 = \mathrm{d}|(b_{22}, p_1)|/\mathrm{d}p_0 = 0$; for $l_h = (v_{l_h1}, v_{l_h2}) = (b_{22}, p_1) \in L^{+0}$, $\mathrm{d}|d_S(b_{22}, p_0)| = +1$; for $l_h = (v_{l_h1}, v_{l_h2}) = (b_{12}, v_2) \in L^{+0}$, $\mathrm{d}|d_S(b_{12}, p_0)|/\mathrm{d}p_0 = -1$.

In equations (11.19) and (11.20), we must compute $d_S(v_{l_h1}, p_0)$ for every $l_h \in L^{+0}$, which varies as p_0 moves. To do so, we separate the variable part and the constant part of $d_S(v_{l_h1}, p_0)$ and rewrite $d_S(v_{l_h1}, p_0)$ with a parametric variable s as:

$$\begin{aligned} d_S(v_{l_h1}, p_0) &= d_S(v_{l_h1}, v_{l_m1}) + d_S(v_{l_m1}, p_0) \\ &= d_S(v_{l_h1}, v_{l_m1}) + \delta_{h0}\, s \quad 0 \le s \le |l_m|, p \in l_m, \end{aligned} \tag{11.23}$$

where:

$$\delta_{h0} = \begin{cases} +1 \text{ if } d_S(v_{l_h1}, v_{l_m1}) \le d_S(v_{l_h1}, v_{l_m2}), \\ -1 \text{ if } d_S(v_{l_h1}, v_{l_m1}) > d_S(v_{l_h1}, v_{l_m2}), \end{cases} \tag{11.24}$$

and $d_S(v_{l_h1}, v_{l_m1})$ is a constant (Figure 11.5b), which is known once $N^{+0} = (V^{+0}, L^{+0})$ and $N^* = (V^*, L^*)$ are constructed.

In equations (11.19) and (11.20), we must also compute the length $|l_h(p_0)|$ for every $l_h \in L^{+0}$. This length can also be written in terms of the parametric variable s. Let v_{l_h1} and v_{l_h2} be the end nodes of $l_h \in L^{+0}$, and v_{l_m1} and v_{l_m2} be those of $l_m \in L^*$. Then, $|l_h(p_0)|$ is written as:

$$|l_h(p_0)| = \begin{cases} d_S(v_{l_h1}, v_{l_h2}) & \text{if } l_h \text{ is not incident to } p_0, \\ d_S(v_{l_h1}, v_{l_m1}) + s & \text{if } l_h \text{ is incident to } p_0 \text{ and } d_S(v_{l_h1}, v_{l_m1}) \le d_S(v_{l_h1}, v_{l_m2}), \\ d_S(v_{l_h1}, v_{l_m1}) - s & \text{if } l_h \text{ is incident to } p_0 \text{ and } d_S(v_{l_h1}, v_{l_m1}) > d_S(v_{l_h1}, v_{l_m2}). \end{cases} \tag{11.25}$$

For example, as Figure 11.5 shows, for $l_h(p_0) = (v_{l_h1}, v_{l_h2}) = (p_1, b_{22})$, which is not incident to p_0, $|l_h(p_0)| = d_S(v_{l_h1}, v_{l_h2}) = d_S(p_1, b_{22})$; for $l_h(p_0) = (v_{l_h1}, v_{l_h2}) = (p_1, p_0) \in L^{+0}$, which is incident to p_0, $|l_h(p_0)| = d_S(p_1, b_{212}) + s = d_S(v_{l_h1}, v_{l_m1}) + s$ (note that b_{212} in Figure 11.4d is v_{l_m1} in Figure 11.5); for $l_h(p_0) = (v_{l_h1}, v_{lh2}) = (v_2, p_0) \in L^{+0}$, which is incident to p_0, $|l_h(p_0)| = d_S(p_1, b_{212}) - s = d_S(v_{l_h1}, v_{l_m1}) - s$.

In almost the same way, $|l_h(p_0)|$ is written in terms of s if l_h is incident to a breakpoint of B_0. Therefore, substituting equation (11.23) with equations (11.24) and (11.25) into equations (11.18)–(11.20), we can write $\mathrm{d}D(p_0)/\mathrm{d}p_0$ as a function of s; i.e., the variable in $\mathrm{d}D(p_0)/\mathrm{d}p_0$ is only s, denoted by $\mathrm{d}D(p_0)/\mathrm{d}p_0 = G(s)$.

Having obtained the derivative $\mathrm{d}D(p_0)/\mathrm{d}p_0$ analytically, we next solve $\mathrm{d}D(p_0)/\mathrm{d}p_0 = G(s) = 0$. The function $G(s)$ is nonlinear and nonconvex. Methods for solving this type of equation have been extensively developed in mathematical optimization, and many textbooks are available. The reader should consult such a book; for example, Gill, Murray, and Wright (1981) or Nocedal and Wright (2006). It should be noted that the solution found may be a local optimum. Let $D(p^*_{0i})$ be an optimal solution for a link l_i of L^*. Then, an optimal location is obtained from:

$$\max\{D(p^*_{0i}), i = 1, \ldots, n_{L^*}, D(p), p \in V_{\geq 3} \cup B_{\geq 3}\}. \quad (11.26)$$

11.3.5 Computational tasks and their time complexities for the Huff-based locational optimization

To find the optimal location of a new store, we must construct $\mathrm{d}D(p_0)/\mathrm{d}p_0$ of equation (11.18) and solve $\mathrm{d}D(p_0)/\mathrm{d}p_0 = G(s) = 0$, where the construction of the equation means to compute all the coefficients and the constants in equations (11.18)–(11.20).

Suppose that p_0 lies on a link l_m. Then, from Property 11.2 (the topologically invariant link set shown in Section 11.3.2), all the coefficients are fixed uniquely. Computations of those coefficients are almost the same as in the computation of $D(p_i)$; the only difference is that the $(n+1)$-th store at p_0 is inserted, and the number of stores is changed from n to $n+1$, which does not affect the order of the computational time. Therefore, the constants c_{hi2} and c_{hi3} can be computed in $\mathrm{O}(n)$ time. Because we have $|L^{+0}| = \mathrm{O}(n\, n_L)$, all the coefficients in equations (11.18)–(11.20) can be computed in $\mathrm{O}(n^2\, n_L)$ time.

It is difficult to solve $G(s) = 0$ analytically, and hence we require some numerical method to improve an approximation of the solution iteratively until we have sufficient accuracy. Let us denote the time complexity to solve $G(s) = 0$ by $T(n)$. Then, constructing $\mathrm{d}D(p_0)/\mathrm{d}p_0$ and solving $G(s) = 0$ requires $\mathrm{O}(n^2\, n_L T(n))$ time for the point p_0 on one link l_m in L^*.

Next, let us evaluate the size of L^*. Recall that $V^* = V^+ \cup B_{\geq 3}$. We have already seen that $|V^+| = \mathrm{O}(n_L\, n)$. On the other hand, $|V_{\geq 3}| = \mathrm{O}(n_L)$ because it happens that most of the nodes of the original network $N = (V, L)$ are of degree three. Each node in $V_{\geq 3}$ generates $\mathrm{O}(n_L)$ breakpoints, and hence the total number $|B_{\geq 3}|$ of the breakpoints in N^* is of $\mathrm{O}(n_L^2)$. Thus, we obtain $|V^*| = \mathrm{O}(n_L\, n + n_L^2)$, and hence $|L^*| = \mathrm{O}(n_L\, n + n_L^2)$ for the planar network N^*.

We must repeat the task of constructing $\mathrm{d}D(p_0)/\mathrm{d}p_0$ and solving $G(s) = \mathrm{O}(|L^*|)$ times. Consequently, we require $\mathrm{O}(n^2\, n_L^2(n + n_L)T(n))$ time to find the optimal location of p_0.

12

GIS-based tools for spatial analysis along networks and their application

In earlier chapters, we presented statistical methods for analyzing events occurring on or alongside a network, together with the computational techniques required for applying these methods. In principle, the interested reader can implement these methods with software following the procedures described in the computational sections of this volume. In practice, however, implementation demands a considerable amount of time and expense, particularly for researchers not very skilled in computer programming. Fortunately, we can reduce this burden to some extent with geographical information systems (GIS), as GIS provide user-friendly tools for efficiently analyzing spatial data in Cartesian space. However, the tools available for analyzing spatial data in network space remain rather limited.

Nonetheless, standard GIS software packages provide some tools that we can employ at least partly for spatial analysis on networks. For example, ArcGIS, one of the main GIS software packages, distributed by the Environmental Systems Research Institute (ESRI), incorporates the *Network Analysts* toolbox, which includes tools for computing the shortest-path distances between nodes and the ordinary network Voronoi diagram (Section 7.4 in Wong and Lee (2005)). Another software package, *Vector Network Analysis*, an application from the Geographic Resources Analysis Support System (GRASS), includes similar functions

Spatial Analysis along Networks: Statistical and Computational Methods, First Edition.
Atsuyuki Okabe and Kokichi Sugihara.

(Section 6.6 in Neteler and Mitasova (2007)). Recently, the City Form Research Group at MIT released a GIS-based toolbox, *Urban Network Analysis*, which provides tools for computing graph-theoretic measures including 'betweenness,' 'closeness,' and 'straightness' (Sevtsuk, 2010). However, the main function of these analytical tools is to analyze the characteristics of a network itself, such as its centrality (Kansky, 1963; Haggett and Chorley, 1969); they are not intended for analyzing the spatial events taking place on and alongside networks ('along' is used for both 'on' and 'alongside' hereafter).

Few software toolboxes deal exclusively with spatial analysis along networks insofar as the GIS community is concerned. The exception is *SANET* (*S*patial *A*nalysis along *NET*works), a software package that plugs into ArcGIS (for useful reviews of the SANET functions, see Okabe, Okunuki, and Shiode (2006a,b), and Okabe and Satoh (2009)). This software package provides the necessary tools for performing the spatial methods presented in preceding chapters. In this chapter, we provide a practical application of these tools. As is often the case, the features of software packages change from version to version as updates become available. Therefore, we present only an outline of the SANET functions. Anyone wishing to understand the detailed operation of these tools should consult the SANET manual, downloadable from the SANET homepage at http://sanet.csis.u-tokyo.ac.jp/.

This chapter consists of two sections. The first section presents the tools for preprocessing, including a test for network connectedness, the assignment of access points to a network, computation of shortest-path distances, and random points generation on a network. The second section outlines the six tools required for spatial analysis on networks: network Voronoi diagrams (as in Chapter 4), the network nearest-neighbor distance method (as in Chapter 5), the network K function method (as in Chapter 6), network cluster analysis (as in Chapter 8), the network kernel density estimation method (as in Chapter 9), and the network spatial interpolation method (as in Chapter 10), along with their application to actual data. Note that the most recent version of SANET available (Version 4) does not explicitly provide tools for the network spatial autocorrelation method (discussed in Chapter 7) or the network Huff model (featured in Chapter 11). However, we can perform these methods using the basic tools shown in Section 12.1. Note that, for brevity, hereafter we omit 'network,' as in 'network...method,' unless specified otherwise and except for headings and subheadings.

12.1 Preprocessing tools in SANET

Before undertaking spatial data analysis, we must examine whether a given dataset satisfies the conditions required by SANET, namely, the 'network connectedness' condition and the 'points on network' condition. We first illustrate the tools for testing these conditions, and then describe the basic tools utilized in developing methods and approaches not available in SANET Version 4.

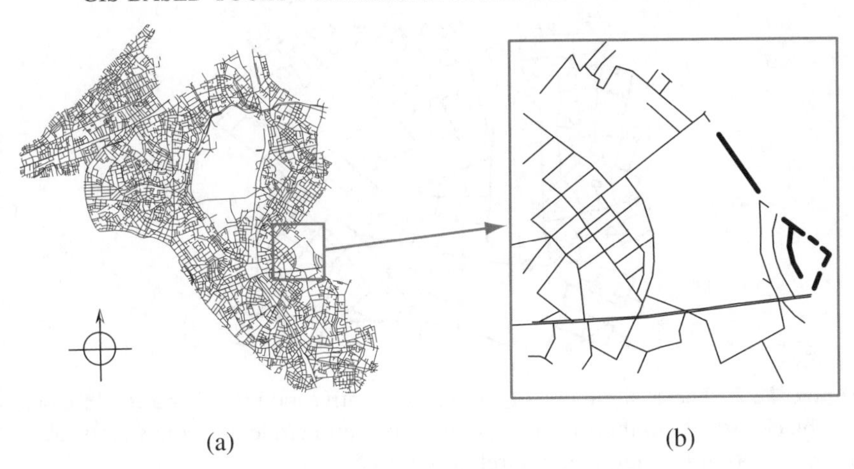

Figure 12.1 The street network in Shibuya ward, Tokyo: (a) the whole network, (b) part of the network, in which the disconnected roads are indicated by the heavy line segments.

12.1.1 Tools for testing network connectedness

In theory, the methods used for network spatial analysis in this volume assume that a network is *connected*, i.e., for any pair of points on the network, at least one path embedded in the network exists. In practice, it often happens that when we view a network (e.g., Figure 12.1a) from a bird's eye view, the network appears connected, but when we consider it more closely, it is actually disconnected (Figure 12.1b). What is worse, the disconnected parts are often difficult to locate by sight. SANET then helps test whether or not a given network is connected (note that SANET regards the two parts connected if the distance between two disconnected parts is less than a given tolerance level, considering human digitizing errors). If not, SANET regards the largest connected part as the main part and indicates the disconnected parts by colors (panel (b) in Figure 12.1 is an enlarged segment of panel (a), where the disconnected line segments are indicated by the heavy line segments). If a user considers that the disconnected parts are really meant to be connected to the main part, the user can choose a larger tolerance level for connectedness. However, if the disconnected parts are too far apart to adjust in SANET, the user may alternatively employ a tool in ArcGIS to edit the network.

12.1.2 Tool for assigning points to the nearest points on a network

The methods for network spatial analysis in preceding chapters all assume that the points representing the events that occur on or alongside a network lie exactly on the network. In practice, many facilities are represented by their access points

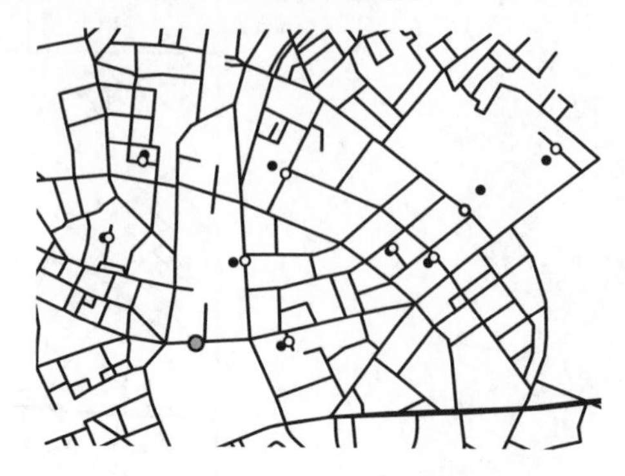

Figure 12.2 The assignment of the representative points of sports clubhouses (the black circles) to their nearest points (the white circles) on the streets around Shibuya station (the large gray circle) in Tokyo.

(entrances or gates), which are close to but do not exactly coincide with streets. Even if events physically take place on streets, such as traffic accidents, they are not always located on the corresponding streets on a map, because different mapmakers usually create accident incidence and street data independently. SANET automatically assigns every representative point to its nearest point on a network. An actual example is shown in Figure 12.2, in which the line segments are the streets around Shibuya station (the gray circle) in Tokyo, the small black circles are the representative points of sports clubhouses, and the white circles are the access points for these sports club houses as assigned by SANET to the nearest streets.

12.1.3 Tools for computing the shortest-path distances between points

All of the tools in SANET are in terms of shortest-path distances. We could also develop new tools for network spatial analysis using these distances. To allow for this potential development, SANET provides two tools. The first is for computing the shortest-path distances between every pair of points in a point set, and the second is for computing the shortest-path distances between every point in a point set and every point in another point set. Figure 12.3 depicts an example, in which panel (a) provides part of the table of the shortest-path distances between each pair of the 14 railway stations in Shibuya ward, Tokyo, and panel (b) depicts the shortest-path distances between the 14 railway stations and the 21 sports clubhouses in Shibuya ward.

12.1.4 Tool for generating random points on a network

One of the most fundamental hypotheses in spatial analysis is, as mentioned in Chapter 2, *complete spatial randomness* (CSR), i.e., points are independently and

FromPntID	ToPntID	Distance
0	1	719.26859
0	2	2121.9828
0	3	1776.2179
0	4	4094.1104
0	5	3251.5453
0	6	3181.7202
0	7	5820.1031
0	8	4848.6889
0	9	4391.4463
0	10	4479.6372
0	11	4774.0600
0	12	4915.6128
0	13	4737.3579
1	0	719.26859
1	2	1450.4300
1	3	1594.7621
1	4	3555.4114
1	5	2924.4992
1	6	2951.4451
1	7	5238.2610
1	8	4521.6428
1	9	4121.1414
1	10	4369.7602
1	11	4447.0139

(a)

FromPntID	ToPntID	Distance
0	0	543.79873
0	1	1769.0238
0	2	1313.7807
0	3	4915.8782
0	4	4620.9697
0	5	1827.2560
0	6	1794.7847
0	7	2171.7148
0	8	1904.2523
0	9	2057.8436
0	10	2974.1677
0	11	1056.1795
0	12	1845.4252
0	13	1189.0212
0	14	2054.2424
0	15	4586.6580
0	16	977.30252
0	17	4910.6330
0	18	5222.2631
0	19	2080.2294
0	20	1624.5902
1	0	1152.2247
1	1	1659.1468
1	2	911.55415

(b)

Figure 12.3 Shortest-path distances: (a) the shortest-path distances between each pair of 14 stations in Shibuya ward, Tokyo (partial), (b) the shortest-path distances between the 14 stations and 21 sports clubhouses in the same region (partial).

identically generated according to the same uniform distribution given by equation (2.4) in Chapter 2 (for the computational method, see Section 3.4.4 in Chapter 3). In testing the CSR hypothesis, we sometimes encounter difficulties in computing test statistics analytically. In such a case, we employ a numerical method, known as *Monte Carlo simulation*, for estimating a function of a statistic from a set of random variables by simulating a stochastic process a large number of times. To carry out this simulation, SANET provides a tool for generating random points on a network following the CSR hypothesis. Figure 12.4 provides an example of 500 points generated on the street network in Shibuya ward, Tokyo.

12.2 Statistical tools in SANET and their application

The preceding section describes general-purpose tools for spatial analysis on networks. This section illustrates the tools needed for specific analyses, and shows their applications to actual examples in Shibuya ward. Shibuya ward is a subcenter district in Tokyo, with an area of about 15.11 square kilometers. At night, some 200 000 inhabitants are resident in Shibuya ward; in the day, the number of people in the ward almost triples. Parks dominate the central part of the ward (the empty areas in Figure 12.4) with the ward street network comprising 7858 links and 5905 nodes.

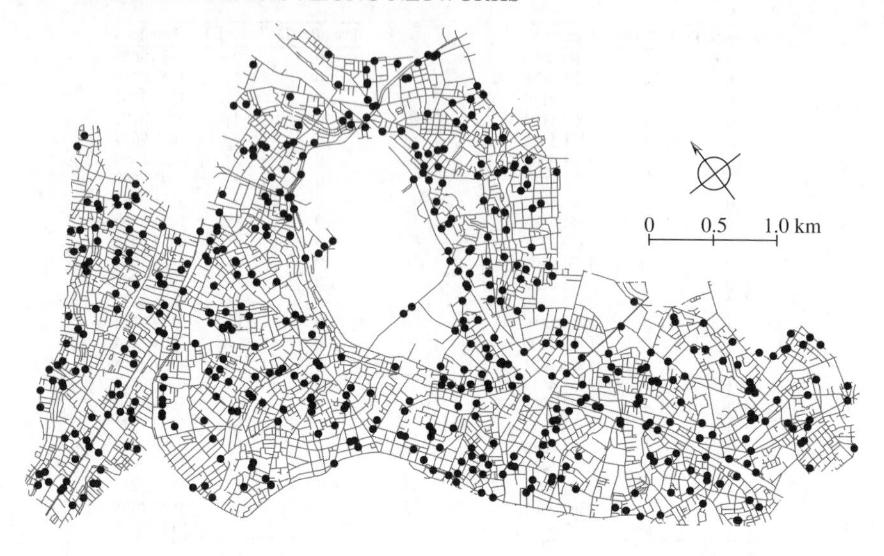

Figure 12.4 Five hundred random points on the street network in Shibuya ward, Tokyo.

The street network data are from 1:2500 scale detailed maps distributed by the National Land Survey Agency of Japan.

We analyzed the spatial patterns of eight kinds of facilities in this ward (number in 2007 in brackets): beauty parlors (793), Japanese restaurants (385), coffee shops (189), preparatory schools (96), aromatherapy houses (56), churches (47), sports clubhouses (21), and railway stations (14). Note that we include subway and aboveground railways stations separately, but if a subway station is close to an aboveground railway station, we represent them with a single point. We obtained the addresses of these facilities from the Nippon Telegraph and Telephone (NTT) Town Pages, and converted them to latitude and longitude coordinates using the online address matching system provided by the Center for Spatial Information Science at the University of Tokyo. Using these datasets, in the following sections we demonstrate how to apply the six tools presented in the introduction of this chapter.

12.2.1 Tools for network Voronoi diagrams and their application

The first tool in SANET is designed to deal with questions Q1 or Q1′ in Chapter 1: How to estimate the approximate market areas (subnetworks) or the dominant service areas (defined in Section 11.1.2 in Chapter 11) of retail stores located alongside streets in a downtown area? The tool responds to such questions using the Voronoi diagrams presented in Chapter 4. SANET generates two types of Voronoi diagram. The first type is the ordinary network Voronoi diagram defined in Section 4.1. Figure 12.5 depicts the ordinary Voronoi diagram generated by the railway stations (the white circles) in Shibuya ward, Tokyo.

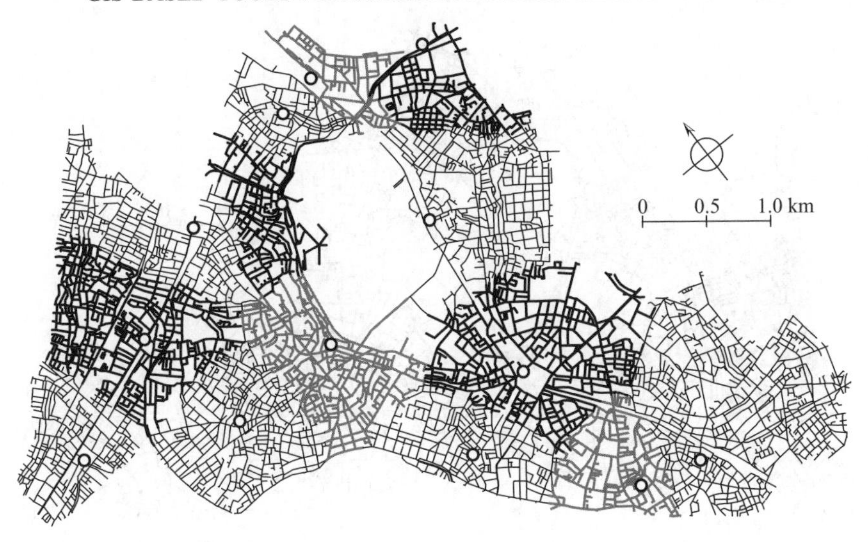

Figure 12.5 The ordinary network Voronoi diagram generated by the railway stations (the white circles) in Shibuya ward, Tokyo.

The second type of Voronoi diagram is the additively weighted network Voronoi diagram defined in Chapter 4. We can use this, for instance, for estimating market areas as determined as a function of not only distribution costs (shortest-path distances) but also the prices of the goods sold at stores (for details, see Section 4.2.2 in this volume, or Section 3.1 in Okabe *et al.* (2000)). We can also use this diagram to estimate the catchment areas of stations when commuters choose their stations to minimize their total commuting time comprising both the walking time to a station and the train-riding time to the central business district (CBD) where their workplace is located. Figure 12.6 illustrates an example, in which the CBD is located around Shibuya station (indicated by the double circles). Note that because every train-riding time between two adjacent stations is less than three minutes, the difference between the ordinary Voronoi diagram in Figure 12.5 and the additively weighted Voronoi diagram in Figure 12.6 is not distinct, but a close look at the fringe of the Voronoi subnetwork of Shibya station shows some difference.

12.2.2 Tools for network nearest-neighbor distance methods and their application

In Chapter 1, we referred to questions Q2 and Q3, or the questions of whether, for instance, beauty parlors tend to stand side-by-side alongside streets in a downtown area or Japanese restaurants tend to cluster around railway stations. To respond to these questions, SANET provides two types of tool: a collection of tools for the nearest-neighbor distance (NND) methods as formulated in Chapter 5, and a collection of

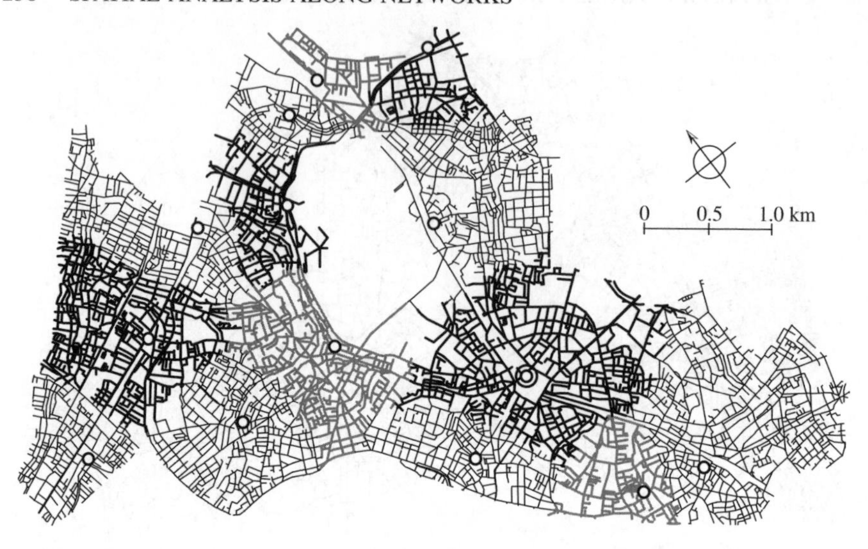

Figure 12.6 The additively weighted network Voronoi diagram generated by the railway stations (the white circles) in Shibuya ward, Tokyo, where the weights are the train-riding time from a station to Shibuya station (the double circle) relative to walking time.

tools for the *K* function methods formulated in Chapter 6. The former includes two methods: the auto NND method (Section 5.1 in Chapter 5) and the cross NND method (Section 5.2). In what follows, we provide applications of these methods.

12.2.2.1 Network global auto nearest-neighbor distance method

We applied a tool in SANET designed for the global auto NND method formulated in Section 5.1.2 in Chapter 5 to the distribution of beauty parlors (the black points) in Figure 12.7.

SANET provides two test statistics: I_G defined by equation (5.4) and the G function defined in Section 5.1.2. The index I_G shows the ratio of the observed value of the average nearest-neighbor distance to the expected value of that distance under the CSR hypothesis. The data indicate that $I_G = 0.62 < 1.00$. This implies that the observed average nearest-neighbor distance is shorter than the expected value obtained under the CSR hypothesis. We statistically tested this result using a normal distribution with $\mathrm{E}(I_G) = 1.00$ and $\sqrt{\mathrm{Var}(I_G)} = 0.023$ calculated by SANET. The test confirms at the 0.95 confidence level that beauty parlors tend to cluster together in Shibuya ward, Tokyo.

An alternative statistic is the G function, $G(t)$, which is the ratio of the number $n(t)$ of points satisfying that the nearest-neighbor distance is less than or equal to t to the total number n of points (i.e., $G(t) = n(t)/n$)). Figure 12.8 illustrates the results of $n(t)$, in which the black curve is the observed function and the two gray curves are the upper and lower envelope curves at the 0.05 significance level under the CSR hypothesis. Noting that the observed curve is above the upper envelope curve for

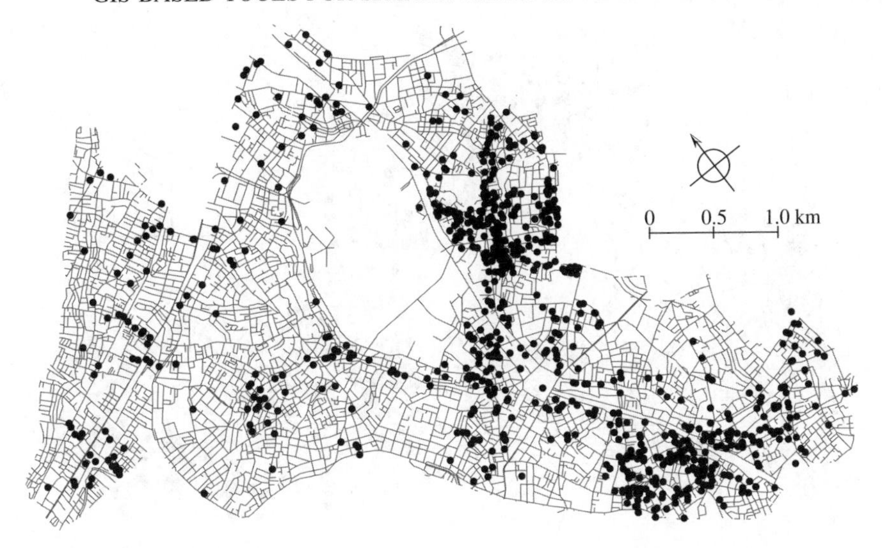

Figure 12.7 Beauty parlors alongside the streets in Shibuya ward, Tokyo.

0–200 meters, we can conclude at the 0.95 confidence level that beauty parlors tend to cluster significantly within 0–200 meters in this region.

12.2.2.2 Network global cross nearest-neighbor distance method

Spatial clusters may also form around a certain type of facility, for example, Japanese restaurants potentially gather around railway stations. We can statistically examine this location tendency using the global cross NND method presented in Section 5.2.2 in Chapter 5. We applied a tool in SANET for performing this method to the dataset illustrated in Figure 12.9, where the black points indicate Japanese restaurants and the white circles indicate railway stations in Shibuya ward, Tokyo.

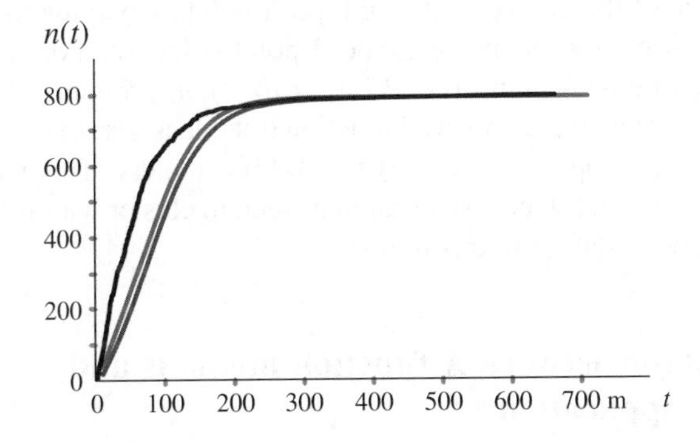

Figure 12.8 The number $n(t)$ of beauty parlors satisfying that the nearest-neighbor distance between beauty parlors is less than or equal to t in Shibuya ward (the black curve is the observed curve and the gray curves are the upper and lower envelope curves at the 0.05 significance level under the CSR hypothesis).

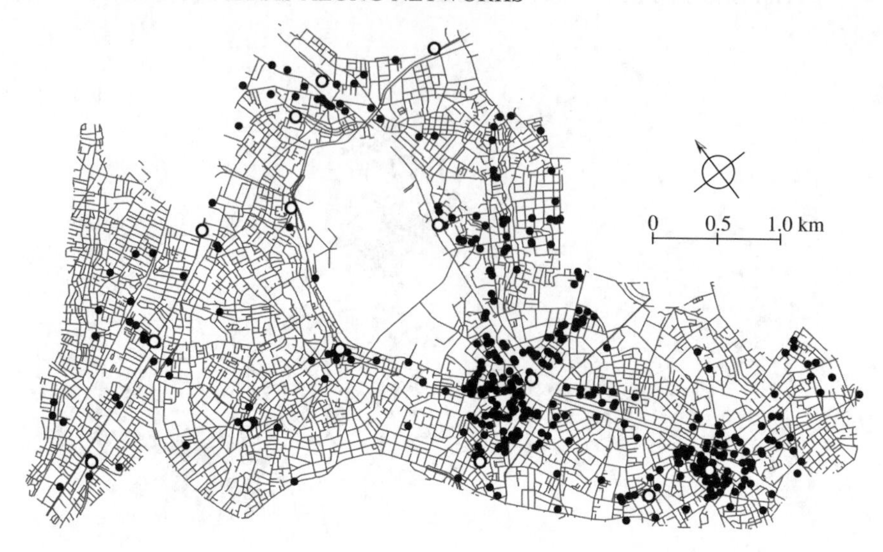

Figure 12.9 Japanese restaurants (the black points) alongside the streets and railway stations (the white circles) in Shibuya ward, Tokyo.

SANET provides two test statistics: I_B defined by equation (5.14) and the F^* function defined in Section 5.2.2 in Chapter 5. The index I_B indicates the ratio of the observed value of the average distance from every type A point to its nearest type B point to the expected value of that distance under the CSR hypothesis. The data show that $I_B = 0.64 < 1.00$. This implies that the observed average nearest-neighbor distance is shorter than the expected value obtained under the CSR hypothesis. We tested this result using a normal distribution with $E(I_B) = 1.00$ and $\sqrt{\mathrm{Var}(I_B)} = 0.20$ calculated by SANET. We can conclude at the 0.95 confidence level that Japanese restaurants tend to locate near railway stations in Shibuya ward, Tokyo.

To examine this global trend more closely, we can use the F^* function, $F^*(t)$, which is the ratio of the number $n_A(t)$ of type A points satisfying that the distance from each type A point to its nearest type B point is less than or equal to t to the total number n_A of type A points (i.e., $F^*(t) = n_A(t)/n_A$). Figure 12.10 depicts the results of $n_A(t)$, in which the observed function (the black curve) is above the upper envelope curve (the upper gray curve) for 0–1100 meters. This suggests that at the 0.95 confidence level, Japanese restaurants tend to cluster within 1100 meters of each nearest railway station in this region.

12.2.3 Tools for network *K* function methods and their application

In the preceding section, we examined questions Q2 and Q3 in terms of the first nearest-neighbor distance. We can alternatively discuss these questions in terms of not only the first nearest-neighbor distance, but also the second nearest, third

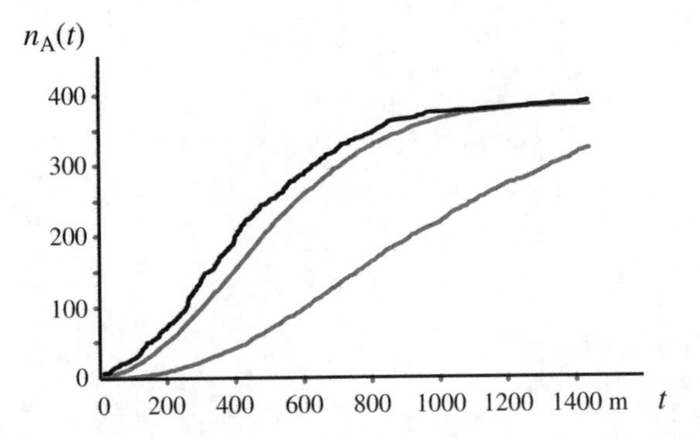

Figure 12.10 The observed number of Japanese restaurants satisfying that the distance from each Japanese restaurant to its nearest railway station is less than or equal to t in Shibuya ward, Tokyo (the black curve is the observed curve and the gray curves are the upper and lower envelope curves at the 0.05 significance level under the CSR hypothesis).

nearest, fourth nearest, up to the farthest nearest-neighbor distance. The family of these methods is the *K* function method (Chapter 6). This family includes the local auto *K* function, the global auto *K* function, the local cross *K* function, the global cross *K* function, and the global Voronoi cross *K* function. We present applications of some of these methods in the following subsections.

12.2.3.1 Network global auto *K* function method

Figure 12.11 illustrates the distribution of preparatory schools (the black points) alongside the streets in Shibuya ward. We applied a tool in SANET designed for the global auto *K* function method formulated in Section 6.1.2 in Chapter 6 to this distribution.

The results are given in Figure 12.12, in which the black curve indicates the observed global auto *K* function and the gray curves indicate the upper and lower envelope curves obtained under the CSR hypothesis at the 0.05 significance level. We note in this figure that the observed global auto *K* function is above the upper envelope curve when the shortest-path distance is less than 2700 meters. Therefore, we conclude at the 0.95 confidence level that preparatory schools tend to cluster when their neighbors are within 2700 meters.

12.2.3.2 Network global cross *K* function method

Some other types of facility tend to cluster around a certain type of facility, for instance, aromatherapy houses (the black points in Figure 12.13) may cluster around railway stations (the white circles). As demonstrated in Section 12.2.2.2, we can use the global cross NND method to examine this tendency. Alternatively, we could employ the global cross *K* function method presented in Section 6.2.2.

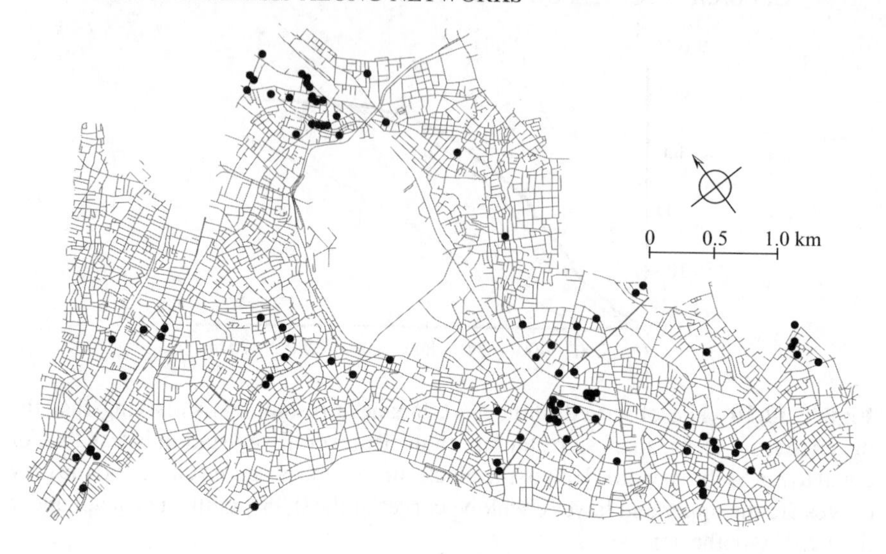

Figure 12.11 Preparatory schools (black points) alongside the streets in Shibuya ward, Tokyo.

We applied a tool in SANET to this method to test the location tendency of aromatherapy houses in relation to railway stations in Shibuya ward, Tokyo.

The resulting observed global cross *K* function, $K_{AB}(t)$, (the black curve) and the upper and lower envelope curves at the 0.05 significance level (the gray curves) are shown in Figure 12.14. This figure shows that the observed global cross *K* function is above the upper envelope curve in the range of 0–700 meters; between the upper and lower envelope curves in the range of 700–2100 meters; below the lower envelope curve in the range of 2100–4800 meters; and so on.

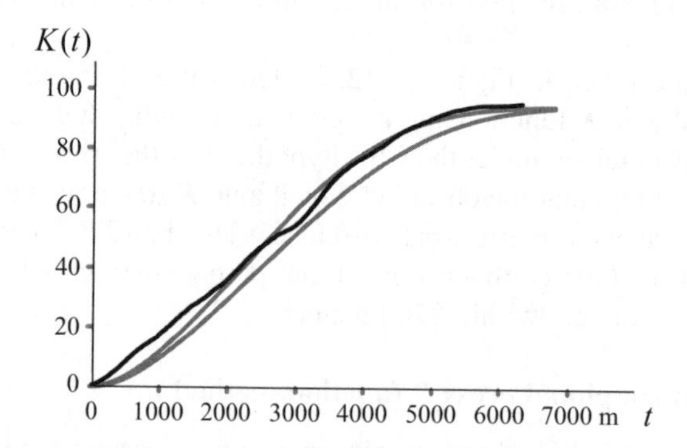

Figure 12.12 The network global auto *K* function of preparatory schools alongside streets in Shibuya ward, Tokyo (the black curve is the observed curve and the gray curves are the upper and lower envelope curves at the 0.05 significance level under the CSR hypothesis).

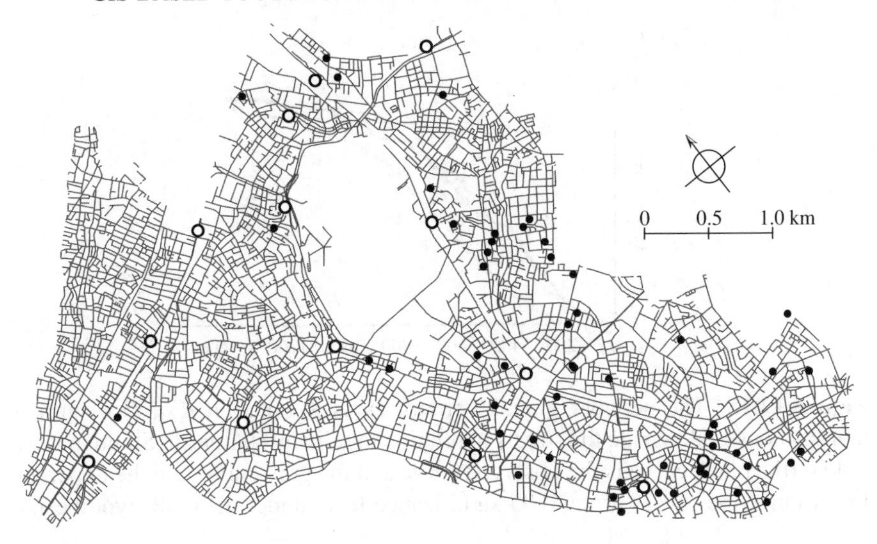

Figure 12.13 Aromatherapy houses (the black points) and railway stations (the white circles) alongside the streets in Shibuya ward, Tokyo.

12.2.3.3 Network global Voronoi cross K function method

To investigate more closely the above location tendency in the 0–1500 meter range, we applied the global Voronoi cross K function method formulated in Section 6.2.3 in Chapter 6 to the same dataset. Figure 12.15 illustrates the results. Because the observed function (the black curve) lies above the upper envelope curve at the 0.05 significance level (the upper gray curve) in the range of 180–700 meters, we

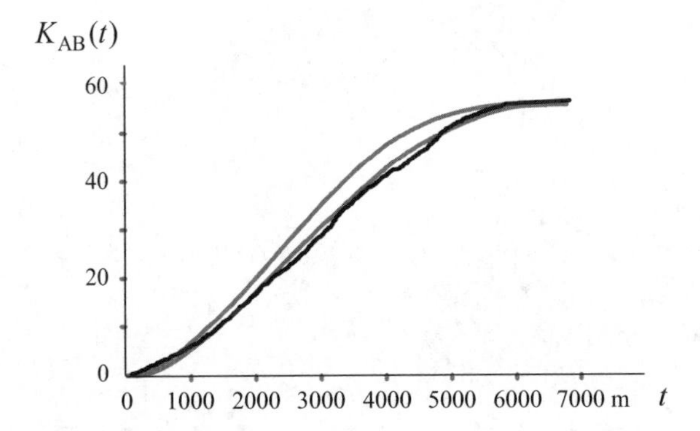

Figure 12.14 The network global cross K function $K_{AB}(t)$ of aromatherapy houses alongside the streets in relation to railway stations in Shibuya ward, Tokyo (the black curve is the observed curve and the gray curves are the upper and lower envelope curves at the 0.05 significance level under the CSR hypothesis).

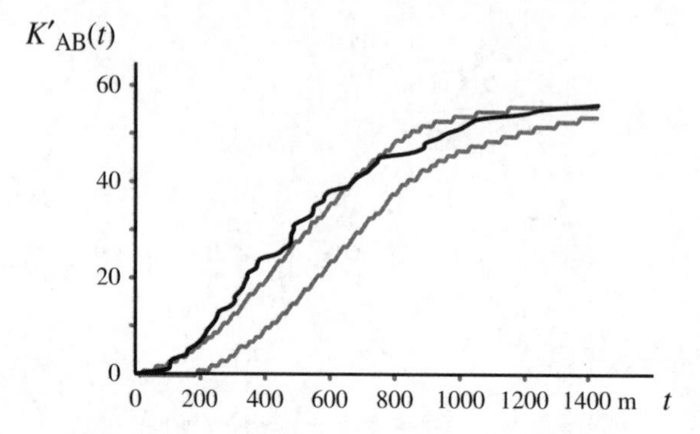

Figure 12.15 The network global Voronoi cross K function $K'_{AB}(t)$ of aromatherapy houses alongside the streets in relation to railway stations in Shibuya ward, Tokyo (the black curve is the observed curve and the gray curves are the upper and lower envelope curves at the 0.05 significance level under the CSR hypothesis).

conclude at the 0.95 confidence level that the number of aromatherapy houses within 180–700 meters of each station is significantly greater than the number realized under the CSR hypothesis. This implies that aromatherapy houses strongly cluster within 180–700 meters of each station.

12.2.3.4 Network local cross *K* function method

When we wish to examine whether facilities, say churches (the black points in Figure 12.16), are clustered around a specific station, say, Shibuya station (the white

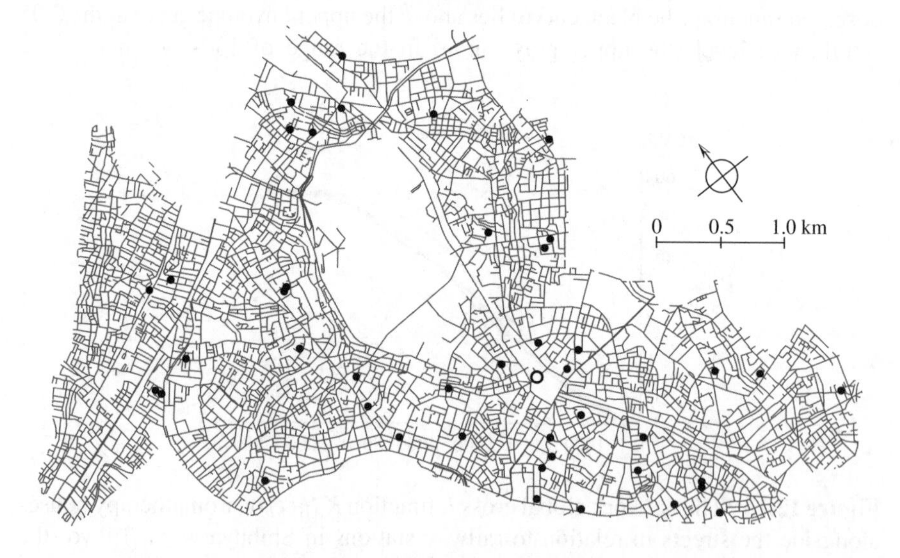

Figure 12.16 Churches (the black points) alongside the streets and Shibuya station (the white circle) in Shibuya ward, Tokyo.

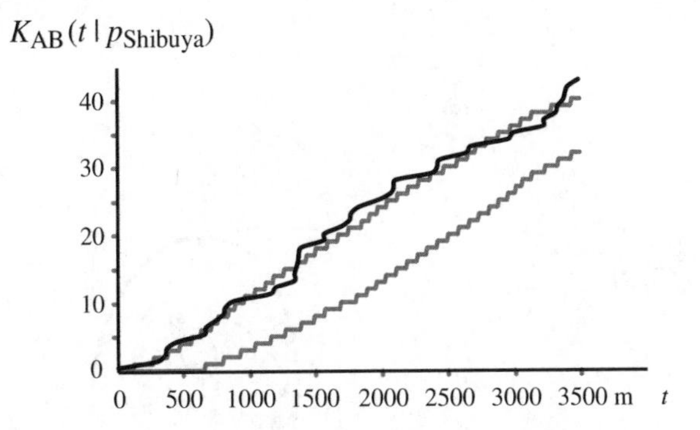

Figure 12.17 The network local cross K function $K_{AB}(t|p_{Shibuya})$ of churches alongside the streets in relation to Shibuya station in Shibuya ward, Tokyo (the black curve is the observed curve and the gray curves are the upper and lower envelope curves at the 0.05 significance level under the CSR hypothesis).

circle) using the first, second, and so on to the farthest distance from the station, then we can use a tool in SANET designed for the local cross K function method presented in Section 6.2.1 in Chapter 6.

Figure 12.17 illustrates the results, showing that the observed local cross K function (the black curve) and the upper envelope curve at the 0.05 significance level (the upper gray curve) intertwine in the range of 0–1000 meters. This implies that churches do not strongly cluster around Shibuya station.

12.2.4 Tools for network point cluster analysis and their application

When we wish to obtain answers to questions Q5 or Q5′ in Chapter 1, i.e., how to find sets of clusters within which points are close to each other (closeness as measured by the shortest-path distance) but between which the points are apart, we can use the cluster analysis developed in Chapter 8. We applied a tool in SANET for the closest-pair point clustering method defined in Section 8.1.2.1 to the distribution of sports clubhouses in Shibuya ward, as in Figure 12.18. The resulting clusters are given by the closed curves in Figure 12.19, where the solid thin closed curves are clusters where the shortest-path distances between sports clubhouses are less than 200 meters. The broken thin closed curves are for less than 400 meters; the solid thick closed curves are for less than 800 meters; and the broken thick closed curves are for less than 1200 meters. We can observe the hierarchy of clusters based on the inclusion relations between the closed curves, where the included clusters are lower than the including clusters. Alternatively, we can describe this hierarchy using the dendrogram of the clusters. SANET provides the data for constructing this diagram. Figure 12.19 depicts an example showing the hierarchical structure of sports clubhouse clusters in the region, where the numbers correspond to those in Figure 12.18.

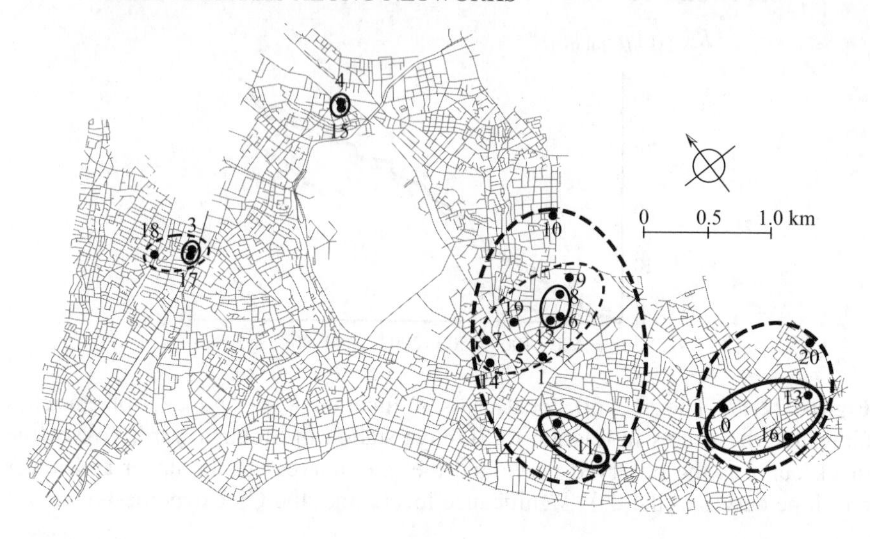

Figure 12.18 Hierarchical clusters of sports clubhouses in Shibuya ward, Tokyo, obtained by the closest-pair point clustering method with the shortest-path distances (the solid thin closed curves are clusters in which the shortest-path distances between sports clubhouses are less than 200 meters; the broken thin closed curves are less than 400 meters; the solid thick closed curves are less than 800 meters; and the broken thick closed curves are less than 1200 meters).

12.2.5 Tools for network kernel density estimation methods and their application

As discussed in Chapter 1, question Q6, or how to estimate the density of points on a network, is often asked in spatial analysis. This is particularly useful in accident and

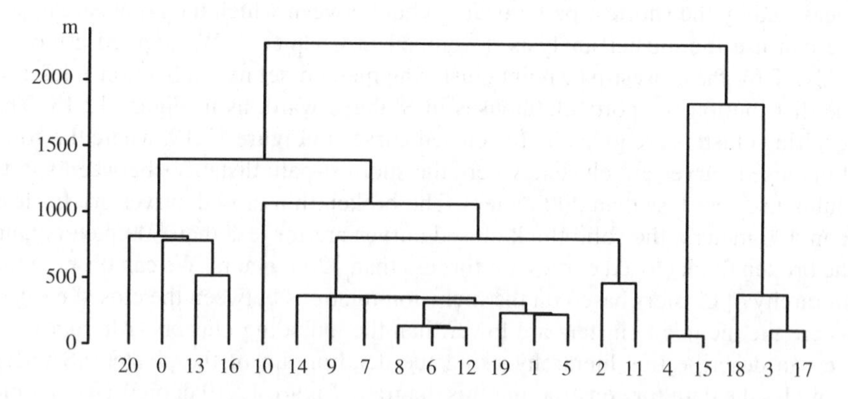

Figure 12.19 Dendrogram of sports clubhouses in Shibuya ward, Tokyo, shown in Figure 12.18.

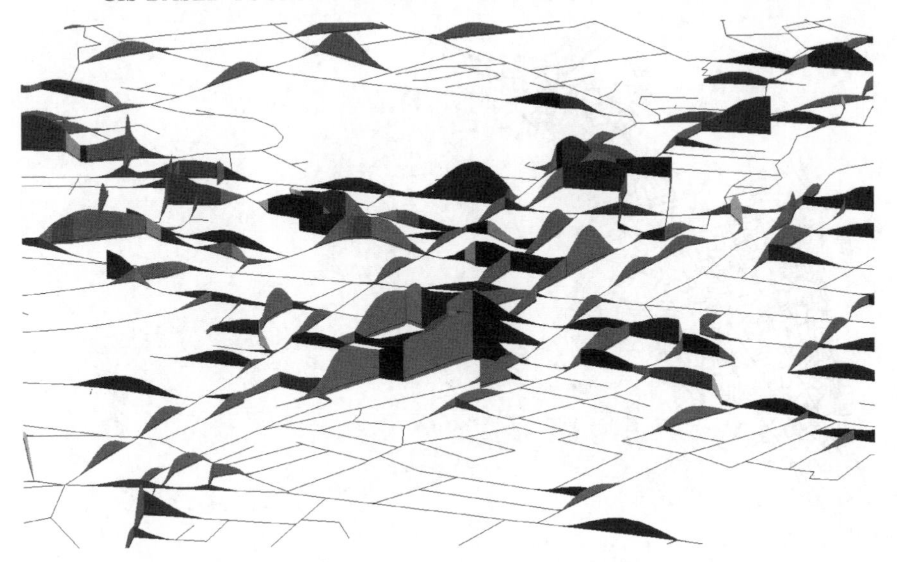

Figure 12.20 The density of beauty parlors around Omotesando subway station in Shibuya ward, Tokyo estimated using the equal-split continuous kernel density function (the distribution of beauty parlors is shown in Figure 12.7.

crime studies where we wish to locate the high-density areas respectively known as *black zones* (Section 1.2.2 in Chapter 1) and *hot spots* (Section 1.2.4). One of the methods for responding to this question is the kernel density estimation method presented in Chapter 9. We applied the equal-split continuous kernel density function defined in Section 9.2.3 to the distribution of beauty parlors in Shibuya ward in Figure 12.7. The estimated density function across the street network is given in Figure 12.20. SANET also provides a tool for the equal-split discontinuous kernel density function defined in Section 9.2.3, which is computationally more efficient than the equal-split continuous kernel density function, particularly where a network includes many short links.

12.2.6 Tools for network spatial interpolation methods and their application

When the attribute value of concern continuously distributes over a network, such as the altitude of the road ground surface, we encounter question Q7 or Q7′ in Chapter 1: How to estimate the unknown altitude of the road ground surface at a point on a road network from the known altitudes of sample points on the network? One of the estimation methods possible is the inverse-distance weighting method presented in Section 10.1 in Chapter 10. We employed a tool in SANET designed for this method to estimate road altitudes across the street network using the known altitudes of the railway stations in Shibuya ward. Figure 12.21 depicts the interpolated altitudes, where darker shades of gray indicate higher altitudes.

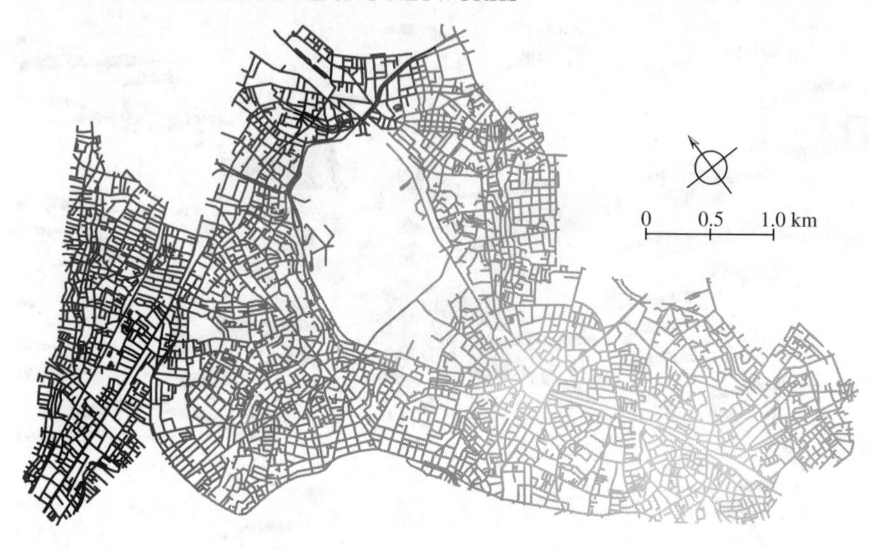

Figure 12.21 The altitude values of the road ground surface interpolated by the inverse-distance weighting method using the known altitude values of railway stations (the white circles in Figure 12.13) in Shibuya ward, Tokyo (the darker the shade, the higher the altitude).

In this last chapter, we overviewed the six tools in SANET for network spatial analysis. At present, SANET does not provide some of the network spatial methods presented in this volume. However, readers can develop their own tools for these methods using the basic tools shown in Section 12.1 and the intermediate data provided by the existing tools. We trust that not only computer-oriented spatial analysts, but also those in fields not always skilled in computer programming, will enjoy and benefit from the network spatial analysis made possible by the user-friendly spatial analysis toolbox available in the form of SANET.

References

Abdel-Aty, M., Cunningham, R., and Gayah, V. (2008) Considering dynamic variable speed limit strategies for real-time crash risk reduction on freeways. *Transportation Research Board*, **2078**, 108–116.

Abellanes, M., Hurtado, F., Sacriston, V., Icking, C., Ma, L., Kline, R., Langetepe, E., and Palop, B. (2003) Voronoi diagrams for services neighboring a highway. *Information Processing Letters*, **86**, 283–288.

Aerts, K., Lathuy, C., Steenberghen, T., and Thomas, I. (2006) Spatial clustering of traffic accidents using distances along the network. *Proceedings 19th Workshop of the International Cooperation on Theories and Concepts in Traffic Safety* (Session 4: Methods for Risk Assessment and of Effect of Complex RSP), pp. 1–10.

Aho, A.V., Hopcroft, J.E., and Ullman, J.D. (1974) *The Design and Analysis of Computer Algorithms*, Addison-Wesley, Reading, MA.

Aichholzer, O., Aurenhammer, F., and Palop, B. (2004) Quickest paths, straight skeletons, and the city Voronoi diagram. *Discrete and Computational Geometry*, **31**, 17–35.

Aldenderfer, M.S. and Blashfield, R.K. (1984) *Cluster Analysis*, SAGE Publications, Newbury Park.

Anderberg, M.R. (1973) *Cluster Analysis for Applications*, Academic Press, New York.

Anderson, R.R., Brown, R.G., and Rappleye, R.D. (1968) Water quality and plant distribution along the upper Patuxent River, Maryland. *Chesapeake Science*, **9** (3), 145–156.

Anselin, L. (1988) *Spatial Econometrics: Methods and Models*, Kluwer Academic Publishers, Dordrecht.

Anselin, L. (1995a) Local indicators of spatial association-LISA. *Geographical Analysis*, **27**(2), 93–115.

Anselin, L. (1995b) *New Directions in Spatial Econometrics – Advances in Spatial Science*, Springer-Verlag, Berlin.

Anselin, L., Cohen, J., Cook, F., Gorr, W., and Tita, G. (2000) Spatial analyses of crime. *Criminal Justice 2000*, **4**, 213–262.

Anselin, L., Griffiths, E., and Tita, G. (2008) Crime mapping and hot spots analysis, in *Environmental Criminology and Crime Analysis*, eds R. Wortley and L. Mazerolle, Willan Publishing, Devon, pp. 97–116.

Arbia, G. (2006) *Spatial Econometrics: Foundations and Applications to Regional Convergence*, Springer-Verlag, Berlin.

Armitage, P. (1949) An overlap problem arising in particle counting. *Biometrika*, **36** (3/4), 257–266.

Armstrong, M. (1998) *Basic Linear Geostatistics*, Springer, Berlin.

Spatial Analysis along Networks: Statistical and Computational Methods, First Edition.
Atsuyuki Okabe and Kokichi Sugihara.

Baddeley, A., Gregori, P., Mateu, J., Stoica, R. and Stoyan, D. (eds) (2006) *Case Studies in Spatial Point Process Modeling*, Lecture Note in Statistics 185, Springer-Verlag, Berlin.

Bailey, T.C. and Gatrell, A.C. (1995) *Interactive Spatial Data Analysis*, Longman Science & Technical, Burnt Mill, Harlow, Essex.

Balaban, I.J. (1995) An optimal algorithms for finding segments intersections. *Proceedings of the 11th ACM Symposium, Computational Geometry*, pp. 211–219.

Ballard, D.H. and Brown, C.M. (1982) *Computer Vision*, Prentice Hall, London.

Barnhill, R.E. and Stead, S.E. (1984) Multistage trivariate surfaces. *Rocky Mountain Journal of Mathematics*, **14** (1), 103–118.

Bartkowiak, M. and Mahan, G.D. (1995) Nonlinear currents in Voronoi networks. *Physical Review B*, **51** (16), 10 825–10 832.

Bartle, R.G. (1964) *Elements of Real Analysis*, John Wiley & Sons, Inc., New York.

Bartle, R.G. (1966) *Elements of Integration*, John Wiley & Sons, Inc., New York.

Bartlett, M.S. (1967) The spectral analysis of line processes. *Proceedings of the 5th Berkley Symposium on Mathematical Statistics and Probability*, **3**, pp. 135–153.

Bartlett, M.S. (1975) *The Statistical Analysis of Spatial Pattern*, Chapman & Hall, London.

Barton, D.E. and David, F.N. (1955) Sums of ordered intervals and distances. *Mathematika*, **2**, 150–159.

Barton, D.E. and David, F.N. (1956a) Some notes on ordered random intervals. *Journal of the Royal Statistical Society. B*, **18** (1), 79–94.

Barton, D.E. and David, F.N. (1956b) Tests for randomness of point on a line. *Biometrika*, **43** (1/2), 104–114.

Bashore, T.L., Tzilkowski, W.M., and Bellis, E.D. (1985) Analysis of deer-vehicle collision sites in Pennsylvania. *Journal of Wildlife Management*, **49**, 769–774.

Bauer, R., Delling, D., Sanders, P., Schieferdecker, D., Schultes, D. and Wagner, D. (2010) Combining hierarchical and goal-directed speed-up techniques for Dijkstra's algorithm. *ACM Journal of Experimental Algorithmics*, **15** (23), 1–31.

Baumgart, B.G. (1972) Winged edge polyhedron representation. *Stanford Artificial Intelligence laboratory Memo*, AIM-179.

Baumgart, B.G. (1974) Geometric modeling for computer vision. *Stanford Artificial Intelligence Laboratory Memo*, AIM 249.

Bellis, E.D. and Graves, H.B. (1971) Deer mortality on a Pennsylvania interstate highway. *Journal of Wildlife Management*, **35**, 232–237.

Ben-Akiva, M. and Lerman, S.R. (1985) *Discrete Choice Analysis*, The MIT Press, Cambridge.

Bentley, J.L. and Maurer, H.A. (1979) A note on Euclidean near neighbor searching in the plane. *Information Processing Letters*, **8** (3), 133–137.

Bentley, J.L. and Ottmann, T.A. (1979) Algorithms for reporting and counting geometric intersections. *IEEE Transactions on Computers*, **C-28**, 643–647.

Bentley, J.L., Weide, B.W., and Yao, A.C. (1980) Optimal expected-time algorithms for closest point problems. *ACM Transactions on Mathematical Software*, **6**, 563–580.

Berg, M.d., Cheoung, O., Kreveld, M.v., and Overmars, M. (2008) *Computational Geometry, Algorithms and Applications*, 3rd edn, Springer-Verlag, Berlin.

Berglund, S. and Karlstrom, A. (1999) Identifying local spatial association in flow data. *Journal of Geographical Systems*, **1**, 219–236.

Bernhardsen, T. (2002) *Geographic Information Systems: An Introduction*, 3rd edn, John Wiley & Sons, Inc., New York.

Berry, B.J. (1967) *Geography of Market Centers and Retail Distribution*, Prentice-Hall, Englewood Cliffs, N.J.

Berry, B.J.L. and Parr, J.B. (1988) Retail location and marketing geography, in *Market Centers and Retail Location*, J.B.L. Berry and J.B Parr, Prentice Hall, Englewood Cliffs, N.J.

Birkin, M., Clarke, G., Clarke, M., and Wilson, A., (1996) *Intelligent GIS Location Decisions and Strategic Planning*, GeoInformation International, Cambridge.

Black, W.R. (1991) Highway accidents: a spatial and temporal analysis. *Transportation Research Record*, **1318**, 75–82.

Black, W.R. (1992) Network autocorrelation in transport network and flow systems. *Geographical Analysis*, **24** (3), 207–222.

Black, W.R. and Thomas, I. (1998) Accidents on Belgium's motorways: a network autocorrelation analysis. *Journal of Transport Geography*, **6** (1), 21–31.

Block, C.R. (1995) STAC hot spots areas: a statistical tool for low enforcement decisions. *Landscape and Urban Planning*, **35**, 193–201.

Block, C.R. (1998) The geoarchive: an information foundation for community policing, in *Crime Mapping and Crime Prevention* (Crime Prevention Studies, Vol. 8), eds D. Weisburd and T. McEwen, Criminal Justice Press, New York, pp. 27–81.

Block, R. and Block, C.R. (1995) Space, place and crime: hot spot areas and hot places of liquor-related crime, in *Crime and Place* (Crime Prevention Studies, Vol. 4) eds J.E. Eck and D. Weisburd, Criminal Justice Press, Monsey, N.Y., pp. 147–185.

Bolduc, D., Laferriere, R., and Santarossa, G. (1992) Spatial autoregressive error components in travel flow models. *Regional Science and Urban Economics*, **22**, 371–385.

Bolduc, D., Laferriere, R., and Santarossa, G. (1995) Spatial autoregressive error components in travel flow models: an application to aggregate mode choice, in *New Directions in Spatial Econometrics*, eds L. Anselin and R.J.G.M. Florax, Springer-Verlag, Berlin, pp. 96–108.

Boots, B.N. (1980) Weighting thiessen polygons. *Economic Geography*, **56** (3), 248–259.

Boots, B.N. and Getis, A. (1988) *Point Pattern Analysis*, SAGE Publications, Newbury Park, CA.

Borruso, G. (2003) Network density and the delimitation of urban areas. *Transaction in GIS*, **7** (2), 177–191.

Borruso, G. (2005) Network density estimation: analysis of point patterns over a network. *Proceedings of Computational Science and Its Applications – ICCSA 2005,* Singapore, May 9–12, 2005, Lecture Note in Computer Science 3482, pp. 126–132.

Borruso, G. (2008) Network density estimation: a GIS approach for analyzing point patterns in a network space. *Transaction in GIS*, **12** (3), 377–402.

Braddock, M., Lapidus, G., Cromley, E., Cromley, R., Burke, G. and Banco L. (1994) Using a geographic information system to understand child pedestrian injury. *American Journal of Public Health*, **84** (7), 1158–1161.

Brandeau, M.L. and Chiu, S.S. (1989) An overview of representative problems in location research. *Management Science*, **35** (6), 645–674.

Brantingham, P.L. and Brantingham, P.J. (1982) Mobility, notoriety, and crime: a study in the crime patterns of urban nodal points. *Journal of Environmental Systems*, **11** (1), 89–99.

Brodsky, H. (2003) Retail area overlap: a case in forensic geography. *The Professional Geographer*, **55** (2), 250–258.

Brodsky, H. and Hakkert, A.S. (1983) Highway accident rates and rural travel densities. *Accidental Analysis and Prevention*, **15** (1), 73–84.

Brown, G.S. (1965) Point density in stems per acre. *New Zealand Forestry Service Research Notes*, **38** (1), 1–11.

Brunell, R.M. and Squibb, B.-M. (1993) An automatic variogram fitting procedure using Cressie's criterion. *Proceedings of the Section on Statistics and the Environment, the Annual Meeting of the American Statistical Association,* San Francisco, California, August 8–12, pp. 135–137.

Bucklin, L.P. (1971) Retail gravity models and consumer choice: a theoretical and empirical critique. *Economic Geography*, **47** (4), 489–497.

Bürger, M.E., Cohn, E.G., and Petrosino, A.J. (1995) Defining the 'hot spots of crime': operationalizing theoretical concepts for field research, in *Crime and Place*, eds J.E. Eck, and D. Weisburd, Criminal Justice Press, New York, pp. 237–257.

Burgess, E.W. (1925) The growth of the city: an introduction to a research project, in *The City*, eds R.E. Park, E. Burgess, and R. McKenzie, University of Chicago Press, Chicago, IL, pp. 47–62.

Burrough, P.A. and McDonnell, R.A. (1998) *Principles of Geographical Information Systems*, Oxford University Press, New York.

Busacker, R.G. and Saaty, T.L. (1965) *Finite Graphs and Networks: An Introduction with Applications*, McGraw-Hill, New York.

Buslik, M. and Maltz, M.D. (1998) Power to the people: Mapping and information sharing in the Chicago Police Department, in *Crime Mapping and Crime Prevention* (Crime Prevention Studies, Vol. 8), eds D. Weisburd and T. McEwan, Criminal Justice Press, New York, pp. 113–130.

Buxton, T. (1992) GIS expected to meet micromarketing challenge. *GIS World*, **February**, 70–72.

Cabrero, J., Lopez-Leon, M.D., Gomez, R., Castro, A. J., Martın-Alganza, A. and Camacho, J. P. M. (1997) Geographical distribution of b chromosomes in the grasshopper *Eyprepocnemis plorans,* along a river basin, is mainly shaped by non-selective historical events. *Chromosome Research*, **5**, 194–198.

Cain, A.J. (1956) The genus in evolutionary taxonomy. *Systematic Zoology*, **5** (3), 97–109.

Camisani-Calzolari, F.A. (2003) A tribute to Prof. D.G. Krige for his contributions over a period of more than half a century compiled by Prof. R.C.M. Minnitt and Dr W. Assibey-Bonsu. *The 31st International Symposium on Application of Computers and Operations Research in the Minerals Industries, APCOM* 2003, pp. 405–408.

Carrothers, G.A.P. (1956) An historical review of gravity and potential models of human interactions. *Journal of American Institute of Planners*, **22**, 94–102.

Case, R.M. (1978) Interstate highway road-killed animals: a data source for biologists. *Wildlife Society Bulletin*, **6** (1), 8–13.

Castle, C.J.E. and Longley, P.A. (2005) A GIS-based spatial decision support system for emergency services: London's King's Cross St. Pancras underground station, in *Geo-Information for Disaster Management*, eds P.V. Oosterom, S. Zlatanova, and E.M. Fendel, Springer, Berlin, pp. 867–881.

Ceder, A. and Livneh, M. (1982) Relationships between road accidents and hourly traffic flow-I Analysis and interpretation. *Accident Analysis and Prevention*, **14** (1), 19–34.

Ceder, A. and Livneh, M. (1978) Further evaluation of the relationships between road accidents and average daily traffic. *Accident Analysis and Prevention*, **10**, 95–109.

Ceder, A. (1982) Relationships between raod accidents and hourly traffic flow-II probabilistic approach. *Accident Analysis and Prevention*, **14** (1), 35–44.

Celik, M., Shekhar, S., George, B., Rogers, J.P. and Shine, J.A. (2007) *Discovering and Quantifying Mean Streets: A Summary of Results,* University of Minnesota – Computer Science and Engineering, Technical Report, 07-25.

Cencov, N.N. (1962) Evaluation of an unknown distribution density from observation. *Soviet Mathematics*, **3**, 1559–1562.

Cevik, E. and Topal, T. (2003) GIS-based landslide susceptibility mapping for a problematic segment of the natural gas pipeline, Hendek (Turkey). *Environmental Geology*, **44**, 949–962.

Chartrand, G. (1984) *Introductory Graph Theory*, Dover Publications, Mineola, N.Y.

Chazelle, B. and Edelsbrunner, H. (1992) An optimal algorithm for intersecting line segments in the plane. *Journal of ACM*, **39**, 1–54.

Chiles, J.P. and Delfiner, P. (1999) *Geostatistics – Modeling Spatial Uncertainty*, John Wiley & Sons, Inc., New York.

Chorley, R. and Haggett, P. (1967) *Models in Geography*, Methuen, London.

Chun, Y. (2008) Modeling network autocorrelation within migration flows by eigenvector spatial filtering. *Journal of Geographical Systems*, **10**, 317–344.

Clark, P.J. (1956) Grouping in spatial distributions. *Science*, **123**, 373–374.

Clark, P.J. and Evans, F.C. (1954) Distance to nearest neighbor as a measure of spatial relationships in populations. *Ecology*, **35** (4), 445–453.

Clauset, A., Moore, C., and Newman, M.E.J. (2008) Hierarchical structure and the prediction of missing links in networks. *Nature*, **453**, 98–101.

Clements, F.E. (1905) *Research Methods in Ecology*, University Publishing Company, Lincoln, NE.

Cliff, A.D. and Haggett, P. (1988) *Atlas of Disease Distributions: Analytical Approaches to Epidemiological Data*, Blackwell, Oxford.

Cliff, A.D. and Ord, J.K. (1973) *Spatial Autocorrelation*, Pion, London.

Cliff, A.D. and Ord, J.K. (1981) *Spatial Processes: Models and Applications*, Pion, London.

Cohen, B. (1980) *Deviant Street Networks*, Lexington Books, Massachusetts.

Converse, P.D. (1945) The development of the science of marketing: an exploratory survey. *The Journal of Marketing*, **10** (1), 14–23.

Cook, D.F. (1998) Topology and Tiger: the Census Bureau's contribution, in *History of Geographic Information Systems: Perspectives from the Pioneers*, ed. T.W. Foreseman, Prentice-Hall, Upper-Saddle River, pp. 47–57.

Cormen, T.H., Leiserson, C.E., Rivert, R.L., and Stein, C. (2001) *Introduction to Algorithms*, 2nd edn, MIT Press, Cambridge.

Cottam, G. and Curtis, J.T. (1949) A method for making rapid surveys of woodlands by means of pairs of randomly selected trees. *Ecology*, **30** (1), 101–104.

Couclelis, H. (1992) People manipulate objects (but cultivate fields): beyond the raster-vector debate in GIS, in *Theories and Methods of Spatio-Temporal Reasoning in Geographic Space*, eds A.U. Frank, I. Campari, and U. Formentini, Springer-Verlag, Berlin, pp. 65–77.

Cova, T.J. and Goodchild, M.F. (2002) Extending geographical representation to include fields of spatial objects. *International Journal of Geographical Information Science*, **16** (6), 509–532.

Cox, D.R. (1955) Some statistical methods connected with series of events. *Journal of the Royal Statistical Society, Series B*, **17** (2), 129–164.

Cox, D.R, and Isham, V. (1980) *Point Process*, Chapman & Hall, London.

Cressie, N. (1985) Fitting variogram models by weighted least squares. *Mathematical Geology*, **17** (5), 563–586.

Cressie, N. (1990) Origin of kriging. *Mathematical Geology*, **22** (3), 239–258.

Cressie, N. (1991) *Statistics for Spatial Data*, John Wiley & Sons, Inc., New York.

Cressie, N., Frey, J., Harck, B., and Smith, M. (2006) Spatial prediction on a river network. *Journal of Agricultural, Biological, and Environmental Statistics*, **11** (2), 127–150.

Cressie, N. and Majure, J.J. (1997) Spatio-temporal statistical modeling of livestock waste in streams. *Journal of Agricultural, Biological, and Environmental Statistics*, **2** (1), 24–47.

Cressie, N., and Wikle, C.K. (2011) *Statistics for Spatio-Temporal Data*, John Wiley & Sons, Inc., New York.

Curriero, F.C. (2006) On the use of non-Euclidean distance measures in geostatistics. *Mathematical Geology*, **38** (8), 907–926.

Dacey, M.F. (1960a) A note on the derivation of nearest neighbor distances. *Journal of Regional Science*, **2** (2), 81–87.

Dacey, M.F. (1960b) The spacing of river towns. *Annals of the Association of American Geographers*, **50**, 59–61.

Dacey, M.F. (1975) Evaluation of the Poisson approximation to measures of the random pattern in the square. *Geographical Analysis*, **12**, 351–367.

Daley, D.J. and Vere-Jones, D. (2003) *An Introduction to the Theory of Point Processes*, 2nd edn, New York, Springer-Verlag.

Daskin, M.S. (1995) *Network and Discrete Location, Models, Algorithms, and Applications*, John Wiley & Sons, Inc., New York.

Date, C.J. (2003) *Introduction to Database Systems*, 8th edn, Addison Wesley, Reading, MA.

Davidson, R. (1974) Construction of line-processes: second-order properties, in *Stochastic Geometry, A Tribute to the Memory of Rollo Davidson*, eds E.F. Harding and D.G. Kendall, John Wiley & Sons, Ltd, London, pp. 55–75.

De Smith, M.J., Goodchild, M.F., and Longley, P. (2007) *Geospatial Analysis: a Comprehensive Guide to Principles, Techniques and Software Tools*, 2nd edn, Troubader Publishing, Leicester.

Dean, R.A. (1966) *Elements of Abstract Algebra*, John Wiley & Sons, Inc., New York.

Deckers, B., Verheyen, K., Hermy, M., and Muys, B. (2005) Effects of landscape structure on the invasive spread of black cherry *Plunus serotina* in an agricultural landscape in Flanders, Belgium. *Ecography*, **28**, 99–109.

DeMers, M.N. (2000) *Fundamentals of Geographic Information Systems*, 2nd edn, John Wiley & Sons, Ltd.

Dent, C.L. and Grimm, N.B. (1999) Spatial heterogeneity of stream water nutrient concentrations over successional time. *Ecology*, **80** (7), 2283–2298.

Devroye, L. and Gyorfi, L. (1984) *Nonparametric Density Estimation: the L_1* View, John Wiley & Sons, Inc., New York.

Devroye, L. and Lugosi, G. (2001) *Combinatorial Methods in Density Estimation*, Springer, New York.

Diggle, P.J. (1983) *Statistical Analysis of Spatial Point Patterns*, Academic Press, London.

Diggle, P.J. (2003) *Statistical Analysis of Spatial Point Patterns*, 2nd edn, Arnold, London.

Dijkstra, E.W. (1959) A note on two problems in connection with graphs. *Numerische Mathematik*, **1**, 269–271.

Diplock, G. and Openshaw, S. (1996) Using simple genetic algorithms to calibrate spatial interaction models. *Geographical Analysis*, **28** (3), 263–279.

Dixon, P.M. (2002) Ripley's *K* function, in *Encyclopedia of Environmetrics*, **3**, eds A.H. El-Shaarawi and W. W. Piegorsch, John Wiley & Sons, Ltd., Chichester, pp. 1796–1803.

Doreian, P. (1990) Network autocorrelation models: problems and prospects, in *Spatial Statistics: Past, Present, and Future*, ed. I.D.A. Griffith, Institute of Mathematical Geography, Ann Arbor, pp. 369–389.

Doreian, P., Teuter, K., and Wang, C.-H. (1984) Network autocorrelation models: some Monte Carlo results. *Sociological Methods and Research*, **13**, 155–200.

Douglas, D.H. and Peucker, T. (1973) Algorithms for the reduction of the number of points required to represent a digitized line or its caricature. *The Canadian Cartographer*, **10** (2), 112–122.

Dow, M.M. (2007) Galton's problem as multiple network autocorrelation effects: cultural trait transmission and ecological constraint. *Cross-Cultural Research*, **41**, 336–363.

Dow, M.M., Burton, M.L., and White, D.R. (1982) Network autocorrelation: a simulation study of a foundational problem in regression and survey research. *Social Networks*, **4**, 169–200.

Dow, M.M., Burton, M.L., White, D.R., and Reitz, K.P. (1984) Galton's problem as network autocorrelation. *American Ethnologist*, **11** (4), 754–770.

Downs, J.A. and Horner, M.W. (2007a) Characterizing linear point patterns. *Proceedings of the Geographical Information Science Research UK Conference,* (GISRUK 07), Maynooth, 11–13th April 2007, pp. 421–425.

Downs, J.A. and Horner, M.W. (2007b) Network-based kernel density estimation for home range analysis. *Proceedings of the 9th International Conference on GeoComputation,* Maynooth, Ireland.

Downs, J.A. and Horner, M.W. (2008) Spatially modelling pathways of migratory birds for nature reserve site selection. *International Journal of Geographical Information Science*, **22** (6), 687–702.

Drezner, T. (1994) Optimal continuous location of a retail facility, facility attractiveness, and market share: an interactive model. *Journal of Retailing*, **70** (1), 49–64.

Drezner, T. and Drezner, Z. (2004) Finding the optimal solution to the Huff based competitive location model. *Computational Management Science*, **1** (2), 193–208.

Drezner, T., Drezner, Z., and Salhi, S. (2002) Solving the multiple competitive facilities location problem. *European Journal of Operational Research*, **142**, 138–151.

Drysdake, R.L. and Lee, D.T. (1978) Generalized Voronoi diagram in the plane. *Proceedings of the 16th Annual Allerton Conference on Communications, Control and Computing*, pp. 833–842.

Drysdale, R.L. (1979) *Generalized Voronoi Diagrams and Geometric Searching*, Ph.D. Thesis, Department of Computer Science, Technical Report STAN-CS-79-705, Stanford University.

Duffala, D.C. (1976) Convenience stores, armed robbery, and physical environment feature. *American Behavioral Scientist*, **20** (2), 227–246.

Eck, J.E. (1998) What do those dots mean? Mapping theories with data, in *Crime Mapping & Crime Prevention* (Crime Prevention Studies, Vol. 8), eds D. Weisburd and T. McEwen, Criminal Justice Press, Monsey, N.Y., pp. 379–406.

Edelsbrunner, H. (1987) *Algorithms in Combinatorial Geometry,* Berlin, Springer-Verlag.

Edelsbrunner, H. and Mucke, E.P. (1988) Simulation of simplicity – a technique to cope with degenerate cases in geometric algorithms. *Proceedings of the 4th ACM Annual Symposium on Computational Geometry,* Urbana-Champaign, pp. 118–133.

Egenhofer, M.J. and Frank, A.U. (1992) Object-oriented modeling for GIS. *URISA Journal*, **4** (2), 3–19.

Eggermont, P.P.B. and LaRiccia, V.N. (2001) *Maximum Penalized Likelihood Estimation: Vol. 1 Density Estimation*, Springer, New York.

Einstadter, W.J. and Henry, S. (2006) *Criminological Theory: An Analysis of Its Underlying Assumptions*, 2nd edn, Rowman & Littlefield Publishers, Lanham, Maryland.

Elbaz-Poulichet, F., Morley, N.H., Cruzado, A., Velasquez, Z., Achterberg, E.P. and Braungardt, C. B. (1999) Trace metal and nutrient distribution in an extremely low pH 2.5 river-estuarine system, the Ria of Huelva, South-West Spain. *The Science of the Total Environment*, **227**, 73–83.

El-Gamal, A., Nasr, S., and El-Taher, A. (2007) Study of the spatial distribution of natural radioactivity in the upper Egypt Nile River sediments. *Radiation Measurements*, **42**, 457–465.

Epanechnikov, V.A. (1969) Nonparametric estimation of a multivariate probability density. *Theory of Probability and Its Applications*, **14** (1), 153–158.

Erdogan, S., Yilmaz, I., Baybura, T., and Gullu, M. (2008) Geographical information systems aided traffic accident analysis system case study: city of Afyonkarahisar. *Accident Analysis in Prevention*, **40** (1), 178–181.

Erwig, M. (2000) The graph Voronoi diagram with applications. *Networks*, **36** (3), 156–163.

Euler, L. (1736) Solutio problematis ad geonatriam situs pertinentis. *Commentarii Academiae Scientiarum Imperialis Petropolitanae*, **8**, 128–140.

Evans, S.G., Hungr, O., and Clague, J.J. (2001) Dynamics of the 1984 rock avalanche and associated distal debris from on mount Cayley, British Columbia, Canada; implications for landslide hazard assesment on dissected volcanoes. *Engineering Geology*, **61**, 29–51.

Everitt, B.S., Landau, S., and Leese, M. (2011) *Cluster Analysis*, 5th edn, Arnold, London.

Faust, N. (1998) Raster based GIS, in *History of Geographic Information Systems: Perspective from the Pioneers*, ed. T.W. Foreseman, Prentice-Hall, Upper-Saddle River, pp. 8–72.

Fernandez, J., Pelegrin, B., Plastria, F., and Toth, B. (2007) Solving a Huff-like competitive location and design model for profit maximization in the plane. *European Journal of Operational Research*, **179**, 1274–1287.

Fetter, F.A. (1924) The economic law of market areas. *Quarterly Journal of Economics*, **38**, 520–529.

Finder, R.A., Roseberry, J.L., and Woolf, A. (1999) Site and landscape conditions at white-tailed deer/vehicle collision locations in Illinois. *Landscape and Urban Planning*, **44**, 44–85.

Fink, K.L. and Krammes, R.A. (1995) Tangent length and sight distance effects on accident rates at horizontal curves on rural two-lane highways. *Transportation Research Record*, **1500**, 162–168.

Fisher, M.M. and Griffith, D.A. (2008) Modeling spatial autocorrelation in spatial interaction data: an application to patent citation data in the European Union. *Journal of Regional Science*, **48** (5), 969–989.

Fisher, R.A. (1932) *Statistical Methods for Research Workers*, 4th edn, Oliver & Boyd, Edinburgh.

Fix, E. and Hodges, J.L. (1951) *Discriminatory Analysis, Nonparametric Discrimination: Consistency Properties*. Report No. 4, Project No. 21-49-004, USAF School of Aviation Medicine, Randolph Field, Texas.

Flahaut, B., Mouchart, M., and Martin, E.S. (2003) The local spatial autocorrelation and the kernel method for identifying black zones a comparative approach. *Accident Analysis and Prevention*, **35**, 991–1004.

Forman, R.T.F. and Alexander, L.E. (1998) Roads and their major ecological effects. *Annual Review of Ecology and Systematics*, **29**, 207–231.

Foster, M.L., and Humphrey, S.R. (1995) Use of highway underpasses by Florida panthers and othe wildlife. *Wildlife Society Bulletin*, **23** (1), 95–100.

Fotheringham, A.S., Brunsdon, C., and Charlton, M. (2002) Local statistics and local models for spatial data, in *Geographically Weighted Regression*, eds A.S. Fotheringham, C. Brunsdon, and M. Charlton, John Wiley & Sons, Ltd, Chichester, pp. 1–25.

Franke, R. (1982) Scattered data interpolation: Test of some methods. *Mathematics of Computation*, **38** (157), 181–200.

Freund, J.E. (1971) *Mathematical Statistics*, Prentice Hall, Englewood, Cliffs.

Furth, P.G. and Rahbee, A.B. (2000) Optimal bus stop spacing through dynamic programming and geographic modeling. *Transportation Research Record*, **1731**, 15–22.

Furuta, T., Suzuki, A., and Okabe, A. (2008) A Voronoi heuristic approach to dividing networks into equal-sized sub-networks. *FORMA*, **23**, 73–79.

Furuta, T., Uchida, M., and Suzuki, A. (2005) An efficient algorithm for the absolute center of a network using the farthest network Voronoi diagram. *The 2nd International Symposium on Voronoi Diagrams in Science and Engineering*, pp. 10–21.

Fustec, J., Cormier, J.-P., and Lode, T. (2003) Beaver lodge location on the upstream Loire river. *Comptes Rendus Biologies*, **326**, 192–199.

Gahegan, M.N. and Roberts, S.A. (1988) An intelligent, object-oriented geographical information system. *International Journal of Geographical Information Science*, **2** (2), 101–110.

Gambini, R., Huff, D.L., and Jenks, G.F. (1967) Geometric properties of market areas. *Papers and Proceedings of Regional Science Association*, **20** (1), 85–92.

Ganio, I.M., Torgersen, C.E., and Gresswell, R.E. (2005) A geostatistical approach for describing spatial pattern in stream networks. *Frontiers in Ecology and the Environment*, **3** (3), 138–144.

Garcia, A.P., Betz, J., Larsen, C., Waitt, R., Rocklage, S., Kellar, D., Arnsberg, B., Milks, D., Varney, M., Burum, D., Key, M., and Groves, P. (2000) *Spawning Distribution of Fall Chinook Salmon in the Snake River*, BPS Report DOE/BP-3776-2, U.S. Department of Energy, Bonneville Power Administration, Division of Fish and Wildlife.

Gardner, B., Sullivan, P.J., and Lembo, A.J., Jr. (2003) Predicting stream temperatures: geostatistical model comparison using alternative distance metrics. *Canadian Journal of Fisheries and Aquatic Sciences*, **60** (3), 344–351.

Garwood, F. (1947) The variance of the overlap of geometrical figures with reference to a bombing problem. *Biometrika*, **34** (1/2), 1–17.

Gasca, M. and Sauer, T. (2000) On the history of multivariate polynomial interpolation. *Journal of Computational and Applied Mathematics*, **122** (1–2), 23–35.

Geary, R.C. (1954) The contiguity ratio and statistical mapping. *The Incorporated Statistician*, **5** (3), 115–145.

Georke, R. and Wolff, A. (2005) Constructing the city Voronoi diagram faster. *Proceedings of the 2nd International Symposium on Voronoi Diagrams in Science and Engineering*, October 10–13, 2005, Seoul, pp. 162–172.

Getis, A. (1991) Spatial interaction and spatial autocorrelation: a crossproduct approach. *Environment and Planning A*, **23**, 1269–1277.

Getis, A. and Boots, B. (1978) *Models of Spatial Processes: An Approach to the Study of Point, Line and Area Patterns*, Cambridge University Press, Cambridge.

Getis, A. and Franklin, J. (1987) Second-order neighborhood analysis of mapped point patterns. *Ecology*, **68** (3), 473–477.

Getis, A. and Ord, J.K. (1992) The analysis of spatial association by use of distance statistics. *Geographical Analysis*, **24** (3), 189–206.

Gilbert, E.W. (1958) Pioneer maps of health and disease in England. *The Geographical Journal*, **124** (2), 172–183.

Gill, P.E., Murray, W., and Wright, M.H. (1981) *Practical Optimization*, Academic Press, New York.

Gilmour, J.S.L. (1937) A taxonomic problem. *Nature*, **139**, 1040–1042.

Golias, J.C. (1992) Establishing relationships between accidents and flows at urban priority road junctions. *Accident Analysis and Prevention*, **24** (6), 689–694.

Golob, T.F., Recker, W.R., and Levine, D.W. (1990) Safety of freeway median high occupancy vehicle lanes: a comparison of aggregate and disaggregate analysis. *Accident Analysis and Prevention*, **22** (1), 19–34.

Goodchild, M. (1992) Geographical data modeling. *Computers and Geoscience*, **18**, 401–408.

Goodchild, M.F. (1986) *Spatial Autocorrelation*, Geo Books, Norwich.

Goodchild, M.F. (1987) A spatial analytical perspective on geographical information systems. *International Journal of Geographical Information Systems*, **1** (4), 327–334.

Gordon, W.J., and Wixom, J.A. (1978) Shepard's method of 'metric interpolation' to bivariate and multivariate interpolation. *Mathematics of Computation*, **32** (141), 253–264.

Gotway, C.A. (1991) Fitting semivariogram models by weighted least squares. *Computers & Geosciences*, **17** (1), 171–172.

Gotway, R.B., Ferguson, R.B., Hergert, G.W., and Peterson, T.A. (1996) Comparison of kriging and inverse-distance methods for mapping soil parameters. *Soil Science Society America Journal*, **60**, 1237–1247.

Gower, J.C. (1967) A comparison of some methods of cluster analysis. *Biometrics*, **23**, 623–637.

Greenhut, M.L. (1952) The size and shape of the market area of a firm. *Southern Economic Journal*, **19** (1), 37–50.

Gregory, K.J., Davis, R.J., and Tooth, R. (1993) Spatial distribution of coarse woody debris dams in the Lymington Basin, Hampshire, UK. *Geomorphology*, **6**, 207–224.

Greig-Smith, P. (1952) The use of random and contiguous quadrats in the study of the structure of plant communities. *Annals of Botany*, **16**, 293–316.

Greig-Smith, P. (1957) *Quantitative Plant Ecology*, Butterworths Publications, London.

Griffith, D.A. (1987) *Spatial Autocorrelation A Premier*, Resource Publications in Geography (publication supported by Association of American Geographers), Washington, D.C.

Griffith, D.A. (2003) *Spatial Autocorrelation and Spatial Filtering, Gaining Understanding through Theory and Scientific Visualization*, Springer-Verlag, Berlin.

Griffith, D.A. and Amrhein, C. (1983) An evaluation of correction techniques for boundary effects in spatial statistical analysis: traditional methods. *Geographical Analysis*, **15**, 352–360.

Griggs, G.B. and McCrory, P.A. (1975) Comparative discharges of fresh and waste water along the California coast. *Environmental Geology*, **1**, 89–95.

Guibas, L. and Stolfi, J. (1985) Primitives for the manipulation of general subdivisions and the computation of Voronoi diagrams. *ACM Transactions on Graphics*, **4** (2), 74–123.

Hagget, T., Cliff, D., and Frey, A. (1965) *Locational Methods*, Edward Arnold, London.

Haggett, P. and Chorley, R.J. (1969) *Network Analysis in Geography*, St. Martin's Press, London.

Haining, R.P. (2003) *Spatial Data Analysis Theory and Practice*, Cambridge University Press, Cambridge.

Hakimi, S.L. (1965) Optimum distribution of switching centers in a communication network and some related graph theoretic problems. *Operations Research*, **13** (3), 462–475.

Hakimi, S.L., Labbe, M., and Schmeichel, E. (1992) The Voronoi partition of a network and its implications in location theory. *ORSA Journal on Computing*, **4** (4), 412–417.

Hale, T.S. and Moberg, C.R. (2003) Location science research: a review. *Annals of Operations Research*, **123**, 21–35.

Handler, G.Y. and Mirchandani, P.B. (1979) *Location on Networks: Theories and Algorithms*, MIT Press, Cambridge.

Harary, F. (1969) *Graph Theory*, Addison-Wesley, Reading, MA.

Härdle, W. (1990) *Applied Nonparametric Regression*, Cambridge University Press, Cambridge.

Harris, C.D. and Ullman, E.L. (1945) The nature of cities. *Annals of the American Academy of Political and Social Science*, **242**, 7–17.

Hayes, J.W., Leathwick, J.R., and Hanchet, S.M. (1989) Fish distribution patterns and their association with environmental factors in the Mokau River catchment, New Zealand. *New Zealand Journal of Marine and Freshwater Research*, **23**, 171–180.

Heise, R.J., Slack, W.T., Ross, S.T., and Dugo, M.A. (2004) Spawning and associated movement patterns of Gulf sturgeon in the Pascagoula River drainage, Mississippi. *Transactions of the American Fisheries Society*, **133**, 221–230.

Hillier, B. (1996) Cities as movement economies, *Space is the Machine: A Configurational Theory of Architecture*, ed. B. Hillier, Cambridge University Press, Cambridge, pp. 149–182.

Hillier, B. and Hanson, J. (1984) *The Social Logic of Space*, Cambridge University Press, Cambridge.

Hiyoshi, H. (2009) Spline interpolation over networks. *Japan Journal of Industrial and Applied Mathematics*, **27** (3), 375–394.

Hiyoshi, H. (2010) Spline interpolation on networks. *Proceedings of the 2009 International Conference on Scientific Computing,* July 13–16, 2009, Las Vegas, pp. 47–52.

Hodder, I. and Orton, C. (1976) *Spatial Analysis in Archaeology*, Cambridge University Press, Cambridge.

Hodgson, M.J. (1981) A location-allocation model maximizing consumers' welfare. *Regional Studies*, **15** (6), 493–506.

Hoef, J.M.V., Peterson, E., and Theobald, D. (2006) Spatial statistical models that use flow and stream distance. *Environmental and Ecological Statistics*, **13** (4), 449–464.

Holzer, M., Shulz, F., and Wagner, D. (2005) Combining speed-up techniques for shortest-path computations. *ACM Journal of Experimental Algorithmics*, **10**, (Article 2. 5), 1–18.

Holzer, M., Schulz, F., and Wagner, D. (2008) Engineering multi-overlay graphs for shortest plan queries. *ACM Journal of Experimental Algorithmics*, **13** (25), 1–26.

Hopkins, B., with appendix by Skellam, J.G. (1954) A new method for determining the type of distribution of plant individuals. *Annals of Botany*, **18** (70), 213–227.

Horton, R.E. (1923) Rainfall interpolation. *Monthly Weather Review*, **51** (6), 291–304.

Hoyt, H. (1939) *The Structure and Growth of Residential Neighborhoods in American Cities*, Federal Housing Administration, Washington, D.C.

Hubbard, M.W., Danielson, B.J., and Schmitz, R.A. (2000) Factors influencing the location of deer-vehicle accidents in Iowa. *Journal of Wildlife Management*, **64**, 707–712.

Huff., D.L. (1963) A probabilistic analysis of shopping center trade areas. *Land Economics*, **39** (1), 81–90.

Huff, D.L. (1964) Defining and estimating a trading area. *Journal of Marketing*, **28** (3), 34–38.

Huff, D.L. (1966) A programmed solution for approximating an optimum retail location. *Land Economics*, **42** (3), 293–302.

Hwang, H., Chiu, Y.-H., Chen, W.-Y., and Shih, B.-J. (2004) Analysis of damage to steel gas pipelines caused by ground shaking effects during the Chi-Chi, Taiwan, earthquake. *Earthquake Spectra*, **20** (4), 1095–1110.

Hyson, C.D. and Hyson, W.P. (1950) The economic law of market areas. *Quarterly Journal of Economics*, **64**, 319–327.

Illian, J., Penttinen, A., Stoyan, H., and Stoyan, D. (2008) *Statistical Analysis and Modeling of Spatial Point Patterns*, John Wiley & Sons, Ltd, Chichester.

Ingram, D.R. (1978) An evaluation of procedures utilized in nearest-neighbor analysis. *Geografiska Annaler*, **60**, 65–70.

Irwin, J.O., Armitage, P., and Davies, C.N. (1949) Overlapping of dust particles on a sampling plate. *Nature*, **163**, 809.

Isaaks, E.H. and Srivastava, R.M. (1989) *Applied Geostatistics*, Oxford University Press, Oxford.

Jacquez, C.M. (2008) Spatial cluster analysis, in *Handbook of Geographic Information Science*, eds J.P. Wilson and A.S. Fotheringham, Blackwell, Malden, MA, pp. 395–416.

Jeffreys, H. and Jeffreys, B.S. (1988) *Wallis's Formula for Methods of Mathematical Physics*, 3rd edn, Cambridge University Press, Cambridge, ch. 15.07, p. 468.

Jiang, B. and Claramunt, C. (2002) Integration of space syntax into GIS: new perspectives for urban morphology. *Transactions in GIS*, **6**, 295–309.

Jin, W., Jiang, Y., Qian, W., and Tung, A.K.H. (2006) Mining outliers in spatial networks. *Proceedings of Database Systems for Advanced Applications,* Singapore, April 12–15, 2006. Lecture Notes in Computer Science, 3882, pp. 156–170.

Johnson, N.L., Kotz, S., and Kemp, A.W. (1992) *Univariate Discrete Distribution*, 2nd edn, John Wiley & Sons, Inc., New York.

Johnson, N.L., Kotz, S., and Kemp, A.W. (1995) *Continuous Univariate Distribution*, 2nd edn, John Wiley & Sons, Inc., New York.

Johnson, S. (1967) Hierarchical clustering schemes. *Psychometrika*, **32** (3), 241–254.

Johnson, S. (2006) *Ghost Map: the Story of London's Most Terrifying Epidemic – and How it Changed Science*, Penguin Group, New York.

Johnston, C.A. (1998) *Geographic Information Systems in Ecology*, Blackwell, Oxford.

Jones, A.P., Langford, I.H., and Bentham, G. (1996) The application of K-function analysis to the geographical distribution of road traffic accident outcomes in Norfolk, England. *Social Science Medicine*, **42** (6), 879–885.

Journel, A.G. and Huijbregts, Ch.J. (1991) *Mining Geostatistics*, 5th edn, Academic Press, San Diego.

Jung, A.F. (1959) Is Railly's law of retail gravitation always true? *Journal of Marketing*, **24**, 62–63.

Kansky, K.J. (1963) *Structure of Transportation Networks: Relationships between Network Geometry and Regional Characteristics*, Department of Geography, The University of Chicago Press, Chicago.

Kelejian, H.H. and Robinson, D.P. (1995) Spatial correlation: a suggested alternative to the autoregressive model, in *New Directions in Spatial Econometrics*, eds L. Anselin and R.J.G.M. Florax, Springer-Verlag, Berlin, pp. 75–93.

Kemp, K.K. (1997a) Fields as a framework for integrating GIS and environment process models. Part 1: representing spatial continuity. *Transaction in GIS*, **1** (3), 219–234.

Kemp, K.K. (1997b) Fields as a framework for integrating GIS and environment process models. Part 2: specifying field variables. *Transaction in GIS*, **1** (3), 235–246.

Kendall, M.G. and Moran, P.A.P. (1963) *Geometrical Probability*, Charles Griffin, London.

Kesidis, G. (2007) *An Introduction to Communication Network Analysis*, John Wiley & Sons, Inc., Hoboken, N.J.

Kim, K. and Boski, J. (2001) Finding fault in motorcycle crashes in Hawaii: environmental, temporal, spatial, and human factors. *Transportation Research Record*, **1779**, 182–188.

King, L.J. (1969) Review article: the analysis of spatial form and its relation to geographic theory. *Annals of the Association of American Geographers*, **59**, 573–595.

King, M.L. (1981) A small sample property of the Cliff-Ord test for spatial correlation. *Journal of the Royal Statistical Society Series B*, **43** (2), 263–264.

Kirkpatrick, D.G. (1979) Efficient computation of continuous skeletons. *Proceedings of the 20th Annual IEEE Symposium on Foundations of Computer Science*, pp. 18–27.

Klemelä, J. (2009) *Smoothing of Multivariate Data: Density Estimation and Visualization*, John Wiley & Sons, Inc., Hoboken, N.J.

Klop, J.R. and Khattak, A.J. (1999) Factors influencing bicycle crash severity on two-lane, undivided roadways in North Carolina. *Transportation Research Record*, **1674**, 78–85.

Knoke, D. and Yang, S. (2008) *Social Network Analysis*, 2nd edn, SAGE Publications, Los Angeles.

Knuth, D.E (1973) *The Art of Computer Programming Vol. 3: Sorting and Searching*, Addison-Wesley, Reading, MA.

Koch, T. (2004) The map as internet variations on the theme of John Snow. *Cartographica*, **39** (4), 1–14.

Koch, T. (2005) *Cartographies of Disease, Maps, Mapping, and Medicine*, ESRI Press, California.

Kolchin, V.F., Sevat'yanov, B.A., and Chistyakov, V.P. (1978) *Random Allocation*, V.H. Winston & Sons, Washington D.C.

Krantz, D.H., Luce, R.D., Suppes, P., and Tversky, A. (1971) *Foundation of Measurement*, Academic Press, New York.

Krebs, C.J. (1999) Estimating abundance: quadrat counts, in *Ecological Methodology*, C.J. Krebs, ed. Addison Wesley, Reading, MA, pp. 105–157.

Kresse, W. and Fadaie, K. (2004) Geometry standard, in *ISO Standards for Geographic Information*, eds W. Kresse and K. Fadaie, Springer-Verlag, Berlin, pp. 125–184.

Krige, D.G. (1951) A statistical approach to some basic mine valuation problems of the Witwatersrand. *The Journal of the South African Institute of Mining and Metallurgy*, **52** (6), 119–139.

Krivoruchko, K. and Gribov, A. (2004) Geostatistical interpolation and simulation in the presence of barriers, in *GeoENV IV Geostatistics for Environmental Application: Proceedings of the Second European Conference on Geostatics for Environmental Applications*, Valencia (Spain), November 18–20, 1998, eds X. Sánchez-Vila, J. Carrera, and J. Gómez-Hernández, pp. 331–342.

Kuiper, N.H. (1962) Tests concerning random points on a circle. *Proceedings of the Koninklijke Nederlandse Akademie van Wetenschappen, Series A*, **63**, 38–47.

Kulldorff, M. (1997) A spatial scan statistic. *Communications in Statistics: Theory and Methods*, **26**, 1481–1496.

Kwan, M.-P. and Lee, J. (2005) Emergency response after 9/11: the potential of real-time 3D GIS for quick emergency response in micro-spatial environments. *Computers, Environment and Urban Systems*, **29**, 91–113.

Labbe, M., Peeters, D., and Thisse, J.-F. (1995) Location on networks (ch. 7), in *Network Routing*, eds M. Ball, M. Magnanti, C. Monma, and G. Nemhauser, Elsevier Science, Amsterdam, pp. 551–624.

LaFree, G.D. (1998) *Losing Legitimacy: Street Crime and the Decline of Social Institution in America*, Willan Publishing, Boulder, CO.

Lam, N.S.-N. (1983) Spatial interpolation methods: A review. *The American Cartographer*, **10**, 129–150.

Lassarre, S. (1986) The introduction of the variables 'traffic volume', 'speed' and 'belt-wearing' into a predictive model of the severity of accidents. *Accident Analysis and Prevention*, **18** (2), 129–134.

Launhardt, W. (1882) Die bestinmmung des zweckmaessigsten standortes einer gewerblichen anlage. *Zeitschrift des Vereines Deutscher Ingenieure*, **26** (3), 105–116.

Launhardt, W. (1885) *Mathematische Begründung der Volkswirthschaftslehre*, Verlag von Wilhelm Engelmann, Leipzig.

Laurini, R. and Thompson, D. (1992) *Fundamentals of Spatial Information Systems*, Academic Press, London.

Lawson, A.B. (2001) *Statistical Methods in Spatial Epidemiology*, John Wiley & Sons, Ltd, Chichester.

Leenders, R.Th.A.J. (2002) Modeling social influence through network autocorrelation: constructing the weight matrix. *Social Networks*, **24**, 21–47.

LeSage, J. and Pace, R.K. (2009) *Introduction to Spatial Econometrics*, Chapman & Hall, Boca Raton.

Leung, Y., Mei, C.-L., and Zhang, W.-X. (2003) Statistical test for local patterns of spatial association. *Environment and Planning A*, **35**, 725–744.

Levine, N. (2004) *CrimeStat: A Spatial Statistic Program for the Analysis of Crime Incidence Locations* (version 3), Ned Levine & Associates, Houston Texas and the National Institute of Justice, Washington, D.C.

Levine, N. (2006) Crime mapping and the CrimeStat program. *Geographical Analysis*, **38**, 41–56.

Levine, N., Kim, K.E., and Nitz, L.H. (1995) Spatial analysis of Honolulu motor vehicle crashes: I. spatial patterns. *Accident Analysis and Prevention*, **27** (5), 663–674.

Lewis, P. (1977) *Maps and Statistics*, University Printing House, Cambridge.

Li, H.J., Zhao, X., Qin, F., Li, L. Li, He, X., Chang, X., Li, Z., Liang, K., Xing, F., Chang, W., Wong, R., Yang, I., Li, F., Zhang, T., Tian, R., Webber, B. B., Wilson, J. B. and Huisman, T. H.J. (1990) Abnormal hemoglobins in the Silk Road region of China. *Human Genetics*, **86**, 231–235.

Lieshout, M.N.M. van (2000) *Markov Point Processes and Their Applications*, Imperial College Press, London.

Little, L.S., Edwards, D., and Porter, D.E. (1997) Kriging in estuaries: as the crow flies, or as the fish swims? *Journal of Experimental Marine Biology and Ecology*, **213**, 1–11.

Liu, C.L. (1968) *Introduction to Combinatorial Mathematics*, McGraw-Hill, New York.

Lloyd, C.D. (2005) Assessing the effect of integrating elevation data into the estimation of monthly precipitation in Great Britain. *Journal of Hydrology*, **308**, 128–150.

Lo, C.P. and Yeung, A.K.W. (2002) *Concepts and Techniques of Geographic Information Systems*, Prentice Hall, College Div.

Loftsgaardes, D.O. and Quesenberry, C.P. (1965) A nonparametrics estimate of a multivariate density function. *The Annals of Mathematical Statistics*, **36** (3), 1049–1051.

Longley, P.A., Goodchild, M.F., Maguire, D.J., and Rhind, D.W. (2005) *Geographic Information Systems and Science*, 2nd edn, John Wiley & Sons, Ltd, Chichester.

Lorr, M. (1983) *Cluster Analysis for Social Scientists*, Jossey-Bass, San Francisco.

Loukaitou-Sideris, A. (1999) Hot spots of bus stop crime. *Journal of the American Planning Association*, **65** (4), 395–411.

Lu, G.Y. and Wong, D.W. (2008) An adaptive inverse-distance weighting spatial interpolation technique. *Computers & Geosciences*, **34**, 1044–1055.

Lu, Y. and Chen, X. (2007) On the false alarm of planar *K*-function when analyzing urban crime distributed along streets. *Social Science Research*, **36**, 611–632.

Luce, R. (1959) *Individual Choice Behavior: A Theoretical Analysis*, John Wiley & Sons, Inc., New York.

MacDonald, J.E. and Gifford, R. (1989) Territorial cues and defensible space theory: the burglar's point of view. *Journal of Environmental Psychology*, **9**, 193–205.

Mack, C. (1954) The expected number of clumps when convex laminae are placed at random and with random orientation on a plane area. *Proceedings of the Cambridge Philosophical Society*, **50**, 581–585.

Mack, M.A. (1948) An exact formula for $Q_k(n)$, the probable number of k-aggregates in a random distribution of n points. *Philosophical Magazine and Journal of Science*, **39** (7), 778–790.

McGuigan, D.R.D. (1981) The use of relationships between road accidents and traffic flow in 'black-spot' identification. *Traffic Engineering and Control*, August/September, 448–453.

McKnight, T.L. (1969) Barrier fencing for vermin control in Australia. *Geographical Review*, **59**, 330–347.

McLeod, K.S. (2000) Our sense of Snow: the myth of John Snow in medical geography. *Social Science & Medicine*, **50**, 923–935.

Mahajan, V., Jain, R.K., and Bergier, M. (1977) Parameter estimation in marketing models in the presence of multicolinearity: an application of ridge regression. *Journal of Marketing Research*, **14**, 586–591.

Maher, M.J. and Mountain, L.J. (1988) The identification of accident blackspots: a comparison of current methods. *Accident Analysis and Prevention*, **20** (2), 143–151.

Maheu-Giroux, M. and Blois, S.d. (2006) Landscape ecology of *Phragmites australis* invasion in networks of liner wetlands. *Landscape Ecology*, **22** (2), 281–301.

Mahoney, M.S. (transl.) (1979) *Rene Descartes, The World of Treatise on Light*, Abaris Books, New York.

Maki, N. and Okabe, A. (2005) Spatio-temporal analysis of aged members of a fitness club in suburbs. *Proceedings of the Geographical Information Systems Association*, **14**, 29–34 [in Japanese].

Malo, J., Suarez, F., and Diez, A. (2004) Can we mitigate animal-vehicle accidents using predictive models? *Journal of Applied Ecology*, **41**, 701–710.

Malvic, T. and Durekovic, M. (2003) Application of methods: inverse distance weighting, ordinary kriging and collocated cokriging in porosity evaluation, and comparison of results on the Benicanci and Star Gradac Fields in Croatia. *NAFTA*, **54** (9), 331–340.

Marcon, E. and Puech, F. (2003) Evaluating the geographic concentration of industries using distance-based methods. *Journal of Economic Geography*, **3**, 409–428.

Mase, S., and Takeda, J. (2001) *Spatial Data Modeling*, Kyoritsu, Tokyo, [in Japanese].

Matern, B. (1960) Spatial variation stochastic models and their application to some problems in forest surveys and other sampling investigations. *Meddelanden Fran Statens Skogsforskningstitut*, **49** (5), 1–144.

Matheron, G. (1963) Principles of geostatistics. *Economic Geology*, **58**, 1246–1266.

Matui, I. (1932) Statistical study of the distribution of scattered villages in regions of the Tonami Plane, Toyama Prefecture. *Japanese Journal of Geology and Geography*, **9**, 251–266.

Medvedev, N. and Syta, P. (1998) *Voronoi's Impact on Modern Science*, The Institute of Mathematics of the National Academy of Sciences of Ukraine, Kiev.

Miaou, S.-P. (1994) The relationship between truck accidents and geometric design of road sections: Poisson versus negative binomial regressions. *Accident Analysis and Prevention*, **26** (4), 471–482.

Michener, C.D. (1957) Some bases for higher categories in classification. *Systematic Zoology*, **6** (4), 160–173.

Michener, C.D. and Sokal, R.R. (1957) A quantitative approach to a problem in classification. *Evolution*, **11**, 130–162.

Miles, R.E. (1964) Random polygons determined by random lines in a plane, II. *Proceedings of the National Academy of Sciences*, **52** (4), 901–907.

Miles, R.E. and Neyman, J. (1970) On the homogeneous planar Poisson point process. *Mathematical Biosciences*, **6**, 85–127.

Miller, H.J. (1994) Market area delimitation within networks using geographic information systems. *Geographical Systems*, **1**, 157–173.

Miller, H.J. (2004) Tobler's first law and spatial analysis. *Annals of the Association of American Geographers*, **94** (2), 284–289.

Millman, R.S., and Parker, G.D. (1977) *Elements of Differential Geometry*, Prentice Hall, Englewood Cliffs.

Milovanovic, M. (2007) Water quality assessment and determination of pollution sources along the Axios/Vardar River, Southeastern Europe. *Desalination*, **213**, 159–173.

Mirchandani, P.B. and Francis, R.L. (1990) *Discrete Location Theory*, John Wiley & Sons, New York.

Mitas, L. and Mitasova, H. (1999) Spatial interpolation, in *Geographical Information Systems*, 2nd edn, **2**, eds P. Longley, M. Goodchild, D. Maguire, and D. Rhind, John Wiley & Sons, Inc., Hoboken, N.J, pp. 481–492.

Mok, J.-H., Landphair, H.C., and Naderi, J.R. (2006) Landscape improvement impacts on roadside safety in Texas. *Landscape and Urban Planning*, **78**, 263–274.

Møller, J. and Waagepetersen, R.P. (2002) *Statistical Inference and Simulation for Spatial Point Processes*, Chapman & Hall/CRC Press, Boca Raton.

Moore, P.G. (1954) Spacing in plant populations. *Ecology*, **35** (2), 222–226.

Moran, P.A.P. (1948) The interpretation of statistical maps. *Journal of Royal Statistical Society. B*, **10** (2), 243–251.

Morisita, M. (1954) Estimation of population density by spacing method. *Memoirs of the Faculty of Science, Kyushu University, Series E*, **1**, 187–197.

Morita, M. and Okunuki, K. (2006) A Study on edge effects with the *K*-function method on a network. *Geographical Review of Japan*, **79** (2), 82–96 [in Japanese].

Morita, M. (2008) A study of *K*-functions at different levels of detail of network data. *Geographical Review of Japan*, **81** (4), 179–196 [in Japanese].

Moss, B., Booker, I., Balls, H., and Manson, K. (1989) Phytoplankton distribution in a temperate floodplain lake and river system. I. Hydrology, nutrient sources and phytoplankton biomass. *Journal of Plankton Research*, **11** (4), 813–838.

Munguira, M.L., and Thomas, J.A. (1992) Use of road verges by butterfly and burnet populations, and the effect of roads on adult dispersal and mortality. *Journal of Applied Ecology*, **29**, 316–329.

Myers, D.E. (1994) Spatial interpolation: an overview. *Geoderma*, **62**, 17–28.

Myint, S.W. (2008) An exploration of spatial dispersion, pattern, and association of socio-economic functional units in an urban system. *Applied Geography*, **28**, 168–188.

Nakanishi, M. and Cooper, L.G. (1974) Parameter estimation for a multiplicative competitive interaction model-least squares approach. *Journal of Marketing Research*, **11**, 303–311.

Nakaya, T. (2001) Geomedical approaches based on geographical information science GIS and spatial analysis for health researches. *Proceedings of Symposium on ASIA GIS* 2001, 20–22 June, Tokyo (CD-Rom).

Nasar, J.L. and Fisher, B. (1993) 'Hot spots' of fear and crime: a multi-method investigation. *Environmental Psychology*, **13**, 187–206.

Neteler, M. and Mitasova, H. (2007) *Open Source GIS: A GRASS GIS Approach*, 3rd edn, Springer-Verlag, New York.

Ng, J.C.N. and Hauer, E. (1989) Accidents on rural two-lane roads: Differences between seven states. *Transportation Research Record*, **1238**, 1–9.

Nicholson, A.J. (1989) Accident clustering: some simple measures. *Traffic Engineering and Control*, **30**, 241–246.

Nicola, G.G. and Almodovar, A. (2004) Growth pattern of stream-dwelling brown trout under contrasting thermal conditions. *Transactions of the American Fisheries Society*, **133**, 66–78.

Nocedal, J. and Wright, S.J. (2006) *Numerical Optimization*, Springer-Verlag, New York.

Oden, N. (1995) Adjusting Moran's I for population density. *Statistics in Medicine*, **14**, 17–26.

Odland, J. (1988) *Spatial Autocorrelation*, SAGE Publications, Beverly Hills, California.

O'Driscoll, R.L. (1998) Description of spatial pattern in seabird distributions along line transects using neighbour *k* statistics. *Marine Ecology Progress Series*, **165**, 81–94.

Okabe, A. (2004) *Islamic Area Studies with Geographic Information Systems*, RoutledgeCurzon, London.

Okabe, A. (ed.) (2005) *GIS-based Studies in the Humanities and Social Sciences*, Taylor & Francis/CRC, Boca Raton.

Okabe, A. (2009a) Future directions of spatial analysis, in *New Frontiers in Urban Analysis*, eds Y. Asami, Y. Sadahiro, and T. Ishikawa, CRC Press, Boca Raton, ix-x.

Okabe, A. (2009b) Preface. *Journal of Geographical Systems*, **11** (2), 107–112.

Okabe, A., Boots, B., and Satoh, T. (2010) A class of local and global *K* functions and their exact statistical methods, in *Perspectives on Spatial Data Analysis*, eds L. Anselin and S.J. Rey, Springer-Verlag, Berlin, pp. 101–112.

Okabe, A., Boots, B., and Sugihara, K. (1994) Nearest neighborhood operations with generalized Voronoi Diagrams: a review. *International Journal of Geographical Information Systems*, **8** (1), 43–71.

Okabe, A., Boots, B., Sugihara, K., and Chiu, S.N. (2000) *Spatial Tessellations: Concepts and Applications of Voronoi Diagrams*, 2nd edn, John Wiley & Sons, Ltd, Chichester.

Okabe, A. and Funamoto, S. (2000) An exploratory method for detecting multi-level clumps in the distribution of points – a computational tool, VCM (variable clumping method). *Journal of Geographical Systems*, **2** (2), 111–120.

Okabe, A. and Kitamura, M. (1996) A computational method for market area analysis on a network. *Geographical Analysis*, **28** (4), 330–349.

Okabe, A. and Okunuki, K. (2001) A computational method for estimating the demand of retail stores on a street network and its implementation in GIS. *Transactions in GIS*, **5** (3), 209–220.

Okabe, A., Okunuki, K., and Shiode, S. (2006a) SANET: A toolbox for spatial snalysis on a network. *Geographical Analysis*, **38** (1), 57–66.

Okabe, A., Okunuki, K., and Shiode, S. (2006b) The SANET toolbox: new methods for network spatial analysis. *Transaction in GIS*, **10** (4), 535–550.

Okabe, A. and Satoh, T. (2005) Uniform network transformation for points pattern analysis on a non-uniform network. *Journal of Geographical Systems*, **8**, 25–27.

Okabe, A. and Satoh, T. (2009) Spatial analysis on a network, in *SAGE Handbook of Spatial Analysis*, eds A.S. Fotheringham and P.A. Rogerson, SAGE Publications, London, pp. 443–464.

Okabe, A., Satoh, T., Furuta, T., Suzuki, A. and Okano, K. (2008) Generalized network Voronoi diagrams: concepts, computational methods, and applications. *International Journal of Geographical Information Science*, **22** (9), 965–994.

Okabe, A., Satoh, T., and Sugihara, K. (2009) A kernel density estimation method for networks, its computational method and a GIS-based tool. *International Journal of Geographic Information Science*, **23**, 7–23.

Okabe, A., and Yamada, I. (2001) The *K*-function method on a network and its computational implementation. *Geographical Analysis*, **33** (3), 271–290.

Okabe, A., Yomono, H., and Kitamura, M. (1995) Statistical analysis of the distribution of points on a network. *Geographical Analysis*, **27** (2), 152–175.

Okunuki, K. and Okabe, A. (2002) Solving the Huff-based competitive location model on a network with link-based demand. *Annals of Operations Research*, **111**, 239–252.

Oliveira, D.P.de., Soibelman, L., and Garrett, J.H., Jr. (2008) GIS applications for spatial analysis of water distribution pipeline breakage and condition assessment data. *Pipelines 2008: Pipeline Asset Management, Maximizing Performance of our Pipeline Infrastructure* (The ASCE Pipelines Conference 2008), eds S. Gokhale and S. Raman, pp. 1–10.

Oosterom, P.V. and Van den Bos, J. (1989) An object-oriented approach to the design of geographic information systems. *Computers & Graphics*, **13** (4), 409–418.

Openshaw, S. (1984) Ecological fallacies and the analysis of areal census data. *Environment and Planning A*, **16**, 17–31.

Openshaw, S. and Taylor, P.J. (1979) A million or so correlation coefficients: three experiments on the modifiable areal unit problem, in *Statistical Applications in the Spatial Sciences*, ed. N. Wrigley, Pion, London, pp. 127–144.

Openshaw, S. and Taylor, P.J. (1981) The modifiable areal unit problem, in *Quantitative Geography: A British View*, eds N. Wrigley and R. Bennett, Routledge and Kegan Paul, London, pp. 60–69.

Ord, K. (1978) How many trees in a forest. *Mathematical Scientist*, **3**, 23–33.

Orlowski, G. and Nowak, L. (2006) Factors influencing mammal roadkills in the agricultural landscape of South-Western Poland. *Polish Journal of Ecology*, **54** (2), 283–294.

O'Sullivan, D. and Unwin, D.J. (2003) *Geographic Information Analysis*, John Wiley & Sons, Inc., Hoboken, N.J.

Painter, K. (1996) The influence of street lighting improvements on crime, fear and pedestrian street use, after dark. *Landscape and Urban Planning*, **35**, 193–201.

Palander, T.F. (1935) *Beitrage zur Standortstheorie*, Almqvist & Wiksell, Uppsala.

Park, R., Burgess, E. and McKenzie, R. (1925) *The City*, University of Chicago Press, Chicago.

Parzen, E. (1962) On estimation of a probability density function and mode. *The Annals of Mathematical Statistics*, **33** (3), 1065–1076.

Paul, A.C. and Pillai, K.C. (1986) Distribution and transport of radium in a tropical river. *Water, Air, and Soil Pollution*, **29**, 261–272.

Pearson, E.S. (1938) The probability integral transformation for testing goodness of fit and combining independent tests of significance. *Biometrika*, **30** (1/2), 134–148.

Pearson, E.S. (1963) Comparison of test for randomness of points on a line. *Biometrika*, **50** (3/4), 315–325.

Pearson, E.S. and Stephens, M.A. (1962) The goodness-of-fit tests based in W_N^2 and U_N^2. *Biometrika*, **49** (3/4), 397–402.

Pearson, K. (1895) Contribution to the mathematical theory of evolution, II. Skew variation in homogenous material. *Philosophical Transactions of the Royal Society of London, A*, **186**, 343–414.

Peeters, D. and Thomas, I. (2009) Network autocorrelation. *Geographical Analysis*, **41** (4), 436–443.

Penn, A. (2003) Space syntax and spatial cognition or why the axial line? *Environment and Behavior*, **35** (1), 30–65.

Penn, A., Hillier, B., Banister, D., and Xu, J. (1998) Configurational modelling of urban movement networks. *Environment and Planning B*, **25** (1), 59–84.

Perkins, D.D., Meeks, J.W., and Taylor, R.B. (1992) The physical environment of street blocks and resident perceptions of crime and disorder: implications for theory and measurement. *Journal of Environmental Psychology*, **12**, 21–34.

Perkins, D.D., Wandersman, A., Rich, R.C., and Taylor, R.B. (1993) The physical environment of street crime: defensible space, territoriality and incivilities. *Environmental Psychology*, **13**, 29–49.

Perona, E., Bonilla, I., and Mateo, P. (1999) Spatial and temporal changes in water quality in a Spanish river. *The Science of the Total Environment*, **241**, 75–90.

Pervin, W.J. (1964) *Foundation of General Topology*, Academic Press, New York.

Pfundt, K. (1969) Three difficulties in the comparison of accident rates. *Accident Analysis and Prevention*, **1**, 253–259.

Philcox, C.K., Grogan, A.L., and Macdonald, D.W. (1999) Patterns of otter *Lutra lutra* road mortality in Britain. *Journal of Applied Ecology*, **36**, 748–762.

Pielou, E.C. (1959) The use of point-to-plant distances in the study of the pattern of plant populations. *Journal of Ecology*, **47**, 607–613.

Pielou, E.C. (1977) *Mathematical Ecology*, John Wiley & Sons, Inc., New York.

Pinder, D.A. and Witherick, M.E. (1973) Nearest-neighbour analysis of linear point patterns. *Tijdschrift voor Economiche en Social Geografie*, **64** (3), 160–163.

Plastria, F. (2001) Static competitive facility location: An overview of optimization approaches. *European Journal of Operational Research*, **129**, 461–470.

Porta, S., Crucitti, P., and Latora, V. (2006a) The network analysis of urban streets: a dual approach. *Physical A: Statistical Mechanics and Its Applications*, **369** (2), 853–866.

Porta, S., Crucitti, P., and Latora, V. (2006b) The network analysis of urban streets: a primal approach. *Environment and Planning B*, **33**, 705–725.

Porta, S. and Latora, V. (2008) Centrality and cities: multiple centrality assessment as a tool for urban analysis and design, in *New Urbanism and Beyond: Designing Cities for Future*, ed. T. Haas, Rizzoli, New York, pp. 140–145.

Porta, S., Strano, E., Iacoviello, V., Messora, R., Latora, V., Cardillo, A., Wang, F. and Scellato, S. (2009) Street centrality and densities of retail and services in Bologna, Italy. *Environment and Planning B*, **36**, 450–465.

Preparata, F.P. and Shamos, M.I. (1985) *Computational Geometry: An Introduction*, Springer, New York.

Pu, S. and Zlatanova, S. (2005) Evacuation route calculation of inner buildings, in *Geo-information for Disaster Management*, eds P.V. Oosterom, S. Zlatanova, and M. Fendel, Springer-Verlag, Berlin Heidelberg, pp. 1143–1161.

Puglisi, M.J., Lindzey, J.S., and Bellis, E.D. (1974) Factors associated with highway morality of white-tailed deer. *Journal of Wildlife Management*, **38**, 799–807.

Raper, J.F. (1999) Spatial representation: the scientist's perspective, in *Geographical Information Systems: Principles and Technical Issues*, 2nd edn, eds P.A. Longley, M.F. Goodchild, D.J. Maguire, and D.W. Rhind, John Wiley & Sons, Inc., New York, pp. 61–70.

Rathbun, S.L. (1998) Spatial modelling in irregularly shaped regions: Kriging estuaries. *Environmetrics*, **9**, 109–129.

Rau, K. (1841) Report to the bulletin of the Belgian Royal Society, in *Precursors in Mathematical Economics: an Anthology*, eds W. Baumol, and S. Goldfeld, London School of Economics and Political Science, London, pp. 181–182.

Reed, D.F., Beck, T.D.I., and Woodard, T.N. (1982) Methods of reducing deer-vehicle accidents: benefit-cost analysis. *Wildlife Society Bulletin*, **10** (4), 349–354.

Reeve, A.F. and Anderson, S.H. (1993) Ineffectiveness of swareflex reflectors at reducing deer-vehicle collisions. *Wildlife Society Bulletin*, **21** (2), 127–132.

Reilly, W.J. (1931) *The Law of Retail Gravitation*, Knickerbocker Press, New York.

ReVelle, C.S. and Eiselt, H.A. (2005) Location analysis: a synthesis and survey. *European Journal of Operational Research*, **165**, 1–19.

Reynolds, R.B. (1953) A test of the law of retail gravitation. *Journal of Marketing*, **17** (3), 273–277.

Richter, B.D., Baumgartner, J.V., Braun, D.P., and Powell, J. (1998) A spatial assessment of hydrologic alteration within a river network. *Regulated Rivers: Research & Management*, **14**, 329–340.

Ries, K.G. III, Steeves, P.A., Gurthrie, J.D., Rea, A.H. and Stewart, D.W. (2008) Stream network navigation in the U.S. geological survey StreamStats web application. *2009 International Conference on Advanced Geographic Information Systems & Web Services*, pp. 80–84.

Rigaux, P., Scholl, M., and Voisard, A. (2002) *Spatial Databases with Application to GIS*, Morgan Kaufmann Publishers, San Francisco.

Ripley, B.D. (1976) The second-order analysis of stationary point processes. *Journal of Applied Probability*, **13**, 255–266.

Ripley, B.D. (1977) Modelling spatial patterns. *Journal of the Royal Statistical Society. B*, **39** (2), 172–212.

Ripley, B.D. (1978) Spectral analysis and the analysis of patterns in plant communities. *Journal of Ecology*, **66**, 965–981.

Ripley, B.D. (1979) Tests of 'randomness' for spatial point patterns. *Journal of the Royal Statistical Society. Series B*, **41** (3), 368–374.

Ripley, B.D. (1981) *Spatial Statistics*, John Wiley & Sons, Inc., New York.

Ripley, B.D. (1988) *Statistical Inference for Spatial Processes*, Cambridge University Press, Cambridge.

Riviere, S. and Schmitt, D. (2007) Two-dimensional line space Voronoi diagram. *International Symposium on Voronoi Diagrams in Science and Engineering*, pp. 168–175.

Roach, S.A. (1968) *The Theory of Random Clumping*, Methuen, London.

Rogers, A. (1974) *Statistical Analysis of Spatial Dispersion*, Pion, London.

Roline, R.A. (1988) The effects of heavy metals pollution of the upper Arkansas river on the distribution of aquatic macroinvertebrates. *Hydrobiologia*, **160**, 3–8.

Romesburg, H.C. (1990) *Cluster Analysis for Researchers*, Robert E. Krieger Publishing Co., Malabar, Florida.

Roncek, D.W. and Maier, P.A. (1991) Bars, blocks, and crimes revisited: Linking the theory of routine activities to the empiricism of "Hot Spots". *Criminology*, **29** (4), 725–753.

Rosenblatt, M. (1956) Remarks on some nonparametric estimates of a density function. *The Annals of Mathematical Statistics*, **27** (3), 832–837.

Sabel, C.E., Bartie, P., Kingham, S., and Nicholson, A. (2006) Kernel density estimation as a spatial-temporal data mining tool: exploring road traffic accident trends. *Proceedings of the GIS Research UK Conference (GISRUK 06)*, Nottingham, 5th–7th April, 2006, pp. 191–196.

Sachs, M. (1973) *The Field Concepts in Contemporary Science*, Charles C. Thomas, Springfield.

Saeki, M. and Macdonald, D.W. (2004) The effects of traffic on the raccoon dog (*Nyctereutes procyonoides viverrinus*) and other mammals in Japan. *Biological Conservation*, **118**, 559–571.

Sakamura, K. and Koshizuka, N. (2005) Ubiquitous computing technologies for ubiquitous learning. *IEEE International Workshop on Wireless and Mobile Technologies in Education (WMTE '05)*, pp. 11–20.

Santalo, L.A. (1976) *Integral Geometry and Geometric Probability*, Addison-Wesley, Reading, MA.

Santalo, L.A., and Yanez, I.L.A. (1972) Averages for polygons formed by random lines in Euclidean and hyperbolic planes. *Journal of Applied Probability*, **9**, 140–157.

Satoh, T. and Okabe, A. (2006) A nearest neighbor distance method for analyzing the distribution of points on a network in relation to subnetworks of the network and its GIS tool. *GIS: Theory and Applications*, **14** (2), 31–39 [in Japanese].

Schaap, E. (2007) *DTFE: The Delaunay Tessellation Field Estimator*, University of Groningen, The Netherlands.

Schafer, J.A. and Penland, S.T. (1985) Effectiveness of swareflex reflectors in reducing deer-vehicle accidents. *Journal of Wildlife Management*, **49** (3), 774–776.

Schwartz, S. (1967) Estimation of probability density by an orthogonal series. *The Annals of Mathematical Statistics*, **38** (4), 1261–1265.

Scott, D.W. (1992) *Multivariate Density Estimation: Theory, Practice, and Visualization*, John Wiley & Sons, Inc., New York.

Scott, J. (1991) *Social Network Analysis, A Handbook*, 2nd edn, SAGE Publications, London.

Seiler, A. (2005) Predicting locations of moose-vehicle collsions in Sweden. *Journal of Applied Ecology*, **42**, 371–382.

Selkirk, K.E., and Neave, H.R. (1984) Nearest neighbor analysis of one-dimensional distribution of points (Signalement/Research note). *Tijdschrift voor Economische en Sociale Geografie*, **75**, 356–362.

Sevtsuk, A. (2010) Path and Place: A Study of Urban Geometry and Retail Activity in Cambridge and Somerville, MA, PhD dissertation in Urban Design and Planning, Massachusetts Institute of Technology.

Shamos, M.I. and Hoey, D. (1975) Closest-point problems. *Proceedings of the 16th Annual IEEE Symposium on Foundations of Computer Science*, pp. 151–162.

Shamos, M.I. (1978) Computational Geometry, Ph.D. Thesis, Yale University.

Shankar, V., Mannering, F., and Barfield, W. (1995) Effect of roadway geometrics and environmental factors on rural freeway accident frequencies. *Accident Analysis and Prevention*, **27** (3), 371–389.

Shepard, D. (1968) A two-dimensional interpolation function for irregularly-spaced data. *Proceedings of the 1968 23rd ACM National Conference*, pp. 517–524.

Shepard, R.N. (1964) Attention and the metric structure of the stimulus space. *Journal of Mathematical Psychology*, **1**, 54–87.

Sherman, L.W., Gartin, P.R., and Bürger, M.F. (1989) Hot spots of predatory crime: routine activities and the criminology of place. *Criminology*, **27** (1), 27–55.

Sherman, L.W. (1995) Hot spots of crime and criminal careers of places, in *Crime and Place* (Crime Prevention Studies, Vol. 4), eds J.E. Eck and D. Weisburd, Criminal Justice Press, Monsey, N.Y., pp. 35–52.

Sherman, L.W. and Weisburd, D. (1995) General deterrent effects of police patrol in crime 'hot spots': a randomized, controlled trial. *Justice Quarterly*, **12** (4), 625–648.

Shieh, Y.N. (1985) K. H. Rau and the economic law of market areas. *Journal of Regional Science*, **25** (2), 191–199.

Shiode, S. (2008) Analysis of a distribution of point events using the network-based quadrat methods. *Geographical Analysis*, **40** (4), 380–400.

Shiode, S. (2011) Street-level spatial scan statistic and STAC for analyzing street crime concentrations. *Transactions in GIS*, **15** (3), 365–383.

Shiode, S. (2012) Revisiting John Snow's map: network-based spatial demarcation of Cholera Area. *International Journal of Geographical Information Science*, (to appear).

Shiode, S. and Okabe, A. (2003) A method for spatial interpolation on a network and comparison with spatial interpolation on a plane. *Papers and Proceedings of the Geographic Information Systems Association*, **12**, 97–100 [in Japaese].

Shiode, S. and Okabe, A. (2004) Analysis of point patterns using the network cell count method. *Theory and Applications of GIS*, **12** (2), 57–66 [in Japanese].

Shiode, S. and Shiode, N. (2009a) Detection of multi-scale clusters in network space. *International Journal of Geographical Information Science*, **23** (1), 75–92.

Shiode, S. and Shiode, N. (2009b) Inverse distance-weighted method for point interpolation on a network, in *New Frontiers in Urban Analysis, In Honor of Atsuyuki Okabe*, eds Y. Asami, Y. Sadahiro, and T. Ishikawa, CRC Press, Boca Raton, pp. 179–196.

Shiode, N. and Shiode, S. (2011) Street-level spatial interpolation using network-based IDW and ordinary kriging. *Transaction in GIS*, **15** (4), 457–477.

Sibson, R. (1981) A brief description of natural neighbour interpolation, in *Interpreting Multivariate Data*, ed. V. Barnett, John Wiley & Sons, Ltd, Chichester, pp. 21–36.

Silverman, B.W. (1986) *Density Estimation for Statistics and Data Analysis*, Chapman & Hall/CRC, London.

Sivertsen, K. (1997) Geographic and environmental factors affecting the distribution of kelp beds and barren grounds and changes in biota associated with kelp reduction at sites along the Norwegian coast. *Canadian Journal of Fisheries and Aquatic Sciences*, **54**, 2872–2887.

Smith, C.E. (1982) The Broad Pump revisited. *International Journal of Epidemiology*, **11**, 99–100.

Smith, T. (1975) A choice theory of spatial interaction. *Regional Science and Urban Economics*, **5** (2), 137–176.

Sneath, P.H.A. (1957) The application of computers to taxonomy. *Journal of General Microbiology*, **17**, 201–226.

Snow, J. (1855) *On the Mode of Communication of Cholera*, 2nd edn, John Churchill, London.

Snow, J. (1936) *Snow on Cholera* (reprint of two papers), D.B. Updike, The Merrymount Press, Boston.

Soares, H.M.V.M., Boaventura, R.A.R., Machado, A.A.S.C., and Esteves da Silva, J.C.G. (1999) Sediments as monitors of heavy metal contamination in the Ave River Basin (Portugal): multivariate analysis of data. *Environmental Pollution*, **105**, 311–323.

Sokal, R.R. and Michener, C.D. (1958) A statistical method for evaluating systematic relationships. *University Kansas Science Bulletin*, **38**, 1409–1414.

Sokal, R.R. and Sneath, P.H.A. (1963) *Principles of Numerical Taxonomy*, W.H. Freeman, San Francisco.

Sorensen, T. (1948) A method of establishing groups of equal amplitude in plant sociology based on similarity of species content and its application to analysis of the vegetation on Danish commons. *Biologiske Skrifter*, **5** (4), 1–34.

Spooner, P.G., Lunt, I.A., Okabe, A., and Shiode, S. (2004) Spatial analysis of roadside *Acacia* populations on a road network using the network K-function. *Landscape Ecology*, **19**, 491–499.

Stanley, W.D. (2003) *Network Analysis with Applications*, 4th edn, Prentice-Hall, Upper Saddle River, N.J.

Stark, B.L. and Young, D.L. (1981) Linear nearest neighbor analysis. *American Antiquity*, **46** (2), 284–300.

Steenberghen, T., Dufays, T., Thomas., I., and Flahaut, B. (2004) Intra-urban location and clustering of road accidents using GIS: a Belgian example. *International Journal of Geographical Information Science*, **18** (2), 169–181.

Steffensen, J.F. (1927) *Interpolation*, The Williams & Wilkins Com Baltimore, Baltimore.

Stein, M.L. (1999) *Interpolation of Spatial Data*, Springer-Verlag, New York.

Stern, E. and Zehavi, Y. (1990) Road safety and hot weather: A study in applied transport geography. *Transactions Institute of British Geographers*, **15**, 102–111.

Stewart, J.Q. (1948) Demographic gravitation: evidence and applications. *Sociometry*, **11** (1/2), 31–58.

Storlazzi, C.D. and Field, M.E. (2000) Sediment distribution and transport along a rocky, embayed coast: Monterey Peninsula and Carmel Bay, California. *Marine Geology*, **170**, 289–316.

Stoyan, D. (2006) Fundamentals of point process statistics, in *Case Studies in Spatial Point Process Modeling*, eds A. Baddeley, P. Gregori, J. Mateu, R. Stoica, and D. Stoyan, Springer-Verlag, New York, pp. 5–22.

Stoyan, D., Kendall, W.S., and Mecke, J. (1995) *Stochastic Geometry and its Applications*, 2nd edn, John Wiley & Sons, Chichester.

Stoyan, D. and Stoyan, H. (1994) *Fractals, Random Shapes and Point Fields*, 2nd edn, John Wiley & Sons, Ltd, Chichester.

Sugihara, K., Satoh, T., and Okabe, A. (2010) Simple and unbiased kernel function for network analysis. *Proceedings of 10th International Symposium on Communications and Information Technologies 2010*, Oct. 26–29 2010, Meiji University Japan, pp. 827–832.

Sugihara, K., Okabe, A., and Satoh, T. (2011) Computational method for the point cluster analysis on networks. *Geoinformatica*, **15**, 167–189.

Sui, D.Z. (2004) Tobler's first law of geography: A big idea for a small world? *Annals of the Association of American Geographers*, **94** (2), 269–277.

Suzuki, T. (1987) Optimum locational patterns of bus-stops for many-to-one travel demand. *Papers of the Annual Conference of the City Planning Institute of Japan*, 22, pp. 247–252 [in Japanese].

Tango, T. (1995) A class of tests for detecting 'general' and 'focused' clustering of rape diseases. *Statistics in Medicine*, **14**, 2323–2334.

Tango, T. (2000) A test for spatial disease clustering adjusted for multiple testing. *Statistics in Medicine*, **19**, 191–204.

Tanner, J.C. (1953) Accidents at rural three-way junctions. *Journal of the Institutions of Highway Engineers*, **2** (11), 56–67.

Tansel, B.C., Francis, R.L., and Lowe, T.J. (1983) Location on networks: A survey. Part I: The c-center and p-median problems. *Management Science*, **29** (4), 482–497.

Tapia, R.A. and Thompson, J.R. (1978) *Nonparametric Probability Density Estimation*, The Johns Hopkins University Press, Baltimore.

Tautu, P. (ed.) (1986) *Stochastic Spatial Processes, Mathematical Theories and Biological Applications Proceedings*, September 10–14, Heidelberg 1984, Springer-Verlag, Heidelberg.

Thomas, I. (1996) Spatial data aggregation: exploratory analysis of road accidents. *Accident Analysis and Prevention*, **28** (2), 251–264.

Tiefelsdorf, M. (1998) Some practical applications of Moran's *I*s exact conditional distribution. *Papers in Regional Science*, **77** (2), 101–129.

Tiefelsdorf, M. (2000) *Modeling Spatial Processes: The Identification and Analysis of Spatial Relationships in Regression Residuals by Means of Moran's I*, Lecture Notes in Earth Sciences 87, Springer-Verlag, Berlin.

Tiefelsdorf, M. and Boots, B. (1995) The exact distribution of Moran's I. *Environment and Planning A*, **27**, 985–999.

Tobler, W.R. (1970) A computer movie simulating urban growth in the Detroit region. *Economic Geography*, **46**, 234–240.

Tobler, W.R. (1979) Smooth pycnophylactic interpolation for geographical regions. *Journal of the American Statistical Association*, **74** (367), 519–530.

Tobler, W.R. (2004) On the first law of geography: a reply. *Annals of the Association of American Geographers*, **94** (2), 304–310.

Tobler, W.R. and Kennedy, S. (1985) Smooth multidimensional interpolation. *Geographical Analysis*, **17** (3), 251–257.

Torgersen, C.E., Gresswell, R.E., and Bateman, D.S. (2004) Pattern detection in stream networks: quantifying spatial variability in fish distribution, in *GIS/Spatial Analysis in Fishery and Aquatic Sciences*, eds T. Nishida, P.J. Kailola, and C.E. Hollingworth, Fishery-Aquatic GIS Research Group, Saitama, Japan, pp. 405–420.

Toth, B., Fernandez, J., Pelegrin, B., and Plastria, F. (2008) Sequential versus simultaneous approach in the location and design of two new facilities using planar huff-like models. *Computers and Operations Research*, **36** (5), 1393–1405.

Tucciarelli, T., Criminisi, A., and Termini, D. (1999) Leak analysis in pipeline systems by means of optimal valve regulation. *Journal of Hydraulic Engineering*, **125** (3), 277–285.

Tufte, E.R. (1997) *Visual Explanations Images and Quantities, Evidence and Narrative*, Graphic Press, Cheshire, Connecticut.

Tuominen, O. (1949) Das Einflussgebiet der Stadt Turku. *Fennia*, **71** (5), 1–138.

Turner, D.J. and Thomas, R. (1986) Motorway accidents: an examination of accident totals, rates and severity and their relationship with traffic flow. *Traffic Engineering and Control*, **27** (7), 377–387.

Unwin, D. (1981) *Introductory Spatial Analysis*, Methuen, London.

Upton, G. and Fingleton, B. (1985) *Spatial Data Analysis by Example Vol. 1: Point Pattern and Quantitative Data*, John Wiley & Sons, Ltd, Chichester.

Van Kirk, D. (2000) *Vehicular Accident Investigation and Reconstruction*, CRC Press, Boca Raton.

Vila, M., Garcés, E., Masó, M., and Camp, J. (2001) Is the distribution of the toxic dinoflagellate Alexandrium catenella expanding along the NW Mediterranean coast? *Marine Ecology Progress Series*, **222**, 73–83.

Voronoi, M.G. (1908) Nouvelles applications des paramètres continus à la théorie des formes quadratiques, deuxième mémoire, recherches sur les parallelloedres primitifs. *Journal fur die Reine und Angewandte Mathematik*, **134**, 198–287.

Wachtel, A. and Lewiston, D. (1994) Risk factors for bicycle-motor vehicle collisions at intersections. *ITE Journal Institute of Transportation Engineers*, **64** (9), 30–35.

Wackernagel, H. (1995) *Multivariate Geostatistics – An Introduction with Applications*, Springer-Verlag, Berlin.

Wagner, D. and Willhalm, T. (2003) Geometric speed-up techniques for finding shortest paths in large sparse graphs. *Proceedings of the 11th Annual European Symposium on Algorithms (ESA) 2003*, Budapest, Hungary, September 16–19, 2003, Lecture Note in Computer Science 2832, pp. 776–787.

Wagner, D. and Willhalm, T. (2007) Speed-up techniques for shortest-path computations. *Proceedings of the 24th Annual Symposium on Theoretical Aspects of Computer Science*, Aachen, Germany, February 22–24, 2007 (STACS 2007), Lecture Note in Computer Science 4393, pp. 23–36.

Wald, A. and Wolfowiz, J. (1940) On a test whether two samples are from the same population. *The Annals of Mathematical Statistics*, **11** (2), 147–162.

Waller, L.A. and Gotway, C.A. (2004) *Applied Spatial Statistics for Public Health Data*, John Wiley & Sons, Ltd, Chichester.

Wand, M.P. and Jones, M.C. (1995) *Kernel Smoothing*, Chapman & Hall/CRC, Hoboken, N.J.

Wang, F.H. (2006) *Quantitative Methods and Applications in GIS*, CRC Press/Taylor & Francis, Boca Raton.

Ward, A.L. (1982) Mule deer behavior in relation to fencing and underpasses on Interstate 80 in Wyoming. *Transportation Research Record*, **859**, 8–13.

Ward, J.H., Jr. (1963) Hierarchical grouping to optimize an objective function. *Journal of the American Statistical Association*, **58** (301), 236–244.

Ward, J.H. and Hook, M.E. (1963) Application of an hierarchical grouping procedure to a problem of grouping profiles. *Educational and Psychological Measurement*, **23** (1), 69–81.

Warden, C.R. (2008) Comparison of Poisson and Bernoulli spatial cluster analysis of pediatric injuries in a fire district. *International Journal of Health Geographics*, **7**, 51.

Wasserman, S. and Faust, K. (1994) *Social Network Analysis: Methods and Applications*, Cambridge University Press, Cambridge.

Watson, D.F. (1992) *CONTOURING: A Guide to the Analysis and Display of Spatial Data*, Pergamon, Oxford.

Watson, D.F. and Philip, G.M. (1985) A refinement of inverse distance weighted interpolation. *Geo-Processing*, **2**, 315–327.

Watson, G.S. (1961) Goodness-of-fit tests on a circle. *Biometrika*, **48** (1/2), 109–114.

Weber, D. and Englund, E. (1992) Evaluation and comparison of spatial interpolators. *Mathematical Geology*, **24** (4), 381–391.

Weber, D.D. and Englund, E.J. (1994) Evaluation and comparison of spatial interpolators II. *Mathematical Geology*, **26** (5), 589–603.

Webster, R., and Oliver, M.A. (2001) *Geostatistics for Environmental Scientists*, John Wiley & Sons, Chichester.

Wee, C.H. and Pearce, M.R. (1985) Patronage behavior toward shopping areas: a proposed model based on Huff's model of retail gravitation. *Advances in Consumer Research*, **12**, 592–597.

Weisbrod, G., Parcells, R., and Kern, C. (1984) A disaggregate model for predicting shopping are market attraction. *Journal of Retailing*, **60** (1), 65–83.

White, D.R., Burton, M.L., and Dow, M.M. (1981) Sexual division of labor in Arfican agriculture: a network autocorrelation analysis. *American Anthropologist*, **83**, 824–849.

Whittaker, E.T. and Watson, G.N. (1990) *A Course in Modern Analysis*, 4th edn, Cambridge University Press, Cambridge.

Whittle, P. (1958) On the smoothing of probability density functions. *Journal of the Royal Statistics Society. B*, **20** (2), 334–343.

Wilks, S.S. (1962) *Mathematical Statistics*, John Wiley & Sons, Inc., New York.

Wilson, A.G. (1970) *Entropy in Urban and Regional Modeling*, Pion, London.

Wishart, D. (1969) An algorithms for hierarchical classifications. *Biometrics*, **25** (1), 165–170.

Wong, D.W.S., and Lee, J. (2005) *Statistical Analysis of Geographic Information with ArcView GIS and ArcGIS*, John Wiley & Sons, Inc., Hoboken, N.J.

Wonnacott, R.J. and Wonnacott, T.H. (1970) *Econometrics*, John Wiley & Sons, Inc., New York.

Worboys, M.F. (1994) Object-oriented approaches to geo-referenced information, Review article. *International Journal of Geographical Information Science*, **8** (4), 385–399.

Worboys, M.F. (1995) *GIS A Computing Perspective*, Taylor & Francis, London.

Worboys, M.F. and Duckham, M. (2004) *GIS A Computing Perspective*, 2nd edn, CRC Press, Boca Raton.

Xie, Y., Ward, R., Fang, C., and Qiao, B. (2007) The urban system in West China: a case study along the midsection of the ancient Silk Road – He-Xi Corridor. *Cities*, **24** (1), 60–73.

Xie, Z.X. and Yan, J. (2008) Kernel density estimation of traffic accidents in a network space. *Computer, Environment and Urban Systems*, **32** (5), 396–406.

Xu, Z. and Sui, D.Z. (2007) Small-worked characteristics on transportation networks: a perspective from network autocorrelation. *Journal of Geographical Systems*, **9**, 189–205.

Yamada, I. (2009) Edge effects, in *International Encyclopedia of Human Geography*, eds R. Kitchin and N. Thrift, Elsevier, Oxford, pp. 381–388.

Yamada, I. and Thill, J.-C. (2004) Comparison of planar and network *K*-functions in traffic accident analysis. *Journal of Transport Geography*, **12**, 149–158.

Yamada, I. and Thill, J.-C. (2007) Local indicators of network-constrained clusters in spatial point patterns. *Geographical Analysis*, **39**, 268–292.

Yamada, I. and Thill, J.-C. (2010) Local indicators of network-constrained clusters in spatial patterns represented by a link attribute. *Annals of the Association of American Geographers*, **100** (2), 269–285.

Yamaguchi, F. (2002) *Computer-Aided Geometric Design*, Springer, Tokyo.

Yap, C.-K. (1988) A geometric consistency theorem for a symbolic perturbation scheme. *Proceedings of the 4th ACM Annual Symposium on Computational Geometry*, Urbana-Champaign, pp. 134–142.

Yiu, M.L. and Mamoulis, N. (2004) Clustering objects on a spatial network. *Proceedings of the ACM Conference on Management of Data (SIGMOD)*, Paris, France, June 2004, pp. 443–454.

Yomono, H. (1993) The computability of the distribution of the nearest neighbor distance on a network. *Theory and Applications of GIS*, **12** (2), 47–56 [in Japanese].

Young, D.L. (1982) The linear nearest neighbour statistic. *Biometrika*, **69** (2), 477–480.

Zahn, C.T. (1971) Graph-theoretical methods for detecting and describing gestalt clusters. *IEEE Transactions on Computers*, **c-20** (1), IEEE, 68–86.

Zimmerman, D., Pavlik, C., Ruggles, A., and Armstrong, M.P. (1999) An experimental comparison of ordinary and universal kriging and inverse distance weighting. *Mathematical Geology*, **31** (4), 375–390.

Index

Spatial Analysis along Networks: Statistical and Computational Methods, First Edition.
Atsuyuki Okabe and Kokichi Sugihara.